Emergent Brain Dynamics

Prebirth to Adolescence

Strüngmann Forum Reports

Julia R. Lupp, series editor

The Ernst Strüngmann Forum is made possible through the generous support of the Ernst Strüngmann Foundation, inaugurated by Dr. Andreas and Dr. Thomas Strüngmann.

This Forum was supported by the
Deutsche Forschungsgemeinschaft

Emergent Brain Dynamics

Prebirth to Adolescence

Edited by

April A. Benasich and Urs Ribary

Program Advisory Committee:

Yehezkel Ben-Ari, April A. Benasich, Julia R. Lupp,
Charles A. Nelson III, Urs Ribary, Wolf Singer,
Terrence J. Sejnowski

The MIT Press

Cambridge, Massachusetts
London, England

Series Editor: J. R. Lupp
Editorial Assistance: M. Turner, A. Ducey-Gessner, C. Stephen
Photographs: N. Miguletz
Lektorat: BerlinScienceWorks

The book was set in TimesNewRoman and Arial.

Library of Congress Cataloging-in-Publication Data
Names: Benasich, April, editor. | Ribary, Urs, editor.
Title: Emergent brain dynamics : prebirth to adolescence / edited by April A.
    Benasich and Urs Ribary.
Other titles: Strüngmann Forum reports.
Description: Cambridge, MA : The MIT Press, [2018] | Series: Strüngmann
    Forum reports | Includes bibliographical references and index.
Identifiers: LCCN 2018005125 | ISBN 9780262038638 (hardcover : alk.
    paper), 9780262545723 (paperback)
Subjects: | MESH: Brain--growth & development | Nervous System
    Physiological Phenomena | Neuronal Plasticity
Classification: LCC QP376 | NLM WL 300 | DDC 612.8/2--dc23 LC record
available at https://lccn.loc.gov/2018005125

Ernst Strüngmann Forum (25th: 2017 : Frankfurt am Main, Germany)

Contents

The Ernst Strüngmann Forum

Science is a highly specialized enterprise—one that enables areas of enquiry to be minutely pursued, establishes working paradigms and normative standards, and supports rigor in experimental research. Some issues, however, do not fall neatly into the purview of a single disciplinary field and for these areas, specialization can actually hinder conceptualization and limit the generation of potential problem-solving approaches. The Ernst Strüngmann Forum was created to address such topics.

Founded on the tenets of scientific independence and the inquisitive nature of the human mind, the Ernst Strüngmann Forum is dedicated to the continual expansion of knowledge. Its activities promote interdisciplinary communication on high-priority issues encountered in basic science. Through its innovative communication process, the Ernst Strüngmann Forum provides an intellectual haven within which experts scrutinize high-priority issues from multiple vantage points.

This process begins with the identification of themes. By nature, a theme constitutes a problem area that transcends classic disciplinary boundaries— a topic of high-priority interest that requires concentrated, multidisciplinary perusal. Proposals are received from leading scientists active in their field and reviewed by an independent Scientific Advisory Board. Once approved, a steering committee is convened to refine the scientific parameters of the proposal and select participants. Approximately one year later, a central gathering, or Forum, is held to which circa forty experts are invited. Expansive discourse is employed to address the problem. Often, this necessitates reexamining long-established ideas and relinquishing previously held perspectives, yet when accomplished, novel insights begin to emerge. Resultant ideas and newly gained perspectives from the entire process are disseminated to the scientific community for further consideration and implementation.

Preliminary discussion on this theme began in 2015, when April Benasich approached me with the wish to extend exploration into dynamic brain coordination and synchrony by utilizing a developmental perspective. Urs Ribary joined us in preparing the proposal and from April 15–17, 2016, the Program Advisory Committee (Yehezkel Ben-Ari, April Benasich, Julia Lupp, Charles Nelson, Urs Ribary, Wolf Singer, and Terry Sejnowski) met to refine the scientific framework for this Forum, which was held in Frankfurt am Main from March 5–10, 2017.

This volume synthesizes the resulting discourse that took place between a diverse group of experts and is comprised of two types of contributions. Background information is provided on specific aspects of the overall theme. These chapters, drafted before the Forum, initiated the discussion at the Forum; they have been peer-reviewed and subsequently revised to provide an up-to-date assessment of these topics. In addition, Chapters 4, 7, 11, and 15 provide an overview of the working groups. These chapters are not consensus documents: their intent was to summarize the discourse, to expose diverging opinions, and to highlight areas where future enquiry is needed.

An endeavor of this kind creates its own unique group dynamics and puts demands on everyone who participates. Each invitee played an active role and for their efforts, I am grateful to all. A special word of thanks goes to the Program Advisory Committee, to the authors and reviewers of the background papers, as well as to the moderators of the individual working groups (Terry Sejnowski, Charles Nelson, Tomáš Paus, and Sylvain Baillet). The rapporteurs of the working groups (Nick Spitzer, Matthias Kaschube, Marina Bedny, and Jennifer Gelinas) deserve special recognition, for to draft a report during the Forum and finalize it in the months thereafter is no simple matter. Finally, I extend my appreciation to April Benasich and Urs Ribary, whose commitment and congenial personalities were essential to this 25th Ernst Strüngmann Forum.

A communication process of this nature relies on institutional stability and an environment that encourages free thought. The generous support of the Ernst Strüngmann Foundation, established by Dr. Andreas and Dr. Thomas Strüngmann in honor of their father, enables the Ernst Strüngmann Forum to pursue its work in the service of science. In addition, valuable partnerships with the following groups are gratefully acknowledged: the Scientific Advisory Board, which ensures the scientific independence of the Forum; the German Science Foundation, for its supplemental financial support; and the Frankfurt Institute for Advanced Studies, which shares its intellectual setting with the Forum. Long-held views are never easy to put aside. Yet, when this is achieved, when the edges of the unknown begin to appear and the resulting gaps in knowledge are able to be identified, the act of formulating strategies to fill such gaps becomes a most invigorating activity. On behalf of everyone involved, I hope this volume will inspire future research focused on understanding the mechanisms by which the brain develops into a mature, fully functioning organ.

Julia R. Lupp, Director
Ernst Strüngmann Forum
Frankfurt Institute for Advanced Studies (FIAS)
Ruth-Moufang-Str. 1, 60438 Frankfurt am Main, Germany
https://esforum.de/

List of Contributors

Baillet, Sylvain   Neurology and Neurosurgery, Biomedical Engineering and Computer Science, Montreal Neurological Institute, McGill University, Montreal QC H2J 2Z3, Canada

Bedny, Marina   Psychological and Brain Sciences, Johns Hopkins University, Baltimore, MD 21218, U.S.A.

Ben-Ari, Yehezkel   Neurochlore, Institut de Neurobiologie de la Mediterranée (Inmed), 13273 Marseille Cedex 09, France

Benasich, April A.   Center for Molecular and Behavioral Neuroscience (CMBN), Rutgers University, Newark, NJ 07102, U.S.A.

Bertrand, Olivier   Lyon Neuroscience Research Center, INSERM, CNRS, Univ. Lyon 1, 69500 Bron, France

Buzsáki, Gyorgy   NYU Neuroscience Institute, New York University, Langone Medical Center, New York, NY 10016, U.S.A.

Chédotal, Alain   Sorbonne Université, INSERM, CNRS, Institut de la Vision, 75012 Paris, France

Doesburg, Sam M.   Biomedical Physiology and Kinesiology, Simon Fraser University, Burnaby BC V5A 1S6, Canada

Fishell, Gordon   NYU Neuroscience Institute, New York University, Langone Medical Center, New York, NY 10016, U.S.A.

Galván, Adriana   Department of Psychology, Brain Research Institute, University of California, Los Angeles, CA 90095-1563, U.S.A.

Gelinas, Jennifer N.   Department of Neurology, Institute for Genomic Medicine and the Gertrude H. Sergievsky Center, Columbia University Medical Center, New York, NY, 10032, U.S.A.

Giedd, Jay   Division of Child and Adolescent Psychiatry, University of California, San Diego, La Jolla, CA 92093, U.S.A.

Gressens, Pierre   Robert Debre Hospital, INSERM U1141, 75019 Paris, France

Hanganu-Opatz, Ileana L.   Developmental Neurophysiology, Institute of Neuroanatomy, University Medical Center Hamburg-Eppendorf, 20251 Hamburg, Germany

Hashemiyoon, Rowshanak   Stereotactic and Functional Neurosurgery, University Hospital of Cologne, 50937 Köln, Germany

Hensch, Takao K.   Molecular and Cellular Biology, Harvard University, Cambridge, MA 02138, U.S.A.

Herculano-Houzel, Suzana Department of Psychology, Vanderbilt University, Nashville, TN 37240, U.S.A.

Hübener, Mark Max Planck Institute of Neurobiology, Synapses – Circuits – Plasticity, 82152 Martinsried, Germany

Kaschube, Matthias Frankfurt Institute for Advanced Studies and Faculty of Computer Science and Mathematics, Goethe University Frankfurt, 60438 Frankfurt am Main, Germany

Kobor, Michael S. Centre for Molecular Medicine and Therapeutics, British Columbia Children's Hospital Research Institute, Vancouver; Department of Medical Genetics, University of British Columbia, Vancouver ; Human Early Learning Partnership, Vancouver; and Canadian Institute for Advanced Research, Toronto, ON, Canada

Kolb, Bryan E. Department of Neuroscience, University of Lethbridge, Lethbridge AB T1K 3M4, Canada

Kolling, Thorsten Developmental Psychology, Goethe University, 60629 Frankfurt am Main, Germany

Lachaux, Jean-Philippe Lyon Neuroscience Research Center–DYCOG Team, INSERM, 69500 Bron, France

Lindenberger, Ulman Center for Lifespan Psychology, Max Planck Institute for Human Development, 14195 Berlin, Germany

Luhmann, Heiko J. Institute of Physiology, University Medical Center Mainz, 55128 Mainz, Germany

Monyer, Hannah Clinical Neurobiology, University Hospital Heidelberg and DKFZ, 69120 Heidelberg, Germany

Moore, Sarah R. Centre for Molecular Medicine and Therapeutics, British Columbia Children's Hospital Research Institute, Vancouver; and Department of Medical Genetics, University of British Columbia, Vancouver, Canada

Nelson, Charles A., III Cognitive Neuroscience, Harvard/Boston Children's Hospital, Boston, MA 02115, U.S.A.

Paus, Tomáš Rotman Research Institute, University of Toronto, Toronto ON M6A 2E1, Canada

Purdon, Patrick L. Anesthesia, Critical Care and Pain Medicine, Massachusetts General Hospital/Harvard Medical School, Charlestown, MA 02129, U.S.A.

Rakic, Pasko Neuroscience, Yale University Medical School, New Haven, CT 06520-8001, U.S.A.

Ribary, Urs Psychology, Simon Fraser University, Burnaby BC V5A 1S6, Canada

Sawa, Akira Johns Hopkins Schizophrenia Center, Department of Psychiatry, Johns Hopkins University, Baltimore, MD 21287, U.S.A.

Sejnowski, Terrence J.   Salk Institute for Biological Studies, La Jolla, CA 92037, U.S.A.

Singer, Wolf   Ernst Strüngmann Institute for Neuroscience in Cooperation with Max Planck Society; Max Planck Institute for Brain Research; and Frankfurt Institute for Advanced Studies, Frankfurt am Main, Germany

Sisk, Cheryl L.   Neuroscience Program, Michigan State University, East Lansing, MI 48824, U.S.A.

Spitzer, Nicholas C.   Neurobiology, UC San Diego, La Jolla, CA 92093, U.S.A.

Stryker, Michael P.   Department of Physiology, University of California, San Francisco, CA 94158, U.S.A.

Sur, Mriganka   Brain and Cognitive Sciences, Massachusetts Institute of Technology, Cambridge, MA 02139, U.S.A.

Uhlhaas, Peter J.   Institute of Neuroscience and Psychology, University of Glasgow, Glasgow G12 8QB, U.K.

1

Exploring Emergent Brain Dynamics

April A. Benasich and Urs Ribary

Abstract

Across development, the coupling and synchronization of oscillations among brain areas are thought to mediate network assembly, coordination, and plasticity, and to play supporting roles in sensorimotor functions and cognition, including perception, language, learning, and memory. Ongoing neural plasticity and sensitivity to environmental cues contribute crucially to this process. To explore the complex mechanisms by which the brain evolves and matures, the Ernst Strüngmann Forum gathered experts to examine collectively how dynamic brain coordination emerges during development. This chapter introduces the extended discourse that transpired and summarizes the perspectives that emerged from multiple disciplines. Background information is provided on the key areas of enquiry (prebirth, childhood, early adolescence, and emerging adulthood), remaining gaps in knowledge are highlighted, and innovative ways forward are proposed to further this evolving area of cross-disciplinary study.

Introduction

Throughout the ages, the developing brain has constituted one of the most fascinating areas for study and reflection. Many efforts have been devoted to the delineation and conceptualization of the remarkably dynamic processes by which the brain evolves and matures over time. Yet despite this long history of inquiry, we still do not completely understand how the developing brain is able to cope with myriad challenges over the life span and emerge as a fully functional adult brain.

The developing brain is an exemplar of an emergent, complex (nonlinear) dynamic system (Freeman 1994). Continual, ongoing changes in brain activity throughout organized neural tissue (i.e., brain dynamics) enable cognition, perception, and language to emerge over the course of development. These processes depend crucially on neural plasticity and are exquisitely sensitive to myriad environmental cues present during early brain development. However, the mechanisms which enable such transformation—from the establishment of early networks during the initial periods of brain growth and organization, to the reorganization and remodeling in early to late adolescence—have long been a matter of debate.

To explore these processes and underlying mechanisms, both event-related and spontaneous dynamics must be investigated. This, in turn, requires input from diverse research domains: nonlinear dynamics, artificial intelligence, and neural networks as well as all of the basic research and clinical fields that monitor and analyze the brain's activity, including genetic and biochemical parameters, using established and emerging techniques.

This Ernst Strüngmann Forum provided an opportunity to increase our understanding of the key components of brain dynamics at local and large-scale networks across development. To this end, experts from wide-ranging fields were invited to participate in a collective examination of how dynamic brain coordination and synchrony emerge during development. This included the possibility that abnormalities in neuronal synchronization and dynamic integration might be causal in developmental disorders (e.g., attention deficit hyperactivity disorder, language learning impairments, schizophrenia, autism spectrum disorder).

If you are unfamiliar with this institute, we offer a brief overview to lend understanding to this resultant volume. Dedicated to the continual expansion of knowledge in basic science, the Ernst Strüngmann Forum is an independent, science-driven entity that convenes "intellectual retreats" that are carefully crafted to permit in-depth scrutiny of problem areas in research. These are not meetings where one fluxes in and out of lectures or presentations: they are integral parts of an interactive process that creates synergy between experts from multiple areas—a process reliant on active engagement, geared toward the (re) conceptualization of pressing problems and the delineation of approaches to address and ultimately resolve these.

From March 5–10, 2017, forty experts participated in the 25th Ernst Strüngmann Forum. Each invitee brought to the discussion table the expertise needed to extend the exploration of dynamic brain coordination and synchrony, and their underlying mechanisms, from a developmental perspective. Each participant contributed to a specific working group, formed around the developmental stages of (a) fetal to birth, (b) childhood, (c) early adolescence, and (d) emerging adulthood. Interactions within and between groups ensured a cross-fertilization of perspectives and ideas. A set of overarching questions, formulated by the organizing committee, provided the starting point for discussion:

- What is the role of dynamic coordination in the establishment and maintenance of brain networks and of structural and functional connectivity?
- How are local and global functional networks assembled and transformed over normative development?

- To what degree do oscillatory patterns vary across development? Are there age-linked "oscillatory signatures" that can serve as maturational biomarkers?
- Do early developmental circuit vulnerability deficits in oscillatory domains have relevance for later developmental disorders?
- Is there a way to quantify the multiple intrinsic and extrinsic factors over development and how they may differentially impact early brain mechanisms including plasticity?
- How might emerging technology enhance early identification, diagnosis, prognosis, and remediation of developmental and neuropsychiatric disorders that may reflect early disruption in dynamic coordination and/or a failure to establish structural and functional connectivity and synchronization between cortical areas?

Specific topics were introduced by background papers. Written, read, and commented on in advance, these papers exposed open issues and posed questions for debate in the working groups. At the Forum, in-depth discussion was cultivated through formal and informal interactions. Throughout, consensus was never forced nor was it necessarily a goal. Instead, long-standing viewpoints were questioned, knowledge gaps were exposed, and novel insights began to emerge. As in any interesting discussion, there is always the danger that emergent, novel ideas might get lost. Thus, the essence of this discourse was captured by the groups in "reports" (see Chapters 4, 7, 11, and 15). These reports are not protocols of the discussions but rather summary statements of the ideas and issues brought forth by each group, from the perspectives of all involved. Their goal is to transfer the ideas, opinions, and remaining contentious issues, along with proposed directions for future research, to a broader audience.

This volume contains the finalized background papers and group reports from this Forum. Our purpose in this introductory chapter is to summarize the main themes covered in this collective examination and to highlight remaining gaps in knowledge and ideas proposed to further this evolving area of cross-disciplinary study.

Emergent Brain Dynamics: A Developmental Perspective

In any living system, effective developmental processes are needed to establish precise relations between the organism's components. In the developing nervous system, functionality depends on highly specific relations among individual neurons, established through biochemical and molecular signaling systems and the electrical activity of neurons.

In his review of the role of oscillations and synchrony in the development of the nervous system (Chapter 2), Wolf Singer discusses the importance of oscillations and their propensity to synchronize in the encoding of relations as

well as the singular effects that synchronization has on synaptic cooperativity
and plasticity. He offers insight into the particular challenges associated with
obtaining causal evidence of synchronous oscillatory activity and discusses
the use of emerging tools to manipulate selectively oscillating network components so as to interfere with synchrony without affecting discharge rates.
His review of the developmental mechanisms that translate temporal relations
among neuronal discharges into functional architectures provides an excellent
starting point for query into emergent brain dynamics across development.

Fetal to Birth

Prior to birth, the brain undergoes a substantive degree of development and
many of these early steps lay the foundation for later maturation. Studying the
developing brain in human embryos and fetuses is laden with challenges. In his
review, Alain Chédotal (Chapter 3) examines the role of axon guidance molecules in the regulation of cell–cell interactions during normative and atypical
development. He points out that although various methods exist to study axon
guidance in animals, technical and ethical issues pose significant challenges in
the developing human. To visualize and track neuronal connectivity successfully in pre-birth humans, Chédotal stresses the need for improved noninvasive
imaging methods (e.g., 3D and 4D obstetrical ultrasonography, 3D power
Doppler ultrasound, *in utero* MRI, and diffusion tensor imaging tractography)
and recommends development of novel methods. Our current ability to correct axon guidance defects or treat neuronal network dysfunction is severely
lacking, and existing surgical methods to improve the effects of monogenic
diseases (e.g., some forms of strabismus caused by congenital cranial dysgenesis) cannot generally be applied to more complex disorders. Chédotal suggests that we need to consider whether aberrant projections should be silenced
or the growth of new connections promoted. He also recommends that existing
observations be expanded to derive general rules of network construction and
developmental sequences.

   In their discussions, Nicholas Spitzer et al. (Chapter 4) asked: What is the
range of normal brain structure and function and how can dynamic changes be
studied over time? How constrained is the specification of cortical cell type
during development? Where does deterministic organization stop and environmental regulation take over? What happens to spatiotemporal patterns of
waves of activity during early cortical development? What are the patterns of
activity and their roles in developmental plasticity? What roles does activity
fulfill, and do changes in activity prefigure development of pathology?

   Spitzer et al. review the difficulties associated with quantifying the range
of normal brain structure and function and stress the value of using multiple dimensions. To determine the range of normality, they find it necessary
to know the range of variation in genes, cells, networks, and oscillations
combined with the impact on normal function and/or pathology. They stress

that brain development requires the maintenance of stability in some states while others are changing and converging, and propose homeostasis as a mechanism by which this is achieved. Further, early patterns of coordinated activity are necessary to assemble the necessary brain networks, whereas subcortical modulatory systems (e.g., cholinergic) seem to shape the activity in all cortical areas. Neurons in the upper layers coordinate the emergence of frequency-specific oscillatory rhythms whereas deeper layers seem to contribute to unspecific activation.

In their discussion of the role of activity, Spitzer et al. discuss transmitter switching: the loss of one transmitter and the gain of another, with corresponding changes in postsynaptic receptors to maintain synaptic function. Transmitter switching involves changes in the levels of transmitters, which seems to occur during psychiatric disorders, thus motivating studies of the role of transmitter switching in depression and schizophrenia. Putting a cell in a new environment may change its properties in ways that allow it to develop normally. Moreover, as Spitzer et al. point out, a lot of descriptive data from development is currently at our disposal, but data on mechanisms is lacking. Thus, inferences drawn from adult physiology and plasticity, while attractive, may be misleading.

With a view toward the future, Spitzer et al. suggest:

1. Using all findings achieved to date to construct conceptual frameworks and directions of experimental work should promote the most rapid progress.
2. Acquiring more data will be an important and necessary step to arrive at general principles by which dynamic brain coordination and dynamics is achieved during development.
3. The formulation of computer models for data analysis and testing circuit mechanisms at different scales should be a component in all research programs.

Further, they recommend that future research focus on

- clarifying the mechanism that governs the transition between discontinuous oscillatory activity during neonatal (mouse) or fetal (human) development and continuous rhythms at juvenile age,
- determining the functional role of different frequency oscillations observed at different ages,
- understanding the basis of developmental changes in cortical wiring,
- identifying the cellular and molecular mechanisms by which these patterns of activity exert their effects,
- exploring clinical diagnostics and treatment protocols that will likely follow from this knowledge, and
- including nonhuman primates and humans (via registry databases) in research, while retaining the mice model.

Early Childhood

After birth, the brain no longer develops solely as the result of internally driven mechanisms. External factors combine with internal mechanisms to shape the brain.

In Chapter 5, Takao Hensch discusses how neural circuitry is shaped by external factors at well-defined periods of time. Using amblyopia as a model of postnatal synaptic plasticity, he reviews the "triggers" and "brakes" that determine the onset and offset of such critical periods. He emphasizes that although the brain may retain the capacity to rewire later in life, adult plasticity may utilize distinct underlying mechanisms. Thus, understanding the differences between developmental and adult plasticity, including differences in how they are measured, is imperative and may enable insights into novel therapies for recovery of visual function from amblyopia in both children and adults. Genetic diversity in mice and humans may provide insight into individual variability and the timing of critical periods and should be pursued. Hensch calls for better models of critical period plasticity across animal species and humans and the identification of biochemical and electrophysiological correlates of these windows.

In Chapter 6, Sarah Moore and Michael Kobor review epigenetic mechanisms involved in the regulation of transcriptional potential. They describe how epigenetic mechanisms shape embryogenesis, neurogenesis and migration, neuronal plasticity, and impact critical windows of development, when neurons and circuitries may be sensitive to external stimulation. They explore the potential role of epigenetic marks as a biological consequence of early forms of social environmental adversity through a review of key animal studies and discuss the existing human literature that links early environments to epigenetic markers and neural structure and function. Despite the limitations in human studies, consistent findings that DNA methylation correlates with early social environment and neural phenotypes suggest that epigenetic marks capture meaningful variation in early environments as well as concurrent neurological measures and mental conditions. Moore and Kobor posit that epigenetic marks in human central and peripheral tissues reflect an important biological substrate of experience-dependent plasticity relevant to current mental health status.

The group discussions by Matthias Kaschube et al. (Chapter 7) focused on the following questions: How does neural activity evolve during early childhood? How does the continuum from internally to externally driven activity shape the brain? What is the role of critical periods? What are the factors that initiate and terminate these periods? How does critical period plasticity differ from adult plasticity? What is the interrelation between neural activity and epigenetics? What measurements and interventions do we have at our disposal? What are the signatures of typical and atypical development?

To navigate in and interact with their environment successfully, infants must construct a comprehensive and predictive internal model of the external world. Kaschube et al. explore the neural bases of this process and how resulting knowledge could be leveraged to treat and prevent neurodevelopmental disorders. They discuss how developing brains form dynamical networks that integrate genetic, epigenetic, and sensory information, and emphasize the interplay between molecules and neural activity. Further, they highlight strategies that the brain uses to tightly control the impact of sensory input onto its developing networks, which are manifest at the molecular, neural activity, and behavioral levels, and which appear pivotal as the brain strives to maintain a fine balance of flexible yet stable configuration. Stressing contributions made by animal models to our understanding of the neural basis of cognitive development, Kaschube et al. point out that in humans, behavioral assays and noninvasive imaging techniques provide only an indirect account of neural activity. Current knowledge of the developing epigenome of the brain is still very limited. They stress the urgent need to link animal and human studies and propose the following:

- Modern data acquisition, analysis, and computing methods should be used to integrate vast amounts of chronic data from a large number of individuals. This data is needed to advance development of high-dimensional formal statistical and dynamical models of typical and atypical neurodevelopmental trajectories.
- Computational efforts are needed to develop better dynamical and statistical neural circuit models, as well as to establish or adapt machine-learning tools to cope with molecular, neural, and behavioral data.
- Size and complexity appear to be key features for typical and atypical development in humans. Representing these features in animal models, beyond rodents, is an important, outstanding task.

Early Adolescence

Adolescence is characterized by maturation of reproductive and other social behaviors and social cognition. Although gonadal steroid hormones are well-known mediators of these behaviors in adulthood, their role in shaping the adolescent brain and behavioral development is not fully understood. Reviewing the impact of pubertal hormones on brain dynamics and maturation, Cheryl Sisk (Chapter 8) describes the organizational effects that pubertal hormones have on sex-specific behaviors during adolescence, as well as the neurobiological mechanisms of structural organization in the adolescent brain by pubertal hormones. To guide further study into the relationship between pubertal hormones, the adolescent brain, and experience, Sisk asks:

- What is the role of social experience in survival, differentiation, and functional incorporation of pubertally born cells?

- Is adolescence a sensitive period distinct from the perinatal period or part of an extended period of postnatal sensitivity for hormone-dependent organization?
- Is adolescent brain development experience expectant or experience dependent?

During adolescence sharp increases in neural plasticity occur as a result of, for example, sensorimotor experiences, stress, diet, drugs, cerebral injury, and the immune system. In Chapter 9, Bryan Kolb reviews different types of neural plasticity: experience-independent, experience-expectant, and experience-dependent. During adolescence he finds that most changes are due to experience-dependent plasticity, although he notes that experience-expectant plasticity (related to gonadal hormones or the increase in socioaffective behaviors) may also occur. Onset of the sensitive period begins around the time of pubertal gonadal hormone production. Offset may be related to the completion of myelination, but likely varies in different cortical regions due to the continuing impact of adolescent socioaffective experiences. Kolb highlights key questions for consideration:

- What is the role of glia in controlling the onset and offset of the sensitive period?
- How do experiences in adolescence vary with exposure age, what are the underlying mechanisms of their effects, and how might they influence brain plasticity later in life?
- What are the sex differences in the timing of the sensitive period and the role of experiences in altering brain plasticity during this period?
- How do changes in gene expression in adolescence influence the duration of the sensitive period?
- What is the role of the immune system in controlling the onset/offset of the sensitive period and synaptic plasticity during this period?

In Chapter 10, Patrick Purdon describes the pivotal role of gamma-aminobutyric acid (GABA) in the maturation of the cerebral cortex. GABAergic inhibition mediates crucial aspects of brain development: the development of structural connections, critical period plasticity, and functional synchronization across large-scale networks. Disturbances in the development of GABAergic circuits are thought to underlie neurodevelopmental disorders (e.g., schizophrenia and autism). Characterizing the developmental trajectory of these circuits in humans is crucial, yet current methods (postmortem studies and noninvasive imaging) provide only indirect links to GABA circuit function. To map the continuous trajectory of GABA circuit function from infancy through adulthood in humans, we ideally need a common set of tools as well as detailed measurements of sensory, cognitive, and language function.

Purdon suggests that studies using anesthesia-induced oscillations provide a way to characterize and track the development of GABAergic oscillatory circuits from childhood through adulthood. In both pediatric and adult practice, positive allosteric modulators of GABA receptors are used most commonly as anesthetic drugs. These drugs induce large, stereotyped oscillations in the unconscious state that are likely generated by the same GABAergic circuits responsible for gamma oscillations in the conscious state. Since these drugs are administered to tens of millions of patients each year, under conditions of both neurotypical and atypical development, Purdon argues this anesthetic experiment of nature could be harnessed and used to develop detailed developmental trajectories of GABAergic circuit function in humans.

In their group discussions, Marina Bedny et al. (Chapter 11) explore the time course and mechanisms of experience-based plasticity in early adolescence. They find that the potential for plasticity in the adolescent brain could theoretically follow one of three types of time courses: (a) the adolescent brain may be no more plastic than the adult brain, (b) adolescence could mark the end of critical periods that began in infancy, or (c) adolescence may constitute its own critical period. In view of current evidence, it may be possible that all three coexist in the human brain.

Each neurocognitive system has its own time course of development: some may be stable over the lifespan, others may begin their sensitive periods early in life and taper off in adolescence, while still others may have a specific critical period of sensitivity that spans adolescence (e.g., sensitivity to social stress). Further research is needed to uncover the time course of plasticity across neurocognitive systems and stages of development.

Bedny et al. also discuss the significant alterations that occur within local and large-scale networks. Changes in neural oscillations that continue throughout childhood and adolescence include

- restructuring neurophysiological synchronization among brain areas,
- reduction in overall power of the oscillation,
- reduction in lower-frequency oscillations ($<\sim 10$ Hz),
- acceleration of the peak frequency of alpha oscillation, and
- increases in higher-frequency oscillations ($>\sim 10$ Hz).

In task-dependent neurophysiological synchronization, these developmental changes may contribute to early cognitive and behavioral maturation; age-dependent increases in inter-regional synchronization during performance of a language task have been shown to correlate with individual differences in language abilities (Doesburg et al. 2016). Bedny et al. discuss the need to link markers of human brain development to changes in experience, and consider the associated challenges. They hold that research into visual cortex function in individuals who are congenitally blind demonstrate that the basic functional properties of cortical networks can change dramatically as a result of developmental experience. In addition, studies in individuals who became blind as adults suggest that the capacity of cortex to respond to changes in experience is qualitatively different in childhood and adulthood.

The Transition to Adulthood

The critical period between adolescence and adulthood marks the final stage of brain development prior to the attainment of the mature state. The neurobiological underpinnings of this transition have been notoriously difficult to characterize, thus creating numerous challenges in the diagnosis and treatment of psychopathologies that commonly emerge during this period (e.g., schizophrenia and affective disorders). Certainly, increased understanding will enable us to address and ultimately prevent these neuropsychiatric diseases. But beyond this, better knowledge of physiological changes, social relationships, and social cognition may lead to a novel conceptualization of this developmental period.

To understand why and how psychopathologies emerge during this period, we need to know the underlying biological vulnerability and mechanisms that confer risk. In Chapter 12, Peter Uhlhaas suggests that important modifications in brain coordination occur during the transition from adolescence to adulthood; these changes involve improved generation of rhythmic activity and its synchronization at low and high frequencies, as well as changes in the functional interactions between brain regions that underlie emotion regulation. Because they coincide with the emergence of brain disorders characterized by profound disruption of reality testing and emotional experience, they provide windows of vulnerability for the expression of dysfunctions that can then lead to behavioral anomalies. Important characteristics of these modifications are (a) the nonlinear trajectory of developmental changes and (b) the close relations to the underlying neurobiological parameters. In the adolescent brain, the transient reduction in large-scale synchronization of cortical networks and the accompanying increase of subcortical input may provide a condition that favors critical fluctuations. If these go beyond the critical threshold during the transition toward the adult state, the brain could remain in a faulty bifurcation and fail to accomplish the final development steps: (a) increase in the precision of synchronized, high-frequency oscillations, (b) integration of frontal and subcortical activity patterns, and (c) shift in the balance between local and global coordinated brain states. Uhlhaas describes the implications for the development of interventions designed to target large networks and brain coordination mechanisms.

Applying insights from research on critical periods in early development, Ulman Lindenberger (Chapter 13) outlines how plasticity can be researched throughout the entire life span. He posits that plasticity is triggered by a mismatch between the current range of functioning and experienced demands, and that it is characterized by inertia. A central nervous system in a permanent state of plasticity-induced renovation would not be able to develop a coordinated set of habits and skills, and would constantly drain a large amount of precious metabolic resources. Thus, mismatches in demand and supply must surpass a threshold of intensity and duration to trade the goal of stability for plasticity. This dynamic equilibrium shifts with age. To guide future research, Lindenberger proposes a set of hypotheses and stresses the following:

- We need to be aware of the almost ubiquitous and often unavoidable confound between age and experience whenever we wish to make claims about age differences in plastic potential.
- We need a better understanding of how age-based changes and between-person differences in large-scale network topography affect the context for local plastic change.
- We need a mechanistic account of the plasticity of higher-order cognition.
- We lack neural theories of generalization and transfer to predict consequences of plastic change.
- We need to better understand the relationship between brain size and neural efficiency.

In Chapter 14, Adriana Galván explores agents of change that impact neurobiological development in the adolescent brain. She begins with a discussion of the concept of "adolescence" and proposes that it be defined using neurobiological criteria, psychosocial responsibilities, and skill-based capabilities. This means that each parameter needs a clear operational definition, but which brain metric should be used, for example, to determine a "mature" versus an "immature" brain? Which skills are necessary for reaching maturation? Resolving these questions requires drawing on input across multiple disciplines. Galván discusses the prevailing neurobiological models of adolescent brain development and considers the impact that physiological changes (e.g., puberty, sleep), social relationships, and risk-taking have on adolescent brain development. She highlights promising areas for future consideration and posits a positive attribute to adolescent brain maturation. Empirical research has shown that the ontogenetic changes in the adolescent brain are adaptive. She encourages a view of the adolescent brain as "a sponge thirsty and receptive for new knowledge"; on the power of the adolescent brain to learn, to engage in prosocial behavior, and to explore the environment in a positive way.

Effective transitioning from adolescence to adulthood is a basic component of a functional society, and increased understanding of this developmental stage has far-ranging benefits—to individuals as well as society. In their group discussions, Jennifer Gelinas et al. (Chapter 15) describe how the numerous nonlinear modifications in the adolescent brain distinguish it from both the child and adult brain. Characterized by a predilection for specific forms of plasticity, these changes predominantly affect neural networks involved in higher cognitive and emotional processes. Gelinas et al. recommend that future research should

- take a multidisciplinary approach focused on the changing patterns of both physiologic and pathologic brain dynamics across adolescence;

- combine multiple research modalities and encourage dedicated and standardized initiatives to collect the relevant longitudinal studies in humans and animal models as well as computational models (e.g., of artificial neural systems); and
- assay the neurophysiologic processes of typical adolescent development and identify neural network level biomarkers and therapeutics for the neuropsychiatric diseases that characteristically emerge during this phase.

Such approaches may enable us to gain a more positive understanding of the adolescent brain as being one that is more adventurous, social, and cognitively mature than a child's, but not yet under the inevitable influences of senescence.

Final Reflections

Dividing a discourse into succinct periods of development (fetal to birth, childhood, early adolescence, transition to adulthood) to give structure to our debate, enabled focused discussion, but it requires us to step back and integrate perspectives. In this final section, we wish to highlight several areas for consideration.

As we think of plasticity and critical periods across brain development, we must bear in mind that it is not optimal to segregate them into specific time windows. Plasticity and critical periods may be active across the entire life span and should be newly defined, enabled by parallel and/or overlapping mechanisms, on smaller or larger timescales, in a nonlinear way, controlled by intrinsic and extrinsic factors. Critical periods previously closed (e.g., in childhood) may reopen at a later stage. Plasticity can be environmentally reopened through a two-step process involving (a) the reactivation of plasticity machinery (the permissive step) and (b) focused sensory experience to stimulate perceptual learning (the instructive step). As Hensch (Chapter 5) states, we need to identify the optimal sensory stimulation that drives change.

As research continues, it will be important to gather multidimensional information on cognition, genomes, epigenomes, molecular biology, neurochemistry, brain structure, and brain function at different points in human development. This information is needed to gain a rich description of the trajectory of development, and could be further enhanced through machine learning and formal statistical and dynamical models. With respect to functional brain dynamics, a focus on nonlinear and multi-scale changes across the life span is needed to determine network stability, network flexibility, and network plasticity.

Clearly many important questions were raised that are as yet unanswered, perhaps because the research has not or cannot be done in humans given the current state of technology and research techniques. Alternatively, we may not yet know the exact research questions to pose to unravel the many layers that still shield us from direct observation of the key neural mechanisms that underlie trajectories of neurophysiological maturation. One intriguing line of inquiry unaddressed here concerns the maintenance of flexibility beyond adolescence: Why do some people remain highly cognitively functional into old age, and what terminates functionality in others at the end of adolescence? Can cognitive flexibility be maintained and, if so, how? Can personality or cultural changes be explained or related to some network shifting or epigenetic factors? These broader questions await well-designed, inclusive longitudinal studies.

By fostering introspection, synergy, and expansion in research foci on dynamic coordination across development, we should be able to move forward to produce a lasting, meaningful impact across disciplines. Such an approach will ultimately facilitate new insights and formulate pragmatic research goals. We hope that both seasoned scientists and the emerging generation of cross-disciplinary minded young scientists will use the issues, collective ideas, and opinions raised here as a guide to further elucidate how the developing brain manifests across development as a mature, healthy, and fully functioning adult human brain.

2

The Role of Oscillations and Synchrony in the Development of the Nervous System

Wolf Singer

Abstract

To orchestrate the stepwise assemblage of building blocks in a living system, effective developmental processes are required to establish precise relations among the organism's components. Particular challenges exist for the development of nervous systems, as their functionality depends critically on highly specific relations among individual neurons. To establish precise connections among neurons, these challenges are met by using both molecular signaling systems and the electrical activity of neurons. Exploiting the exquisite sensitivity of synaptic modification rules for the precise timing of discharge patterns, the temporal correlation structure of both self-generated and environmentally induced activity is used to encode relations, thereby specifying the functional architecture of neuronal networks. Among the multiple mechanisms implemented to generate temporally structured activity, the propensity of microcircuit networks to engage in oscillatory activity plays a prominent role: network oscillations permit precise timing relations between discharges of distributed neurons to be established through synchronization, systematic phase shifts, and cross-frequency coupling. Developmental mechanisms are reviewed that translate temporal relations among neuronal discharges into functional architectures.

Relations Matter

Advances in the identification of genes and their products have revealed that the building blocks of living systems are strikingly similar. This implies that the often marked differences in the organisms' organization are essentially due to differences in the arrangement of components. A considerable amount of information specifying the idiosyncratic nature of organisms is thus contained in the relations between their rather stereotyped components. This raises the question of how information about these relations is encoded and read out during the organism's development. With the advent of whole genome sequencing methods, it soon became evident that the information which defines the future relations between the building blocks is stored in complex interaction networks constituted by coding and, in particular, noncoding genes. It is the extremely complex and still poorly understood dynamics of these reciprocal gene–gene interactions, together with epigenetic signals from their environment, that orchestrate the sequence of gene expression and ultimately define how the building blocks will be assembled. As development of the organism unfolds, the complexity and the origin of the environmental signals modulating this self-organizing process change. Initially, these signals are provided essentially by the constituents of the egg, but as differentiation proceeds, the environment-influencing gene expression expands to cell assemblies and ultimately the whole environment in which the developing organism is embedded. During early developmental stages, signals are mainly conveyed by physical contact among cells, membrane-bound recognition molecules, and diffusible messengers. Later, when neuronal systems come into play, electrochemical signals assure fast distribution of orchestrating signals throughout the developing organism over large distances and with high topological specificity. Ultimately, with the maturation of exteroceptive sensory systems, even distant environmental factors become influential for the developmental process. Thus, the information required for the development of an organism resides in an exceedingly complex network of relations that extends across multiple scales.

Definition of Relations in the Developing Nervous System

For obvious reasons, the precise definition of relations among components is particularly important and challenging for the development of the nervous system, because functions depend crucially on extremely complex and specific interactions among neurons. In addition, the functional architecture of neuronal networks provides the storage space for data and programs. Knowledge about the conditions of the world, acquired through evolutionary selection, is stored in the genome and expressed during development in the functional architecture of nervous systems. This architecture is subsequently refined by experience: after structural development has come to an end, network interactions continue to be modified by learning to modulate the efficiency of connections. Thus, the internal model of the world required for predictive coding and the programs for adapted behavior reside in the functional architecture of the brain, and hence in the idiosyncratic relations between neurons.

During the early stages of development, the signaling systems that control cell differentiation, migration, and contact formation resemble those supporting the formation of other organs: diffusible molecules and cell-specific surface markers. These signaling systems suffice to support the differentiation of the various brain structures, the coarse specification of the connectome, and the formation of protomaps. However, once neurons become electrically excitable and sensitive to synaptic input, the ability of nerve cells to convey signals over large distances with high speed and spatial selectivity is used to further support the development of the nervous system. This not only enables establishment of relations among distant maturation processes but also permits the unique computational abilities of neuronal networks to be exploited for the specification of developmental steps. This process has profound implications, as it permits (a) the selection of connections and the formation of maps with a precision that goes well beyond that attainable with the other signaling systems, (b) the optimization of developing circuits according to functional criteria, and (c) the adaptation of functional architectures to the actual conditions of the embedding environment.

The Role of Neuronal Activity in Development

Neuronal activity fulfills several functions in the context of the development and maintenance of neuronal architectures:

- Neuronal discharges and related synaptic activity are associated with the release, uptake, and transport of trophic signals necessary for the survival of neurons, the motility of cells and their processes, and the maintenance of synaptic connections.
- Activity has a normalizing function by regulating the number and efficiency of excitatory and inhibitory synapses that converge on a particular neuron, so that the average activity of the cell is kept within the optimal dynamic range.
- Temporal relations between discharge patterns of converging inputs are evaluated and used for the selective stabilization and disruption of connections.
- Neuronal activity controls the degree of myelination and thus conduction velocity of axons, as suggested by recent evidence (Elbaz 2016). Thus, activity plays a crucial role in shaping not only the topology of neuronal networks but also the temporal dimension of interactions.

In this chapter, I will focus on the third mechanism—the activity-dependent stabilization and disruption of connections—as this mechanism is crucial for the development of complex nervous systems and the realization of higher cognitive and executive functions. The fourth mechanism—the activity-dependent regulation of conduction velocities—may be equally important for the specification of network dynamics and ensuing functions, but research in this domain is still at the very beginning. Excluded from this review are also the numerous other developmental changes that target critical functions of neuronal networks and have no direct relation with synaptic plasticity and circuit formation: developmental changes in the subunit composition and spatial distribution of transmitter and voltage-gated membrane channels, as well as the extensive structural modifications of developing neurons (differentiation of dendritic and axonal ramifications). These variables play a crucial role in the spatial and temporal integration of signals and the homeostasis of cell excitability.

My reason for focusing on activity-dependent synaptic modifications is that the underlying mechanisms exploit network dynamics for the detection and encoding of relations and their translation into lasting changes of the functional architecture of the brain.

Hebbian Mechanisms: Evaluation of Relations and Their Translation in Network Architectures

Mechanisms which support activity-dependent circuit selection during development and use-dependent long-term modifications of synaptic gain, thought to underlie learning in the adult, share numerous similarities. Major differences are that during development, functionally weakened synaptic connections eventually get physically and irreversibly removed while the pool of connections available for selection is permanently replenished by newly formed connections. The initial steps, however, that serve the evaluation of relations and their translation in selective modifications of synaptic gain seem to be based on very similar molecular mechanisms and to follow closely the rules proposed by Donald Hebb for establishing permanent relations between frequently co-occurring and hence statistically related events. Hebb postulated that connections among neurons should strengthen if the coupled neurons are repeatedly active in temporal contiguity (Hebb 1949). This prediction received experimental support through the seminal discovery of long-term potentiation (LTP) in the hippocampus by Bliss and Lomo (1973). These authors found that tetanic stimulation of excitatory pathways led to a long-lasting enhancement of the efficacy of the synapses between the activated fibers and the respective postsynaptic target cells. Later, it was shown in multiple studies that this increase in synaptic efficacy occurred only if the postsynaptic cells were actually responding with action potentials to the tetanic stimuli, thus fulfilling the criterion of contingent pre- and postsynaptic activation. If postsynaptic cells were prevented from responding, modifications either did not occur or had opposite polarity; that is, they consisted of a reduction of synaptic efficacy. This phenomenon has become known as long-term depression (LTD). It is now well established that both modifications depend on a surge of calcium in the subsynaptic space of the postsynaptic dendrites and that the polarity of the modifications depends on the rate of rise, amplitude, and sources of this Ca increase (Bröcher et al. 1992b; Hansel et al. 1996, 1997). Fast and strong increases lead to LTP, whereas slow and smaller increases trigger LTD. Moreover, the source of the Ca increase is of importance. Calcium entering through N-methyl-D-aspartate (NMDA) receptor-associated channels favors the induction of LTP, whereas Ca entering through voltage-dependent Ca channels is more likely to trigger LTD. Moreover, secondary release of Ca from intracellular stores plays an important role in gating polarity and duration of gain changes (Cho et al. 2012). Both modifications, however, can be obtained merely by raising intracellular Ca concentrations through the liberation of caged Ca in a concentration-dependent manner (Neveu and Zucker 1996). A vast number of studies have been performed to elucidate the site of change (pre- or postsynaptic) and the molecular cascades that mediate the respective changes. It is now well established that modifications involve changes of both transmitter release and the number and sensitivity of postsynaptic receptors. These molecular approaches have led to a deep understanding of the extremely complex regulatory processes that translate neuronal activity into lasting changes of synaptic transmission (for a review, see Morishita et al. 2005).

It has long been held that the polarity of use-dependent synaptic gain changes depends on the extent to which pre- and postsynaptic activity is cor-related in time. The evidence that activation of NMDA receptors was one of the decisive variables agreed with this notion, because these channels function as coincidence detectors, becoming permeable for Ca ions only if glutamate is bound to the receptor and if the postsynaptic cell is sufficiently depolarized to remove the magnesium block (Nowak et al. 1984; Artola and Singer 1987; Kleinschmidt et al. 1987; Artola et al. 1990; Bear et al. 1990; Collingridge and Singer 1990). Since the level of depolarization of the postsynaptic membrane does not only depend on the activity of the local excitatory synapses, but also on all the other excitatory and inhibitory inputs, this mechanism also accounts for the cooperativity that characterizes use-dependent synaptic modifications. Even weak inputs can increase their gain if they are active in synchrony with other nearby excitatory inputs that contribute to depolarization and the removal of the magnesium block. With the advent of two-photon imaging technology it became possible to demonstrate *in vivo* that contingent activation of weak inputs converging onto the same dendritic branch could induce sufficient depo-larization to activate regenerative dendritic responses (Na and Ca spikes) and to induce LTP (Grienberger et al. 2015). Conversely, concomitant activation of inhibitory inputs can prevent even strongly activated inputs from depolarizing the postsynaptic dendrite above LTP threshold. In this case, presynaptic activ-ity that would normally induce LTP may either induce LTD or no change at all (Artola et al. 1990).

In conclusion, the net effect of these use-dependent synaptic modifications of excitatory connections include

- a strengthening of (reciprocal) connections among pairs of cells that are frequently activated in temporal contiguity,
- a strengthening of the gain of converging inputs that are frequently ac-tive in temporal contiguity,

- a weakening of connections among pairs of cells whose activity is un- or anti-correlated,
- a weakening of inputs active in contiguity with inhibition of the postsynaptic cell, and
- a weakening of connections that are inactive while the postsynaptic cell is strongly activated by other inputs (heterosynaptic depression).

In this conceptual framework, the crucial variable that determines the occurrence and polarity of synaptic gain changes is the temporal coherence (contiguity) of the activity of converging presynaptic inputs and/or the activity of presynaptic afferents and the depolarization of the postsynaptic neuron.

The notion that the occurrence and polarity of use-dependent synaptic modifications depend crucially on precise timing relations between the discharges of converging inputs received further support from the demonstration that postsynaptic spikes can backpropagate into dendrites, and that the ensuing depolarization also contributes to the gating of synaptic plasticity (Markram et al. 1997; Bi and Poo 1998; Stuart and Häusser 2001; but see Stiefel et al. 2005). Varying the timing between a single excitatory postsynaptic potential (EPSP) and the backpropagating spike revealed that small changes in the temporal relation have a massive impact on synaptic modifications. No changes occurred when the interval between the EPSP and the back-propagating spike was longer than about 50 ms. When the EPSP preceded the backpropagating action potential, the probability of obtaining LTP increased with decreasing delays; once the EPSP occurred after the backpropagating spike, there was a sharp transition toward LTD. The underlying mechanism is the same as detailed above. If the backpropagating spike occurs shortly *before* the EPSP, it can contribute to lifting the Mg block, allowing LTP to occur; if it arrives *after* the EPSP, the repolarizing currents prevent NMDA receptor activation, and LTD is the likely result. This special case of a use-dependent synaptic modification, known as spike timing-dependent plasticity, has an important implication: It suggests that synaptic changes may not only be sensitive to the coherence of converging activity but also to causal relations. The gain of excitatory connections increases if their activity can be causally related to the activation of the postsynaptic neuron and weakens when this is not the case.

Temporal Relations among Neuronal Discharges Signal the Degree of Relatedness

Consistent temporal relations between events signal relatedness. Simultaneously occurring events usually have a common cause or are interdependent because of reciprocal interactions. If one event consistently precedes the other, the first is likely the cause of the latter; if there are no temporal correlations between the events, they are most likely unrelated. The learning rules adopted by evolution exploit these relations, thereby permitting internal models of the world to be generated that have considerable predictive power—a likely reason for the striking conservation of the mechanisms supporting use-dependent modifications of synaptic transmission.

The fact that the *learning rules* are exquisitely sensitive to *temporal relations* has several important consequences for the way nervous systems process information and attribute significance to temporal relations. One requirement is that the precise timing relations between events in the environment are reliably encoded in neuronal responses to permit learning of correct associations. This requirement is met in all sensory modalities by the implementation of transmission chains, commonly referred to as "phasic systems," which operate with high temporal resolution and accuracy. *In vivo* recordings from higher visual areas as well as the auditory and the somatosensory cortex revealed that the discharges of individual neurons signal the temporal structure of stimuli with extreme precision in the millisecond range. This proves that precise timing of discharges can be preserved despite numerous intervening synaptic transmission steps (Buracas et al. 1998; Reinagel and Reid 2002). Simulation studies, partly based on the concept of synfire chains proposed by Moshe Abeles (1991), confirmed that conventional integrate-and-fire neurons are capable of transmitting temporal information with the required precision (Mainen and Sejnowski 1995; Diesmann et al. 1999).

Additional mechanisms are required, however, when selective associations have to be established between neurons that represent features which lack temporal structure and just give rise to sustained responses, or when a particular set of neurons (out of many simultaneously active neurons) have to become selectively associated because they code for contents that should be bound together. These types of mechanisms are most likely required for the segmentation of stationary scenes or when associations have to be formed among contents stored in memory, as is the case during reasoning. To accommodate this, one possibility would be to implement mechanisms of synaptic plasticity that are insensitive to the relative timing of inputs but establish associations according to other than temporal criteria. To the best of my knowledge, such mechanisms have not yet been described. An alternative and parsimonious solution is to impose a temporal structure on neuronal responses that satisfies the contingency requirements of the classical plasticity rules and to utilize the existing mechanisms also for the association of signals that initially lack temporal structure. Responses that become associated would have to be made temporally coherent, whereas those that do not should remain uncorrelated.

Neuronal mechanisms responsible for the generation of temporally structured activity are diverse, abundant, and already implemented in simple nervous systems. A common and highly conserved strategy is the oscillatory patterning of activity, the basic principle of parsing time, used in virtually all clocks. Certain neurons are endowed with pacemaker currents that support oscillatory discharge patterns (Heyer and Lux 1976; Gray and McCormick

1996). Networks which function as central pattern generators produce rhythmic activity for the coordination and synchronization of a large variety of effector systems (Grillner 2006). Neuronal networks endowed with reciprocal connections among their nodes engage spontaneously in temporally coordinated activity that manifests itself in traveling waves (Meister et al. 1991; Ermentrout and Kleinfeld 2001; Plenz and Thiagarajan 2007), especially during development when neurons often interact synaptically as well as through gap junctions. Finally, most neuronal networks share the motive of recurrent inhibition, which endows them with the propensity to engage in oscillatory activity, whereby the frequency of these oscillations is determined by the various time constants of the interacting elements and the conduction velocity of the coupling connections. Prominent examples are the septo-hippocampal circuits which generate the theta rhythm (Buzsáki 2006), the thalamocortical interactions responsible for the alpha rhythm (Steriade et al. 1993), and the cortical microcircuits which generate the gamma oscillations known as ING and PING circuits (Kopell et al. 2000; Börgers and Kopell 2008; for a review, see Buzsáki et al. 2013). These mechanisms provide ample opportunities to impose temporal structure on neuronal activity and to establish precise temporal relations among discharges. These relations can then be converted by the established mechanisms of time-sensitive synaptic plasticity into selective modifications of functional architectures.

The exquisite sensitivity of plasticity mechanisms for precise timing relations also constrains strategies for information processing in general. This implies that information about the relatedness of neuronal responses be expressed in precise temporal relations among the discharges of neurons. Indeed, there is ample evidence for mechanisms that render responses temporally coherent (if they need to be bound together) or which make them uncorrelated (if they represent unrelated contents). Most of these mechanisms are based on an oscillatory patterning of activity, and the resulting option is to synchronize responses with variable phase lags and across different oscillation frequencies for the definition of relations. These mechanisms are considered relevant for signal processing and the dynamic coordination of distributed neuronal processes in the context of feature binding, scene segmentation, the formation of Hebbian assemblies, selective routing of signals, interareal communication, and the specification of functional networks. These aspects of temporal coding have been discussed in numerous reviews (Singer 1999; Buzsáki 2006; Fries 2009; Uhlhaas et al. 2009a; von der Malsburg et al. 2010; Buzsáki et al. 2013) and will thus not be considered further here. Also excluded from review are mechanisms that influence and shape the sequence order of discharges at longer timescales (e.g., synfire chains) and more global synchronized fluctuations of excitability that occur, for example, during alternations between up- and down-states.

The Role of Temporally Structured Activity
in Embryonic Development

As soon as neurons become active, temporal relations between their firing patterns are exploited by developmental mechanisms to guide the selection of connections. A well-examined example involves the refinement of topological maps in the visual system. Because of the cooperative nature of the synaptic modification rules, afferents that convey well-correlated activity mutually support their consolidation on common target cells and repress afferents whose activity is less well correlated—neurons wire together if they fire together. The correlation structure of activity used to select circuits results from widely differing mechanisms, and these change during development. At early stages, before interactions with the environment become important, correlated discharge patterns result from interactions among spontaneously active neurons; that is, from self-generated activity. Slowly oscillating burst activity in thalamic and cortical structures leads to synchronized discharges of local clusters of neurons (Yang et al. 2009b), and in the retina, traveling waves cause sequential, highly synchronized volleys of activity whose sequence order or phase offset reflects precisely the neighborhood relationships among ganglion cells (Meister et al. 1991). While research on the role of the slow burst activity is still in an exploratory phase, evidence indicates that the traveling waves, through the synchronized discharges at the respective wave front, are causally involved in the formation and refinement of topographical maps. First proposed by Kohonen (1982) and Willshaw and von der Malsburg (1976), the neighborhood relations of the array of ganglion cells are encoded in the correlation structure of the activity of axonal projections and then reconstituted by selective stabilization and disruption of connections in the respective target structures such as the optic tectum, the lateral geniculate body, and to some extent the visual cortex (Penn et al. 1998). Although coarse maps have already formed by matching gradients of diffusible and membrane-bound molecules (Bonhoeffer and Gierer 1984), the refinement of topological correspondence required for high-resolution vision is achieved only through this activity-dependent sorting of axons. Self-generated activity patterns in the two eyes are—as far as is known—uncorrelated and thus the clustering of afferents from the same eye and the segregation of afferents from different eyes can be supported by activity-dependent pruning. Blocking retinal activity reduces clustering of afferents in the lateral geniculate nucleus and cortex into eye-specific domains and segregation of the afferents from the two eyes. However, there is a possible confound that is difficult to resolve: blocking activity altogether may also interfere with the transport and uptake of eye-specific marker molecules.

It is likely that temporally structured self-generated activity also plays a role in refining circuits responsible for the coordination of movements, because embryos exhibit coordinated movements at early stages of maturation. Here, however, evidence for causal relations is still sparse. Currently it is unknown to which extent traveling waves and synchronous low-frequency oscillations contribute to map formation in nonvisual sensory systems and the refinement of connections between subcortical nuclei or cortical areas. As demonstrated by Khazipov et al. (2004) and Minlebaev et al. (2011), correlated activity patterns occur in the somatomotor system, already at early stages of development, and likely contribute to the shaping of functional architectures. Because the mechanisms mediating use-dependent modifications of synaptic transmission are ubiquitous, it would be surprising if activity-dependent selection of circuits is restricted to the model systems investigated so far, but the critical experiments have not yet been conducted. Occasional observations indicate, however, that the capacity of central structures to adapt to their respective afferent input is astounding. In a routine scan of one of our subjects—a 14-year-old girl—we discovered that the subject lacked one cerebral hemisphere, including the striatum and the entire thalamus (Muckli et al. 2009). To our great surprise, the girl exhibited only minor distal apraxia of the hand, contralateral to the missing hemisphere, but had a close to normal visual field. Closer examination with structural scans and fMRI mapping of retinotopic representations revealed, however, that both the nasal and temporal retina of the ipsilateral eye projected to the intact hemisphere, and that two complementary retinotopic maps had developed in the primary visual cortex: one for the ipsilateral, the other for the contralateral hemifield. These maps met at the representation of the vertical meridian. Apparently, the normally crossing axons from the nasal retina were rerouted to the ipsilateral hemisphere because their natural target was missing, and the nasal and temporal afferents from the ipsilateral eye got sorted to form two independent maps that covered the two hemifields, respectively. To which extent these drastic rearrangements were mediated by molecular markers or structured activity patterns from the retina is unfortunately unresolved.

A recent study on the development of the entorhinal-hippocampal circuitry provides direct evidence for the crucial function of activity in promoting sequential maturation of functional networks. If activity in the respective feeding structures is blocked, the maturation of all elements constituting the subsequent processing stages is jeopardized (Donato et al. 2017).

The Role of Correlated Activity Patterns in Postnatal Development

Once sensory systems become responsive to stimulation from the outer world, the correlation structure of activity in sensory pathways reflects the correlation structure of (a) events in the embedding environment and (b) characteristic features of perceptual objects. The developmental mechanisms described above utilize these additional sources of information to refine microcircuits and maps and improve the inherited internal model of the world that is stored in the brain's functional architecture.

Results from lesion and deprivation experiments clearly indicate that the formation of sensory maps is influenced by the activity conveyed by sensory afferents. This has been shown for the somatosensory system (Jenkins et al. 1990) and the somewhat special variant, the cortical representation of the whiskers in rodents (Jeanmonod et al. 1981; Petersen et al. 2004). Of particular relevance in the present context are findings that artificial synchronization of proprioceptive signals from different fingers or whiskers led to the fusion of the cortical representations that are normally well segregated. This suggests that the correlation structure of afferent activity is used to associate or segregate the territories occupied by the respective axonal arborizations.

Very similar results have been obtained in the visual system. For binocular vision and stereoscopic depth perception, it is imperative to assure that those afferents converge on common cortical target cells that convey signals from precisely corresponding loci in the two eyes. Genetic instructions alone do not suffice, because correspondence depends on variables such as interocular distance and size of the eyes—variables which are subject to epigenetic influences and cannot be anticipated with precision. For this reason, the correlation structure of afferent activity is again used for the refinement of circuitry. By definition, afferents from corresponding retinal loci convey precisely correlated activity when the two eyes are fixating an object. Thus, selective stabilization of these afferents will establish precise retinal correspondence. However, selection of afferents must be confined to epochs when the two eyes are actually fixating. This is one of the reasons why many experience-dependent refinement processes are supervised. In this particular case, synaptic plasticity is gated by nonretinal signals from extraocular muscles (Buisseret and Singer 1983) and additional "now print" systems capable of evaluating the adequacy of retinal activity in a more global behavioral context (for a review, see Singer 1995). This important issue will be addressed in more detail below. If activity from the two eyes is consistently de-correlated, as occurs with strabismus or monocular deprivation, afferents from different eyes compete with one another and a winner-take-all mechanism destroys binocular convergence.

From the visual cortex, causal evidence also indicates that correlated sensory signals play a role in shaping the dense network of recurrent tangential connections, which link feature-selective neurons with one another (Löwel and Singer 1992). In the mature cortex, these connections are particularly dense between columns that share similar functional properties, such as eye dominance and orientation preference (Gilbert and Wiesel 1989). However, when visual experience is withheld, this bias does not develop and connections distribute randomly across functional domains (Löwel and Singer 1992). This observation can be accounted for if one assumes that (a) connections stabilize selectively between columns that have a high probability of being active simultaneously and (b) the structure of the visual world favors coherent activation of columns responding to similar features. Both prerequisites are fulfilled. As demonstrated in numerous slice experiments, tangential connections are endowed with Hebbian synapses that obey the "fire and wire together" rule and "orientation" columns with similar preference—in particular if they have collinearly aligned receptive fields—have a high probability of being simultaneously activated, because elongated contours are prominent in natural scenes (Kayser et al. 2003). In essence, these findings indicate that the contingencies of features in the outside world get translated during development into the weight distributions of the myriads of connections that link neurons with corresponding feature preferences. Because these graded and selective weight distributions reflect the statistics of the outer world, it has been proposed that they serve as priors in predictive coding and, in fact, correspond to the Gestalt criteria applied for figure-ground segregation, perceptual grouping, and feature binding (Singer 2013; Singer and Lazar 2016). Strongly coupled neurons engage more likely in synchronous firing than weakly interacting neurons (Schillen and König 1991). Thus, as previously proposed (Singer and Gray 1995; Singer 1999), readout of the "learned" binding criteria would occur through enhanced synchronization of neurons coding for features that are related (e.g., features that co-occur in frequently observed objects). If experience is withheld during the critical period of development in which these architectural changes are induced, cognitive functions become irreversibly impeded. The likely reason is that sensory signals cannot be interpreted adequately if the priors reflecting meaningful relations among features have not been installed by experience-dependent shaping of intercolumnar connections. This interpretation agrees with reports of patients suffering from early-life visual deprivation, who have problems with figure-ground segregation and feature binding.

Although other sensory systems have been studied less extensively, it is very likely that their development follows the same principle and that the priors representing the statistical contingencies in the respective sensory environment become internalized by use-dependent adaptation of cortical circuits. Obvious examples are provided by the development of auditory functions, such as the acquisition of kin-specific songs in songbirds and the learning of the mother tongue in children (see Kaschube et al., this volume). In both cases, priors are installed that permit automatic parsing of sound streams into syllables, and there are critical periods for the development of these abilities (Tchernichovski et al. 2001; Ortiz-Mantilla et al. 2016). This suggests that the acquisition of the respective priors goes along with circuit changes that are only possible during development. In the case of auditory stimuli, the temporal patterning of activity is already inherent in the rhythmicity of the utterances, but it is unclear how exactly this information is used to fine-tune circuits for appropriate chunking of sounds.

It is not too surprising that activity-dependent development and learning rely on similar mechanisms because in both cases, relations between distributed neurons must be established, specified, and modified. For memory formation in the adult, the modifications of the interaction architecture seem to be confined to gain changes of synaptic transmission, whereas during development, such gain changes lead to either consolidation or disruption of connections, and thus to irreversible changes of anatomical architecture. In both cases, however, the critical variable determining the polarity of the respective changes appears to be the correlation structure of the activity displayed by the interacting elements—not only the level of activation but also the precise temporal relations among activation patterns.

The Need for Supervision

As mentioned above, in the context of the development of precise binocular correspondence, selection processes that rely on temporal correlations are often supervised. Activity-dependent changes of connectivity are permitted only when certain additional conditions are fulfilled. Experimental observations and theoretical arguments suggest that such gating of activity-dependent developmental processes may be a general strategy. Evidence from the visual system indicates that activity, even when it is strong and well structured, may alone not be sufficient to support use-dependent plasticity and circuit changes. The seminal experiment by Held and Hein (1963) demonstrated that development of normal visual abilities requires active exploration; passive exposure to a visual environment is insufficient. Likewise, when animals are anesthetized or eye movements abolished, ocular dominance changes do not occur in response to monocular stimulation, even when cortical neurons are strongly driven by the visual stimuli (Buisseret et al. 1978). Circuit modifications also fail to occur in freely behaving animals when the retinal signals are in conflict with inbuilt visuomotor reflexes; for example, when the retinal coordinates are rotated (Singer et al. 1982; for a review, see Singer 1995). These findings suggest that central gating systems can permit or veto activity-dependent circuit modifications. Some systems, whose activation is required for the induction of experience-dependent circuit changes, have been identified. These are the cholinergic, noradrenergic, and serotonergic projections originating in the brain stem and basal forebrain (Bear and Singer 1986; Gu and Singer 1995). The dopaminergic projections, prime candidates for the supervision of use-dependent synaptic plasticity, have not been investigated in this context. The more recent findings of a strong enhancement by locomotion of responses, even in primary visual cortex, might account for the early finding of Held and Hein (1963) that exploratory behavior is a necessary prerequisite for experience-dependent maturation of visual functions.

Using neuronal activity to optimize the development of architectures has obvious advantages: it permits the information-processing capacities of nervous systems to be exploited for their own development. This option is used not only to adapt sensory and motor systems to the constraints of the environment but also to coordinate developmental processes that assure cooperativity between the various subdivisions of the brain. One example is the adaptation of the auditory map to the visual map in the tectum (Knudsen and Knudsen 1989). Whether this strategy is applied ubiquitously to establish correspondence, and to which extent self-generated patterned activity may play a role, is still little explored.

Despite the obvious advantages of using experience (i.e., signals) from the environment to guide developmental processes, there is a price to be paid: abnormal activity patterns can have deleterious effects on development, as demonstrated by the often severe consequences of deprivation. One way to minimize these risks is to permit circuit changes to take place only when the respective activity patterns have been identified as consistent or appropriate in a more general context. Studies on the experience-dependent development of the visual system suggest that consistency criteria for sensory signals could be correct predictions about the effects of eye movements or locomotion on retinal image slip or the congruence of signals conveyed by different sensory systems.

How distinctions are made between internally generated activity patterns which should or should not induce circuit changes is unknown. This question is intimately related to the equally unresolved riddle of how the brain "knows" when it has arrived at a solution—or in other words, how activity patterns which result from computations during the search for a result differ from those representing the result. For development as well as for signal processing and learning, the brain needs to prevent spurious activity from changing neuronal architectures. As mentioned above, gating functions that prevent inappropriate activity from inducing changes in circuitry seem to be realized by globally organized modulatory systems whose activity facilitates synaptic plasticity, and thereby serves as a "now print" signal. These systems are activated as a function of arousal, attention, and reward expectancy (Tobler et al. 2005) and have been shown to facilitate activity-dependent synaptic modifications. They either control dendritic depolarization directly, by modulating the conductance of ion channels, or indirectly, by regulating the excitability of inhibitory networks. In addition, they act through metabotropic receptors on the second messenger cascades that mediate long-term changes of synaptic transmission. Through these complex actions, modulatory systems can veto or permit synaptic modifications as well as adjust the set points of synaptic plasticity. As *in vitro* and *in vivo* studies suggest, a given pattern of activity can lead to transient adaptation, long-term depression or potentiation, depending on the state of the modulatory systems. However, just how these modulatory systems are informed about activity constellations that warrant synaptic changes is by and large unknown.

Gating of Synaptic Plasticity by Local Circuit Dynamics

Considering the activity requirements for the induction of synaptic modifications, it is likely that gating of plasticity is not only mediated by globally organized modulatory systems but also by local computations. As summarized above, a favorable condition for the induction of circuit changes is strong dendritic depolarization: it results from cooperativity among converging excitatory inputs and/or coincidence between pre- and postsynaptic discharges. These two conditions are particularly well fulfilled if neuronal activity is synchronous. Since synchrony is, in turn, enhanced when neuronal groups engage in oscillations, there could be a relation between the occurrence of an oscillatory patterning of activity and plasticity, in particular in high-frequency oscillations, as these lead to high-precision synchrony. The prediction is that circuits which engage in synchronous high-frequency oscillations should be particularly susceptible to undergo use-dependent modifications of synaptic transmission. In actuality, there is a close correlation between the state of plasticity-enhancing modulatory systems, the propensity of microcircuits to engage in synchronous oscillations, and the occurrence of synaptic modifications. Synchronized high-frequency oscillations in the gamma-frequency range are facilitated by acetylcholine (Munk et al. 1996; Herculano-Houzel et al. 1999)—one of the neuromodulators proven to enhance synaptic plasticity *in vitro* (Bröcher et al. 1992a) and memory formation *in vivo* (Letzkus et al. 2011). High-frequency oscillations are also enhanced by attention (Fries et al. 2001), an important variable that controls learning processes. Thus, it is likely that synchronous oscillations contribute to the gating of synaptic plasticity, by means of enhancing the coincidence of discharges (cooperativity).

Direct support for this conjecture comes from experiments that relate the occurrence of oscillations to use-dependent modifications of neuronal response properties. The receptive fields of neurons in the visual cortex can be modified by appropriate visual stimulation, even in anesthetized preparations, if the brain is concomitantly activated by electrical stimulation of the mesencephalic reticular formation (Singer and Rauschecker 1982). Reticular stimulation increases the release of plasticity-enhancing neuromodulators while, at the same time, favoring the occurrence of gamma oscillations in response to the applied stimuli. Post hoc analysis of the neuronal responses to change-inducing light stimuli revealed that lasting changes in receptive field properties (in this case, orientation preference) occurred only in response to stimuli associated with a strong oscillatory modulation and synchronization of neuronal responses in the gamma band. Changes consisted in a shift of orientation tuning toward the stimuli used for conditioning. In the absence of gamma oscillations, the cells preserved their initial tuning properties although still vigorously driven by the light stimulus. However, in this case, the cells became less responsive to the change-inducing stimuli: they showed adaptation (Galuske, pers. comm.). A dependence of synaptic plasticity on oscillations, albeit in the theta-frequency range, has also been found in the hippocampus (Huerta and Lisman 1995). Here, the so-called theta-burst stimulation that entrains the hippocampus in the characteristic theta rhythm turned out to be particularly effective for the induction of long-lasting synaptic modifications. Other evidence for an instrumental role of synchronized oscillations in use-dependent changes of synaptic transmission comes from research on memory formation (Fell et al. 2001; Tallon-Baudry et al. 2001, 2004). In human subjects implanted with depth electrodes for the localization of epileptic foci, it was found that successful formation of episodic memories was accompanied by transient increases in gamma- and theta-oscillatory synchrony between the hippocampus and neighboring entorhinal cortex, structures known to be involved in memory formation. In trials in which memory formation was not successful, these increases in synchronization were not observed (Fell et al. 2001, 2003, 2011). Likewise, simultaneous recordings from limbic structures (amygdala and hippocampus) have shown that fear conditioning is associated with transient synchronization of oscillatory activity between the two structures (Seidenbecher et al. 2003; Narayanan et al. 2007; Liu et al. 2012a; Igarashi et al. 2014; Yamamoto et al. 2014). Finally, studies on memory consolidation during sleep have revealed correlations between an oscillatory patterning of neuronal activity and memory formation. Memory consolidation during sleep has been reported to be enhanced following induction of slow oscillations with direct-current stimulation in human subjects (Marshall et al. 2006).

If synchronous oscillations facilitate use-dependent synaptic modifications, they could indeed serve as local gates and enable activity to induce changes if this activity meets certain criteria of consistency. There are some indications that patterns elicit strong gamma oscillations if they match the prewired response properties (priors) of local cortical networks, if their structure is sufficiently regular to allow for predictions across different parts of the pattern, and if the stimulus is expected (Lima et al. 2011; Vinck and Bosman 2016; for a review, see Singer and Lazar 2016). Thus, synchronous oscillations (or, in other terms, high temporal coherence) could be signatures of consistent states that are worth being reinforced by synaptic modifications.

These considerations of mechanisms that relate neuronal plasticity and oscillatory activity have recently experienced an unexpected twist. Two independent studies indicate that the entrainment of circuits in narrow-band gamma oscillations triggers specific signaling systems that act upon the microglia and the extracellular matrix. Iaccarino et al. (2016) showed that entrainment of neuronal networks in synchronized narrow-band gamma oscillations reduces synthesis and accelerates degradation of Aß, the plaque-generating polypeptide in Alzheimer disease. These effects on metabolic processes were mediated by the activation of microglia and were surprisingly frequency specific. Entrainment of oscillations was ineffective in higher- or lower-frequency bands. As reported by Takao Hensch (see Kaschube et al., this volume), entrainment of cortical circuits in narrow-band gamma oscillations also has the unexpected effect of rendering adult visual cortex susceptible again to use-dependent synaptic plasticity that shares all characteristics of critical period plasticity. This effect appears to be mediated by changes in the extracellular matrix.

These new findings suggest the possibility that abnormalities in the ability of networks to generate well-synchronized narrow-band gamma oscillations impede developmental processes, such as pruning and stabilization of connections, due to the effects that reduced and imprecise synchrony has on Hebbian mechanisms as well as its direct effects on metabolism, glial cells, and the extracellular matrix. At present one can only speculate as to why there seems to be something special about gamma oscillations. When networks engage in well-synchronized narrow-band gamma oscillations, parvalbumin-containing inhibitory interneurons—the basket cells—exhibit a drastic change in their discharge pattern: they emit a burst of spikes in each cycle and show an overall increase in activity. Pyramidal cells, by contrast, show little change in overall activity because they skip cycles and do not burst. Their spikes become concentrated around the depolarizing peak of the oscillations, but their discharge rate does not signal engagement in an oscillatory process. The activity of the basket cells—the pacemakers of the gamma-generating PING circuit (Whittington et al. 2000)—is thus a reliable indicator for the transition of a network in the gamma mode. If, as suggested above, entering the gamma-processing mode is associated with an increased likelihood of ensuing synaptic modifications, entering the gamma mode would predict enhanced metabolic demands. During gamma oscillations, basket cells are likely to experience a massive influx of Ca ions, and this could be the trigger for the generation of metabolically relevant signals. Unlike pyramidal cells, they contain, for example, NO synthase, and nitrous oxide is a very potent diffusible messenger for a host of downstream processes. Support for this admittedly speculative scenario comes from the finding that entrainment of cortical circuits in gamma oscillations is particularly effective in increasing the hemodynamic response (Niessing et al. 2005).

Concluding Remarks and Outlook

Despite intensive research on the functional role of oscillatory activity and the concomitant synchronization of discharges in neuronal processing and its ubiquitous occurrence from the early stages of development throughout the rest of life, there is still only sparse evidence for a causal role of these phenomena in brain development. However, given the importance that relations play in defining mechanisms in development, the central role that oscillations and their propensity to synchronize play in the encoding of relations, and the special effects that synchronization has on synaptic cooperativity and plasticity, it seems highly likely that entrainment of networks into synchronous oscillatory activity plays an important role in the development of the nervous system.

Obtaining causal rather than solely correlative evidence for this conjecture is, however, notoriously difficult. Oscillations are nearly as fundamental a property of neuronal activity as discharge rate, and the two variables are closely intertwined. Hence, manipulating oscillatory patterning without simultaneously interfering with discharge rate is almost impossible. Still, this approach will have to be taken to obtain more direct evidence for a causal role of oscillatory activity in promoting developmental self-organization of neuronal networks.

Tools are now available that should permit us to selectively manipulate oscillating network components and to interfere with synchrony without affecting discharge rates. Still, the epistemic mandate to provide causal evidence may well encounter insurmountable methodological hurdles which may need to be relaxed as we investigate these extremely complex systems that exhibit nonstationary, nonlinear high-dimensional dynamics. The classical approach of manipulating an *independent* variable and investigating the consequences is likely to fail because of inherent circularities. Thus in certain domains of systems neuroscience, we may have to be content with correlative evidence, inductive reasoning, arguments of plausibility, and consistency of simulation results. This epistemic challenge is not unique to brain research. Other domains in the natural sciences face similar problems, either because it is impossible to manipulate the independent variables, as in evolutionary anthropology or cosmology, or because the systems are simply too complex and nonlinear, as in the Earth sciences and climatology.

Fetal to Birth

3

Molecular Guidance and Cell-to-Cell Interactions in Intrauterine Brain Construction during Typical and Atypical Development

Alain Chédotal

Abstract

Our understanding of the etiology of axon guidance disorders as well as our ability to correct axon guidance defects or treat neuronal network dysfunction is limited. Surgical methods currently employed to improve some forms of strabismus cannot, for example, be readily applied to more complex disorders, although experimental neurosurgery for neuropsychiatric disorders can now successfully target thalamocortical networks. Should aberrant projections be silenced or should the growth of new connections be promoted? This chapter examines the role of axon guidance molecules in the regulation of cell–cell interactions during normative and atypical development. It discusses how this affects the formation of neural circuit connections (normal and pathological) and posits what types of experiments and novel tools are needed to explore these processes. It is recommended that these observations be expanded to derive general rules of network construction and developmental sequences.

Introduction

In vertebrate embryos, developing organs undergo dramatic changes in size, shape, and cellular constitution. This is particularly striking in the central nervous system (CNS) where glial and neuronal cells are born at a distance from their final location. After undergoing their final division, postmitotic neurons migrate through a highly complex and changing cellular and molecular environment. Concomitantly, most neurons extend an axon that will have to find its appropriate target cells among the billions of neurons that constitute the CNS. This is a particularly daunting task for axons which form point-to-point connections on specific compartments (e.g., cell bodies, dendrites, spines) of one or a few distant target cell(s). Specificity is lower for aminergic axons that extend and branch throughout the CNS, although with variable density between brain regions (Chédotal and Richards 2010). The task appears easier for most interneurons which synapse with partner neurons located in their vicinity (Ascoli et al. 2008).

Since the end of the nineteenth century, studies have shown that the processes of neuronal migration and axonal elongation are not random but precisely orchestrated by cells and molecules distributed in the developing CNS (Dickson 2002; Valiente and Marín 2010). In this chapter, I address the role of axon guidance molecules in the development of neuronal networks and their possible involvement in neurological diseases.

A Brief History of Axon Guidance Molecules

The existence of axon guidance cues in the developing CNS was postulated by Ramón y Cajal at the end of the nineteenth century. This hypothesis was based on the observation of a polarized growth of dorsal spinal cord axons toward the ventral midline or floor plate (Ramón y Cajal 1892). It was later shown that in all species with bilateral symmetry, one of the first decisions that newborn neurons make is to project their axons to target cells located either on the ipsilateral side or on the opposite (or contralateral) side (Chédotal and Richards 2010; Chédotal 2014). Axons crossing the midline are called commissurals; they represent a paradigm for the analysis of axon guidance mechanisms. Over the last 25 years, genetic and biochemical studies have identified a variety of axon guidance molecules in multiple families of secreted or membrane-bound proteins. These guidance cues either promote or inhibit/repel axon outgrowth by acting on the stabilization of the growth cone, the motile structure found at their distal tip (Tessier-Lavigne and Goodman 1996; Dickson 2002).

The most studied axon guidance proteins belong to four protein families whose structure and function have been reviewed extensively. *Semaphorins* (with more than 20 members in mammals), secreted or membrane bound, bind to neuropilin and plexin receptors, respectively. *Slits* are secreted and bind to Roundabout (Robo) receptors and some proteoglycans. *Ephrins* are membrane-bound ligands of Eph receptor tyrosine kinases, but signaling is bidirectional: Ephs can act as receptors (or co-receptors) for ephrins and vice versa. *Netrins* comprise soluble and membrane-bound proteins related to laminins. The founding member, netrin-1, has diverse receptors such as deleted in colorectal carcinoma (DCC) and Unc5s (Unc5a–Unc5d). Semaphorins, Slits and ephrins/Ephs are primarily repulsive for axons unlike netrins, which can be either attractive or repulsive depending on axon types or developmental stages.

Importantly, many unrelated proteins can also guide axons in addition to the "canonical" ones. Many are immunoglobulin superfamily members, such as Down syndrome cell adhesion molecules (Yamagata and Sanes 2008; Dascenco et al. 2015; Alavi et al. 2016), draxin (Islam et al. 2009; Shinmyo et al. 2015), and L1-related IgCAMs (Castellani et al. 2000; Ango et al. 2004; Chauvet et al. 2007; Huang et al. 2007). They can also be morphogens, such as Sonic hedgehog (Charron et al. 2003; Okada et al. 2006), members of the Wnts/planar cell polarity pathway (Lyuksyutova et al. 2003; Liu et al. 2005; Zhou et al. 2008; Shafer et al. 2011; Chai et al. 2014), and bone morphogenetic proteins (Butler and Dodd 2003). Other notable ones are the repulsive guidance molecules (Monnier et al. 2002; Rajagopalan et al. 2004), some neurotrophins (Lumsden and Davies 1983; O'Connor and Tessier-Lavigne 1999; Park and Poo 2012), homeobox-containing proteins (Brunet et al. 2005; Sugiyama et al. 2008), chemokines (Zhu et al. 2009), and even lipids (Guy et al. 2015).

This is a non-exhaustive list and new molecules are still to be found. Importantly, recent studies indicated that posttranslational modifications, such as glycosylation (with potential sugar codes), modify the activity of axon guidance proteins (Conway et al. 2011; Blockus and Chédotal 2012; Wright et al. 2012). Axon guidance gene splicing has also been described and increases their structural diversity (Chen et al. 2008; Colak et al. 2013). Along this line, up to more than 350,000 combinations of clustered protocadherin ectodomain isoforms might exist (Zipursky and and Sanes 2010; Rubinstein et al. 2015). These proteins, which exhibit isoform-specific homophilic binding, were shown to play a role in neuronal self and nonself recognition (Lefebvre et al. 2012). Protocadherins also control axon guidance (Uemura et al. 2007; Leung et al. 2013; Hayashi et al. 2016) and some have been associated with neurological diseases such as epilepsy (Nabbout et al. 2011; Aran et al. 2016).

Recent studies show that growth cones integrate multiple guidance signals and that this combinatorial action might have a synergistic or antagonistic outcome (Bielle et al. 2011b; Lokmane et al. 2013; Poliak et al. 2015; Sloan et al. 2015; Morales and Kania 2016). A plethora of *in vitro* and *in vivo* data show that axon guidance molecules control the targeting of axons from long projection neurons (including aminergic ones) and interneurons throughout the nervous system. There is also evidence in the neocortex that clonally related excitatory pyramidal neurons within a column are preferentially interconnected, but the underlying guidance mechanism (if any) is unknown (Li et al. 2012). Whether this lineage-driven connectivity pattern applies to cortical interneurons is still under debate (Harwell et al. 2015; He et al. 2015; Mayer et al. 2016; Sultan et al. 2016).

Notably, axon guidance molecules are pleiotropic and control cell–cell interactions during tangential and radial neuronal migration, angiogenesis, and immune response, among others. Therefore, it would be simplistic to expect that axon guidance disorders result only from mutations or risk variants in axon guidance genes.

Cellular Sources of Axon Guidance Molecules

In the mammalian central nervous system, axon guidance cues are produced by a variety of neural and nonneural cell types. Midline glia cells localized at the floor plate in the midbrain, hindbrain, and spinal cord (Tessier-Lavigne and Goodman 1996; Bashaw et al. 2000; Chédotal 2011; Neuhaus-Follini and Bashaw 2015); the indusium griseum and glial wedge in the forebrain (Suárez et al. 2014); or the optic chiasm in the diencephalon (Kuwajima et al. 2012) are all major sources of signals that may attract precrossing commissural axons and repel ipsilateral and postcrossing axons. Radial glial cells, such as in the optic tectum (Drescher et al. 1995; Monnier et al. 2002), play a role in axon guidance.

More recently it has been found that transient corridors for growing axons are established at specific locations, such as the basal forebrain and corpus callosum, by migrating neurons, which express specific guidance cues for thalamocortical and callosal axons, respectively (López-Bendito et al. 2006; Niquille et al. 2009; Bielle et al. 2011a). Axon–axon interactions also play an important role to promote the fasciculation of follower axons and pioneer ones, as well as to interconnect neurons coming from distinct locations. One of the most classic examples is the so-called "handshake" between cortico-thalamic and thalamocortical axons (Molnár et al. 1998; Mandai et al. 2009; Deck et al. 2013). Notably, axons also express cues that guide migrating neurons. This happens, for instance, in the case of (a) olfactory and vomeronasal axons, which are followed by neurons secreting gonadotropin-releasing hormone (Messina et al. 2011; Casoni et al. 2016; Cariboni et al. 2012), and (b) some spinal cord ventral interneurons, which are guided by commissural axons (Laumonnerie et al. 2015).

Other types of cells also influence axon guidance. Meningeal cells produce chemokines such as SDF1/CXCL12 which influence the growth and migration of some hindbrain cortical neurons (Zhu et al. 2002, 2009; Borrell and Marín 2006). Netrin-1 and endothelins are produced by the vasculature and guide sympathetic axons innervating vessels in the periphery (Makita et al. 2008; Brunet et al. 2014), and this might also be the case in the CNS where vascular endothelial growth factor was already shown to pattern commissural projections (Erskine et al. 2011; Ruiz de Almodovar et al. 2011). Finally, microglia which invade the CNS at early embryonic ages (E9.5 in mice) appear to accumulate first at specific choice points for some axonal tracts (Squarzoni et al. 2015). Dopaminergic and callosal axons as well as cortical interneurons are misrouted following microglia depletion (Pont-Lezica et al. 2014; Squarzoni et al. 2014).

Altogether these results suggest that intrinsic and extrinsic factors perturbing the development of cells expressing axon guidance cues could indirectly alter the development of neuronal connectivity.

Axon Guidance, an Intrauterine Process

In mammals, most neuronal networks are built during embryonic development. In the mouse CNS, the first axons are born at embryonic day 8 (E8) (Mastick and Easter 1996) around Carnegie stages 11–12 (CS11–CS12). This is equivalent to 23–30 postconception days (E23–E30) in humans (Rhines and Windle 1941; Humphrey 1944; O'Rahilly and Müller 1987). These pioneer neurons appear in the hindbrain and diencephalon following a developmental sequence that is largely conserved in all vertebrates.

In the human neocortex, cells expressing neuronal markers, called predecessor neurons (Bystron et al. 2006), were described as early as E33, before the initiation of cortical neurogenesis and are therefore suspected to originate from outside the cortical anlage. Interestingly, a recent study shows that in the postnatal mouse, meningeal-derived cells could generate cortical neurons (Bifari et al. 2017). The first postmitotic pyramidal neurons reach the cortical plate around E50, and extrinsic axons, including thalamocortical axons, enter the intermediate zone around the cortical plate (Larroche 1981; Bystron et al. 2006, 2008). The corpus callosum, the largest commissural tract in the CNS, is detectable as of gestation week 11 (GW11) and its size increases until after birth (Rakic and Yakovlev 1968). The corticospinal tracts (CST), the longest axonal tracts in the CNS, reach the spinal cord at CS23 (E56–E60) and their decussation is completed at GW15 (Eyre 2000, 2003; ten Donkelaar et al. 2004). They reach the lumbosacral region at the caudal end of the spinal cord by GW29 and contact motor neurons by GW37. Therefore, in humans, axonal development almost exclusively occurs during intrauterine life, with the noticeable exception of cerebellar granule cell interneurons, two-thirds of which are produced postnatally (Kiessling et al. 2014). This is not the case in rodents. In rats, for example, CSTs just reach the spinal cord at birth and their caudal growth proceeds at least until postnatal day 16 (P16) (ten Donkelaar et al. 2004).

Importantly, in vertebrates, including humans, the size of the CNS continues to expand well after axons have contacted their targets. It is estimated that the weight of the brain increases 40-fold between the end of the embryonic period and birth (O'Rahilly and Müller 2008). This "noncanonical" axonal growth, also known as stretch growth (Weiss 1941), involves mechanical forces (Franze 2013). Experimentally, axons can be forced to elongate at a speed of 400 μm/hr for at least two weeks (Pfister 2004; Heidemann and Bray 2015). Interestingly, in the fish lateral line, some sensory axons are towed by their target cells as they migrate (Gilmour et al. 2004). Mounting evidence suggests that mechanical forces and tension also influence axonal growth and guidance before axons contact their targets (Athamneh and Suter 2015; Polackwich et al. 2015). In the developing *Xenopus* retinotectal pathway, growing ganglion cell axons appear to respond, via piezo1 ion channels, to mechanical signals and probe the stiffness of the surrounding tissue (Koser et al. 2016). However, the molecular mechanisms that control mechanical axon growth and guidance and their possible contribution to neurological diseases are largely unknown (Budday et al. 2014). These observations raise an important question: How can axonogenesis and axonal tract development in human embryos/fetuses be technically studied *in utero*?

How Can We Study Axon Guidance in Humans?

Although a multitude of genetic and imaging methods can be used to study axon guidance in animal models, specific technical and ethical issues make this extremely difficult in human embryos and fetuses. Most studies are based on postmortem brains and incomplete analysis of a limited number of samples and tissue sections, in which axons are labeled with silver staining or immunostaining using only a few axonal markers, such as GAP43.

Moreover, most neuropsychiatric diseases are only diagnosed well after birth, and it is therefore difficult to link them to anomalies of axon guidance. This demonstrates the need for novel or improved imaging methods, in particular noninvasive ones, to visualize and follow the intrauterine development of neuronal connectivity in humans.

Important progress has been made in the noninvasive medical imaging of embryos and fetuses *in utero* during pregnancy to detect congenital anomalies and malformations. This now includes three- and four-dimensional obstetrical ultrasonography, which can generate holographic images of the embryo (Kurjak et al. 2005; Pooh et al. 2011; Baken et al. 2015) but mostly provides information about surface features and cavities. Likewise, 3D power Doppler ultrasound was used to visualize the embryo vasculature (Weisstanner et al. 2015). *In utero* magnetic resonance imaging (MRI) also provides a good appreciation of the development of the CNS (Weisstanner et al. 2015) in the fetus, and diffusion tensor imaging (DTI) tractography is now used as a prenatal diagnostic of callosal dysgenesis as early as GW20 (Jakab et al. 2015).

Validation of these *in utero* 3D data is challenging and problematic, as it currently relies on postmortem evaluation of histological sections.

A variety of tissue clearing techniques, such as Clarity and 3DISCO, have been developed over the past few years, and using them in combination with whole-mount immunostaining and light sheet fluorescence microscopy allows high-resolution three-dimensional images of adult mouse brains and embryos to be generated. This method, now adapted to human embryos and fetuses (Belle et al. 2017), should help us obtain a better understanding of the time course and characteristics of axon development before term in normal and pathological cases. It will also be useful in interpreting and validating the *in utero* images obtained using noninvasive methods.

Current Evidence Supporting the Developmental Origin of Some Neurological Disorders: Intrauterine Axon Guidance and Neuronal Migration Defects in Patients

The developmental origin of various monogenic diseases, with dominant or recessive inheritance, has been demonstrated (Blockus and Chédotal 2015; van Battum et al. 2015). This is the case for congenital cranial dysgeneses (Assaf 2011; Nugent et al. 2012), which are primarily due to a lack or mistargeting of oculomotor nerves and cause strabismus and other eye movement disorders. In albino patients, binocular vision is altered due to a significant reduction of the size of the ipsilateral contingent or retinal ganglion cell axons (Guillery and Kaas 1973; Neveu and Jeffery 2007). Likewise, abnormal corticospinal tract and corpus callosum decussation have been described in patients suffering from congenital mirror movements (Izzi and Charron 2011). *NTN1*, *DCC*, and *RAD51* (involved in DNA repair) are the three known causal genes (Srour et al. 2010; Depienne et al. 2011, 2012; Meneret et al. 2015). Patients suffering from the horizontal gaze palsy with progressive scoliosis syndrome (HGPPS) display a severe loss of commissural connections, including the CST and lateral lemniscus (Jen et al. 2004; Chédotal 2014; Zelina et al. 2014). All HGPPS patients carry autosomal recessive mutations in the *ROBO3* gene which in mammals encodes a transmembrane receptor involved in commissural axon attraction (Marillat et al. 2004; Sabatier et al. 2004; Zelina et al. 2014). Interestingly, Robo3 knockout mice completely lack commissures in the midbrain, hindbrain, and spinal cord, but axons that fail to cross the midline still connect to their proper target, albeit on the wrong side of the brain (Renier et al. 2010; Badura et al. 2013). Other diseases that have a clear axon guidance basis are corpus callosum dysgenesis/agenesis (Paul et al. 2007; Edwards et al. 2014; Suárez et al. 2014).

How then can we demonstrate that abnormal intrauterine neuronal guidance is involved in the etiology of polygenic and complex diseases, such as autism spectrum disorders (ASD), schizophrenia, bipolar disorders, and other psychiatric diseases, which are often linked to multiple genetic risk variants? Some genetic studies have identified mutations or single nucleotide polymorphisms (SNPs) in genes encoding axon guidance molecules (either ligand or receptors), suggesting that abnormal brain wiring might contribute to those diseases. For example, genome-wide association studies have shown that rare mutations in *PLXNA2* (a gene on chromosome 1q32 which encodes a semaphorin receptor) could contribute to schizophrenia in some individuals, although this is still under debate (Mah et al. 2006; Fujii et al. 2007; Allen et al. 2008; Ripke et al. 2013). A few studies show a possible involvement of *ROBO3* and *ROBO4*

(Anitha et al. 2008; Suda et al. 2011) and *SEMA5D* in ASD (Melin et al. 2006; Weiss et al. 2009) and *ROBO1* in dyslexia (Hannula-Jouppi et al. 2005; Lamminmaki et al. 2012).

Interestingly, the *DCC* gene was also associated with schizophrenia and adolescence-related psychiatric diseases and suicidal behavior (Grant et al. 2012; Manitt et al. 2013). Finally, polymorphism and the identification of certain SNPs in axon guidance genes (e.g., *DCC*, *EphB1*, *SEMA5A*, *SLIT3*) might even predispose to Parkinson disease (Lesnick et al. 2007; Lin et al. 2009).

Although these data suggest the existence of a complex and precise genetic program for building neuronal networks, there is also evidence for axon guidance errors (Hutson and Chien 2002; Poulain and Chien 2013), stochastic events, and activity-dependent regulation of axonal development (Mire et al. 2012; Hassan and Hiesinger 2015). Thus the following questions should be addressed:

- How plastic is the system?
- Given evidence of extreme abnormalities in developing systems (e.g., Muckli et al. 2009; Hoffmann et al. 2012; Warner et al. 2015), how does the system accommodate intrauterine axon guidance errors or brain malformations?
- What role does timing play in the ability to reorganize?
- Are some circuits more plastic than others?
- How much interindividual variability exists in axonal connectivity?

Using iPSCs and Organoids to Study Axon Guidance

It will not be easy to obtain direct *in vivo* evidence linking axon guidance and neuronal migration defect and neurological disorders, because most of these developmental processes occur *in utero* and are currently unable to be assessed directly in humans using existing noninvasive imaging techniques. Moreover, neurological and psychiatric disorders are diagnosed postnatally, often years after developmental errors have occurred. News tools that would allow the recapitulation of normal and pathological brain development *in vitro* could provide important insights (Suzuki and Vanderhaeghen 2015; Quadrato et al. 2016).

In less than ten years, two major technical advances have completely revolutionized our ability to study the etiology of complex neurological diseases: somatic cell reprogramming and CRISPR/Cas9-mediate genome editing (Jinek et al. 2012; Cong et al. 2013; Doudna and Charpentier 2014; Hsu et al. 2014; Shi et al. 2017). Human-induced pluripotent stem cells (hiPSCs) have already been derived from normal individuals and patients suffering from various neurological and neuropsychiatric diseases, and a collection of differentiation protocols allow many different types or neuronal and glial cells to be produced (Yoon et al. 2014). Using CRISPR/Cas9, mutations of candidate genes can be introduced in control cells to test their pathogenicity and in patient cells to correct risk alleles.

The ability of hiPSCs-derived neurons to migrate and extend axons on various substrates can be easily studied and compared (Brennand et al. 2011) in classic 2D cultures, but this will not tell us much about the ability of those axons to find their targets in a complex cellular environment, like the embryonic brain. Importantly, each differentiation protocol produces, most often, a limited number of neuronal types, and the usual target cells are likely to be absent from the cultures.

Recently, organoid models have emerged that show the expansion, differentiation, and self-organization of hiPSC-derived cells. Such approaches can generate eye cups containing a pigmented epithelium and a simple retina with multiple neuronal types (Eiraku et al. 2011; Reichman et al. 2014), cerebellar (Muguruma et al. 2015), hippocampal (Sakaguchi et al. 2015), and forebrain organoids (Paşca et al. 2011). Although more complex and self-patterned cerebral 3D organoids have been produced (Li et al. 2017), the reconstitution of long-range projections circuits (e.g., CSTs, thalamocortical or nigrostriatal pathways) in hiPSC-derived miniature brains still appears beyond our reach (not even considering behaviors).

Grafting iPSC-derived neurons or organoids into the brain of animal models appears to be an interesting option, as shown for the dopaminergic system (Hargus et al. 2010; Korecka et al. 2016). Unless this is possible *in utero*, the grafted cells will develop in an environment quite different from their normal one. This disparity, in turn, could influence cellular growth potential and the ability to reach their target cell and integrate in a circuit, although there is evidence that some of the cues are still present. Despite these limitations, some recent results using embryonic derived stem cells (Michelsen et al. 2015) support the potential of this strategy for understanding the role of axon guidance disorders in the etiology of neuropsychiatric diseases.

Disruptors of Axon Guidance during Intrauterine Life

What is the evidence for disruption of axon guidance, and what factors might be involved? As discussed above, in humans most of the axon guidance process occurs *in utero* during the first semester of gestation. However, some deleterious agents (viruses or bacteria) and molecules (e.g., alcohol) that are able to pass through the placental barrier (Syme et al. 2004) can perturb the development of axonal connectivity.

Psychoactive drugs, such cannabinoids, are another candidate disruptor: they can regulate serotonin transporter (SERT) activity in the placenta, thereby influencing the clearance of serotonin (see below). Also, by acting on cannabinoid receptors, they can change Slit/Robo signaling (Alpár et al. 2014). A recent study showed that using SERT inhibitors for treating depression during pregnancy might also perturb the development of the enteric nervous system and contribute to gastrointestinal disturbances that accompany ASD (Margolis et al. 2016). Likewise, valproic acid, which is sometimes used in the treatment of epilepsy, influences axon outgrowth (Tashiro et al. 2011; Lv et al. 2012; Yang et al. 2012).

Conclusion

Our understanding of the etiology of axon guidance disorders is far from complete, and attaining this will not be easy. Currently, our ability to correct axon guidance defects or treat neuronal network dysfunction is severely lacking. Surgical methods being used to improve some forms of strabismus cannot be applied to more complex disorders. Thus we need to consider whether aberrant projections should be silenced or the growth of new connections promoted.

4

Fetal to Birth

Nicholas C. Spitzer, Terrence J. Sejnowski,
Yehezkel Ben-Ari, Alain Chédotal, Gordon Fishell,
Ileana L. Hanganu-Opatz, Suzana Herculano-Houzel,
Heiko J. Luhmann, Hannah Monyer, and Michael P. Stryker

Abstract

Prior to birth, the brain becomes highly developed, and many early events lay the foundations for later maturation. This chapter begins with a focus on the range of normal brain structure and function, with consideration given to how dynamic changes over time can best be studied. It then explores the extent to which the specification of cortical cell types is constrained during development, followed by a review and discussion of what happens to spatiotemporal patterns of waves of activity during early cortical development and their roles in developmental plasticity. Consideration of the central role of activity in organizing the developing nervous system prompted us to ask how changes in activity prefigure development of pathology. Key conclusions and future directions are summarized at the end of this report.

All theories are wrong but some are useful (after George E. P. Box).

If you torture the data long enough, it will always confess (after Ronald H. Coase).

What Is the Range of Variation of Normal Brain Structure and Function, and How Should We Study the Dynamic Changes Over Time?

We often report averages and jump to general conclusions when we analyze brain structure or function, lumping variation in the numbers of neurons, the sizes of structures, and the numbers of fibers and physiological attributes.

Group photos (top left to bottom right) Nick Spitzer, Terry Sejnowski, Suzana Herculano-Houzel, Yehezkel Ben-Ari, Hannah Monyer, Gordon Fishell, Michael Stryker, Heiko Luhmann, Ileana Hanganu-Opatz, Terry Sejnowski, Michael Stryker and Alain Chédotal, Gordon Fishell and Nick Spitzer, Suzana Herculano-Houzel, Nick Spitzer and Terry Sejnowski, Heiko Luhmann, Ileana Hanganu-Opatz, Gordon Fishell, Yehezkel Ben-Ari, Hannah Monyer, Alain Chédotal, Michael Stryker

Does the range of the data have implications for function? Is the range during development correlated with outcomes postnatally? Because some of this variability is likely to be relevant, we need to learn the normal range. The first rigorous quantification of the number of neurons in the adult human brain—86 billion—came from analysis of the number of neurons in four adult males (Azevedo et al. 2009).

For example, mutant mice have different defects in their patterns of axon guidance, visualized using whole animal imaging, that are likely to be functionally significant (Belle et al. 2014). However, individual sensory innervation and branching patterns overlap only to a limited extent between the right and left hands in every human embryo tested (Belle et al. 2017), and in this case, the variability in axon guidance may be functionally irrelevant. Variations across species are also highly relevant: mutation of *Robo3* leads to the absence of commissures between bilateral pre-Bötzinger nuclei, and mice, but not humans, die at birth (Jen et al. 2004; Bouvier et al. 2010). In another case, preconceptual and gestational forms of stress lead to larger and smaller brains, respectively, but performance is poor in both cases (Kolb, this volume); furthermore, initially there are unusually large numbers of neurons in the autistic brain. These findings argue that brain size is not directly correlated with ability. In line with this view, there can be twofold differences in the numbers of neurons in a cortical area without an obvious effect on the overall function of the brain.

Is it useful to recommend specific criteria for determination of normality? Detecting the range of variation in a single dimension is not enough. Comparison of numbers across individuals and within individuals (e.g., on both sides of the brain) will be best achieved using multiple criteria. The number and identity of such criteria are likely to differ when assessing different aspects of normality. Particularly useful are measures of brain structure, brain function, and the gain or loss of particular neurons, since some cells are more impactful than others (e.g., hubs versus outliers or pre-Bötzinger neurons regulating respiration). Unfortunately, causality has frequently been inferred from single, inadequately parameterized events. In China, more education has been associated with a lower incidence of Alzheimer disease (AD) (Zhang et al. 1990). In the United States, in the "Nun Study," cognitive reserve was interpreted to be an important determinant in the onset of AD (Snowdon et al. 1996) even when formal education was modest.

It is crucial to inspire investigators to be more aware of the significance of understanding the range of normal variation. We also need to develop new techniques for stereology to accelerate the rate of quantification of anatomical data. Larger sample sizes will make conclusions more robust, and machine learning will now enable accumulation of these larger samples. The Allen Brain Atlas data are based on only six human donors at this point, although one-third of genes are still consistently expressed across 33 gene areas. Interestingly, some aspects of brain development are stochastic: examples include map formation in the visual system (Owens et al. 2015) and the ratio of red-to-green cone pigments (Wang et al. 1999).

Light sheet microscopy is an important, recently developed technology for rapid analysis of brain structure; using 3DISCO/iDISCO to clarify the tissue, it allows the range of variation in nervous system architecture to be determined (Belle et al. 2014, 2017; Renier et al. 2014; see Figure 4.1). The standard 3×5 cm specimen size is sufficient for analysis of the embryonic and fetal human brain or adult rodent brain. The adult human brain has to be cut into 5 cm slices, imaged, and individual images then stitched together. Resolution is at the level of single axons and spines. One person can process 100 mouse brains per week, so analysis of 3000 lines of mutant mice is feasible. For example, it becomes easy to count the number of neurons expressing a given transcription factor, such as Foxp2, in the whole brain. The ability to see the whole brain allows one to find neurons easily and visualize all their projections in 3D. At present, it is possible to examine three or four markers at one time and it should be possible to do more in human tissues using monoclonal antibodies. This procedure has been successfully applied to birds, reptiles, and other animals. It allows the collection of raw data unhampered by hypotheses or conjectures, thus creating the opportunity for other investigators to analyze the data from their own perspective. One is led to wonder whether serial block face electron microscopy, presently underway in many labs, will be capable of measuring variability that is in the noise? There are now 120 light sheet microscopes

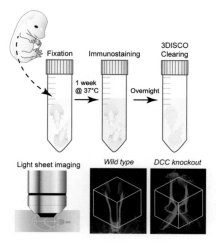

Figure 4.1 A simple procedure combines immunolabeling, solvent-based clearing, and light sheet fluorescence microscopy for rapid high-resolution whole-brain neuroanatomical analysis. This technique allows large-scale screening of axon guidance defects and other developmental disorders in mutant mice (Belle et al. 2014). Reprinted with permission from Alain Chédotal.

around the world, stitching software has developed rapidly, and data compression facilitates handling large files. Activity-dependent markers such as c-fos and Arc can be used as readouts of function. Among other benefits, these observations provide information about where to interrogate cellular or circuit function.

Although stereology is good for small, well-demarcated samples, the isotropic fractionator ("brain soup") technology (Herculano-Houzel and Lent 2005) is another method that facilitates more accurate counts and higher throughput. Dissection and dissociation of as many as a dozen tissue samples per day, collection of nuclei, and staining for nuclear markers make it simple to count cell numbers. Less training is required and the method is useful for developmental studies. Other methods include recording with new calcium (Ca)indicators and large-scale multiple electrode recording over long periods of time, combined with selective manipulation of distinct neuronal populations that can now be accomplished in freely behaving animals. Currently missing and urgently needed is a generalizable method for tracking gene expression in single neurons during development. All these methods create new ways of collecting, studying, and presenting the data. It is now possible to study multiple parameters in the same brain and in the same neurons over time.

Although we are becoming richer in techniques, conceptualization remains poor. *Phenotypic checkpoints* constitute an example of a useful concept (Ben-Ari and Spitzer 2010). The idea is that developmental expression of a gene generates a transient phenotype, which is necessary for further normal gene expression. If the phenotype is not generated, further development is arrested or proceeds along an abnormal pathway. As hard-headed scientists, we are reluctant to engage in overgeneralization of our findings, and this may account for our hesitancy to formulate developmental rules. One may proceed, however, by elaborating concepts on the basis of simple experiments and then test their validity with more complex studies. Even though knockouts can be uninformative and solid information about phenotype is required, it is possible to identify a phenotype that is consistent and work backward. Increasing evidence supports the relevance of phenotypic checkpoints for development (Donato et al. 2017).

The gap between knowledge and understanding arises in part because we do not know the underlying neurobiology. To study and determine the range of normality, we need to know the range of variation in genes, cells, networks oscillations *combined* with impact on normal function, and/or pathology. To this end, the Free University of Amsterdam operates a brain analysis program linked to euthanasia, wherein ca. 100 brains are processed and scans acquired per year. Over time, the program aims to establish relationships between individual experiences and brain structure. In addition, registries in Scandinavian countries attempt to link birth and life experience to brain outcome.

How Constrained Is the Specification of Cortical Cell Type during Development?

Determining Cell Types

How do we determine the different cell types? One view is that all that matters is input, intrinsic properties, and output. In the past, cells were often classified by their receptive field (input) alone. In artificial networks, this is not sufficient: one needs to know the projective field (output); that is, where axons project. Another view is to determine how cells cluster by shared properties. As noted by Wolf Singer, "you have to bundle them somehow." For example, parvalbumin (PV) cells from different regions have more in common than PV cells and pyramidal neurons. However, we may not yet have identified all the cells' properties.

Further discussion led to the conclusion that classification is arbitrary and there is an unlimited number of ways to classify. Classification may be fluid; for example, the bursting properties of thalamic neurons depend on the state of the membrane potential. The same problem arises in discussion of the number of cortical areas in the brain, classified by function, functional potential, and other criteria. Classification should serve a purpose, and different purposes will be served by different classifications. It is important to avoid "theological" approaches—getting stuck on a particular view. There is no inherent ground truth. In the retina, one classifies the same cells into the same groups using a variety of measures. Is this unusual or atypical? Perhaps we should start with this as the null hypothesis until it can be shown that it does not work.

The values of classification include having a common language for discussion, getting genetic access for manipulations (although classification should not be made using a single tool), increasing replicability of experiments, and comparing cell types across animals and within animals.

Determining the Origin of Cell Types

How does cell type specification occur? We considered excitatory pyramidal neurons and inhibitory interneurons, remembering that evolution did not need to have names for cell specification. The source of neuronal properties is a key question. Sydney Brenner famously referred to the American plan (in which fates are determined by the environment) and the European plan (in which fates are determined by lineage). We discussed invertebrate neurons (and included neurons in the sensory and motor peripheral vertebrate nervous system, since they seem to share properties with invertebrate neurons) and contrasted them with vertebrate CNS interneurons and pyramidal cells.

Transcription factor codes are established through genes and checkpoints that provide a constraint on cell type (Lieberam et al. 2005). Lineage is predictable and determines cell fate in invertebrates in some cases. Interestingly, in flies, cuticle progenitors are equipotential and can assume different fates (Lawrence 1973), potentially determined by positional cues. Overproduction and apoptosis appear to create more options in vertebrates. Moreover interneurons projecting to dendrites versus cell body have different synaptic properties, and properties of synapses depend on innervation. Lineages are not determinate in the retina, spinal cord, and cortex. For interneurons, generally, lineage does not seem to determine fate, but it appears to be more important for pyramidal neurons. It is useful to think of cardinal specification of interneurons as the initial program; definitive properties then result from position and activity. Interestingly, single cell transcriptomes from interneurons reveal that 23 classes fall into four different developmental groups: PV, somatostatin, VIP, and neuroglioform cells. Analysis does not allow identification of specific precursors as the basis for differentiation.

Where Does Deterministic Organization Stop and Environmental Regulation Take Over?

The timing of this switch varies in different parts of the nervous system. Specification is not perfect and some axons project to the wrong place. However, neurons that make these mistakes are usually eliminated. Three classes of postnatal interneurons emerge from the ganglionic eminences, and all initially express Nkx2.1. When Nkx2.1 is lost at a specific point in time, postnatal interneurons diverge into the separate classes. Diversity is seeded by a small number of genes. Satb1 and Sox6 are expressed in the medial but not the central and lateral ganglionic eminences. Surprisingly, a small number of genes specify different cell types, much like the programming of induced pluripotent stem cells (iPSCs). The small number of genes creates an attractor network that draws in other genes (Theunissen et al. 2016). This may be why chimeric neurons are not observed. The intrinsic propensity is then acted on at a later stage to become further specified.

Gord Fishell and colleagues have found that while the position of pyramidal neurons is predictable based on the position of their progenitors, the position of interneurons is not. Positionally constrained pyramidal neurons provide positional information to less-constrained interneurons, allowing the latter to be programmed to position at their settling point. This interaction enables the reorganizations that occur during the later critical periods. In ongoing work (Fishell, unpublished), a bar-coded virus is being used to track interneuronal lineages to determine where they go and what they become. These cells are recovered by unique molecular indicators to determine their relationship to lineage.

Pasko Rakic described the protomap concept, which he developed from his work on the reeler mouse in which neurons do not migrate properly (Rakic et al. 1991). The hypothesis is that neurons are specified at an early stage, thus establishing the relationship of the progenitors to the cortex. Even when they go to the wrong layer, these neurons still look like pyramidal cells. Similar observations have been made in macaque monkeys (Herculano-Houzel et al. 2013; Ribeiro et al. 2013). Whether these observations, which concern the formation of the somatosensory cortex, apply to visual, auditory, and other cortical areas remains to be determined.

What is the value of inside-out formation of the cortex? Different cells are located in different layers, and this is very likely to be important for establishing correct connectivity. The protomap hypothesis was controversial because it seemed too deterministic. However the subventricular zone is thick in the human, thinner in the monkey, and thin in the mouse, presumably to reflect the different levels of cortical complexity. If neocortical cells are prevented from migrating and then patched for electrophysiological recordings and dye filling, morphologically they look like neocortical pyramidal cells. Electrophysiologically they can also look normal, but they have altered function. Lineage analysis in cortical progenitors shows that you get a column of neurons above the progenitors and that neuronal birthdate predicts position. One can thus conclude that pyramidal neurons know their position whereas interneurons do not.

Convergence and Stability during Dynamic Changes of the Developing Nervous System: Attractor States and Homeostasis

Neuronal gene expression, cell identity, and circuit structure and function appear to converge on a finite set of phenotypes. The dynamics of gene networks that drive development have much in common with certain types of neural network models that exhibit attractor states (Hopfield 1982). An attractor is a region of the state space of a dynamical system toward which trajectories tend with time. Attractors occur in many dynamical systems with nonlinearly interacting variables. A useful metaphor for an attractor is a valley, and its dynamic behavior is like the flow of water into a basin (Milnor 1985). Attractors may also explain why differentiated cells can be reprogrammed into pluripotency by any 4 of 10 transcription factors (Hochedlinger and Jaenisch 2015). In addition to point attractors, which lead to steady state, dynamical systems can also have limit cycles, driving oscillations that are common during development.

Starting from many different initial states, it is possible to end up in the same attractor state. However, if the initial state is too far away from the basin, the system can end up in a completely different attractor state. This may be a useful metaphor for different stages of development in most cases; however it may also provide the basis for pathological states in cases resulting from a large perturbation early in brain development. For example, chaos is another type of attractor that occurs in nonlinear systems, which happens in the heart during fibrillation.

Maintaining stability of some states while others are changing and converging is another requirement of brain development, and homeostasis is a mechanism by which this is achieved. For a one-dimensional variable, such as temperature, the goal of a control signal is to reduce the difference between the current temperature and a set point. The situation is much more difficult when the set point is multidimensional. For example, a neuron may want to maintain the shape of an action potential as well as the firing rate for a given input, while holding constant some of its other properties. There are dozens of ion channels in neurons that interdependently affect these two variables. Moreover, many different combinations of parameters, such as densities and time constants, may lead to the same outcome (Prinz et al. 2004). Thus it is important to measure the correlations between these parameters as well as their means.

Computational modeling offers an increasingly powerful approach to understand how the brain works and it continually evolves in a reciprocal fashion with increasing neuroscience data. Neuromorphic architectures, for example, have brought several advantages for computer design (Liu et al. 2015). During development of the nervous system there is an overproduction of neurons and synapses followed by pruning. Using this approach to structure neural networks has been shown to produce more robust, distributed computer networks than incremental approaches (Navlakha et al. 2015), with application to improvements in the design of routing networks such as the development of airline networks.

What Happens to Spatiotemporal Patterns of Waves of Activity during Early Cortical Development?

Before we focus on humans, we begin with a discussion of rodents at equivalent stages of development, examining the properties of neurons in the somatosensory cortex. Neurons are born at an early stage, and GABA influences proliferation (LoTurco et al. 1995). Since migration is inside out in rodents at the time of birth, the cerebral cortex consists of subplate, layer VI, layer V, cortical plate, and the marginal zone (future layer I) (Figure 4.2). At this stage, the subplate in rodents is only about three cells thick (although much thicker in primates) and much smaller than the cortical plate.

Lesions of the subplate have identified its contributions to cortical development, including formation of thalamocortical projections (Ghosh et al. 1990), descending axonal connections (McConnell et al. 1989), columnar organization (Ghosh and Shatz 1992; Kanold et al. 2003), and maturation of cortical inhibition (Kanold and Shatz 2006).

There is functionally mature thalamic input to the subplate at birth (Hanganu et al. 2002) along with innervation by neuromodulatory systems (NE, 5-HT and ACh). Electrophysiological signatures are evident: subplate neurons can fire at ~20 Hz when activated by cholinergic input (muscarinic receptors, mostly m1 and m5) (Hanganu et al. 2009), while neurons in layers V and VI fire only single spikes upon depolarizing current injection (Luhmann et al. 2000). EEG and 16-site depth silicon electrodes enable recording of the thalamic input and intracortical processing *in vivo* in newborn rat barrel cortex (Yang et al. 2009b). The thalamic input onto subplate neurons is glutamatergic, via AMPA and NMDA receptors (Hanganu et al. 2002; Hirsch and Luhmann 2008), and targets a heterogeneous cell population of glutamatergic and to a minor extent GABAergic neurons. Intracortical multielectrode recordings reveal spindle bursts (spindle-shaped discharges with 10–20 Hz bursts on top, which are reminiscent of human delta brushes), along with gamma oscillations at 30–40 Hz (Yang et al. 2009b; Figure 4.2).

Substantial research has elucidated the organization and origin of coordinated patterns of electrical activity in the neonatal brain. Most of the available data originate from sensory cortices. In the visual cortex, spindle bursts (less clear for gamma oscillations) arise from the spontaneous activation of the sensory periphery (retinal waves) (Hanganu et al. 2006). In the somatosensory cortex, in the absence of active whisking, the spindle bursts arise from spontaneous muscle twitches triggered by central pattern generators in the spinal cord (Khazipov et al. 2004; Inacio et al. 2016), brainstem (Blumberg et al. 2013),

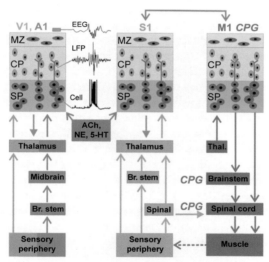

Figure 4.2 Model for the generation of 10–20 Hz spindle burst/delta brush activity in the newborn (P0) rodent and 3rd trimester (preterm) human cerebral cortex. At this stage the cortex consists of the marginal zone (MZ, future layer 1), the cortical plate (CP, future layers 2 to 4), layer 5, layer 6, and the subplate (SP). Thalamocortical input innervates glutamatergic subplate neurons, which also receive neuromodulatory inputs (ACh, NE, 5-HT). Subplate neurons can fire repetitively at 10–20 Hz when activated and transmit this activity to electrically coupled neurons in a local columnar syncytium, thereby generating the spindle burst/delta brush activity. The drive for this cortical activity comes from the sensory periphery (retinal waves, spontaneous activity in the cochlea) or from somatosensory receptors, which are activated following muscle movements. Central pattern generators (CPGs) in the motor cortex, brainstem, and spinal cord trigger these spontaneous movements ("twitches"). Modified after Luhmann et al. (2016).

or motor cortex (An et al. 2014; Luhmann et al. 2016). Multiple electrodes and voltage-sensitive dyes show that some of these events are local (Yang et al. 2013; Luhmann 2017).

The emerging picture is that subplate cells amplify input from the thalamus (Luhmann et al. 2009) and are connected by gap junctions (probably Cx36) to other immature subplate and layer VI–V cells (Dupont et al. 2006). The temporal pattern of activity is different if recordings are made in different areas (e.g., the prefrontal cortex; Bitzenhofer et al. 2015) and is probably also dependent on the developmental stage. In sensory cortices we can think of this as vertical (columnar) organization above input from the thalamus prior to P5. Among us, there was some debate about the patterns and source of patterns. These early patterns of coordinated activity are necessary to assemble the network. Hardwiring the network from time zero, independent of activity, would require a vast amount of genetic information and is, therefore, less probable.

Different cortical areas show distinct temporal dynamics of coordinated patterns of oscillatory activity. While all cortical areas show discontinuous oscillations, the onset of activity reflects different degrees of maturation. For example, coordinated patterns of oscillatory activity are present already at birth in the sensory cortices and hippocampus (Hanganu-Opatz 2010); in contrast, in the prefrontal cortex (PFC) these rhythms emerge toward the middle to end of the first postnatal week (Brockmann et al. 2011). Major differences can be also detected in the frequency organization of rhythms with rather simple organized spindle bursts and gamma oscillations in the primary sensory cortices (Hanganu-Opatz 2010) and with complex nested gamma spindle bursts (theta–alpha ground rhythm superimposed with fast discharges in beta to low gamma range and high-frequency oscillations) (Cichon et al. 2014; Bitzenhofer et al. 2015; Bitzenhofer et al. 2017). These differences reflect the multiple generators involved: endogeneous drive from the sensory periphery (e.g., retinal waves in the V1) versus theta drive from the hippocampus in PFC (Hanganu et al. 2006; Brockmann et al. 2011). Subcortical modulatory systems (e.g., cholinergic) seem to shape the activity in all cortical areas (Hanganu et al. 2009; Janiesch et al. 2011).

In addition to extrinsic generators, neonatal oscillations can emerge within cortical areas through cross talk across layers. For example, in the prelimbic subdivision of the PFC, different neuronal populations coordinate discontinuous patterns of oscillatory activity at the end of the first postnatal week; this corresponds to the third gestational trimester in humans (Figure 4.3). All layers are in place and the neurons in the upper layers coordinate the emergence of frequency-specific oscillatory rhythms whereas the deeper layers seem to contribute to unspecific activation.

The way in which this compares with preterm human circuitry has been discussed (Khazipov and Luhmann 2006). The cortex looks similar. Spindle bursts/delta brushes in preterms are similar, occur spontaneously, or are triggered by light touch (Milh et al. 2007; Colonnese et al. 2010), but currently

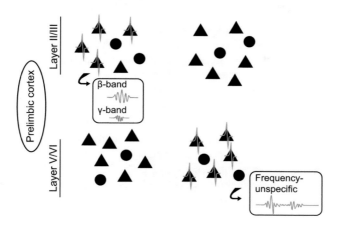

Figure 4.3 Cellular substrate of activity patterns in the neonatal prefrontal cortex. Schematic diagram depicting the contribution of pyramidal neurons in layer II/III of the prelimbic subdivision of the prefrontal cortex to the emergence of fast (beta to low gamma-frequency bands) components of discontinuous network oscillations and the ability of layer V/VI to generate a broad frequency-unspecific activation of the PFC.

there is no evidence for gamma oscillations. Significantly, if this spindle burst activity is blocked in rodent barrel cortex, the barrel field pattern is disturbed (Tolner et al. 2012). One hypothesis is that spindles are signaling the position of coactive muscles to columns of neurons to allow columns to link together, and that this needs to be updated during development as distances change. While initial insights of a functional circuit assembling during development in humans have been obtained for the somatosensory system, sparse knowledge is available for other areas. In line with the rodent data, we hypothesize that driving input in the visual system are the retinal waves; for the auditory system it is likely to be similar activity from the cochlear nuclei (Tritsch et al. 2007; Wang et al. 2015); for the PFC, it is the theta-band hippocampal input (Bitzenhofer et al. 2015).

A different structural and temporal organization of the patterns of coordinated electrical activity is present in the hippocampus. The first signals recorded in the rodent hippocampus are intrinsic Ca currents generated by voltage-gated Ca channels, followed around the time of delivery by synchronized Ca plateaux that interconnect interneurons via connexon-dependent gap junctions. Shortly afterward, the first synapse-driven activities, referred to as giant depolarizing potentials (GDPs), appear (Ben-Ari et al. 1989; Ben-Ari 2014) and have also been observed with some different properties in the neocortex, where they are referred to as early network oscillations (Allene and Cossart 2010). Recording from hippocampal pyramidal neurons at birth and reconstructing them revealed that neurons with no dendrites have no synaptic currents. Neurons with small apical dendrites have only GABA synaptic currents, and neurons with long dendrites extending to the distal lacunosum layer have synaptic currents generated by both GABA and glutamate, both in rodents and in primates *in utero* (Tyzio et al. 1999; Khazipov et al. 2001). A similar developmental sequence also occurs in interneurons at an earlier stage, indicating that GABAergic neurons are the first ones to be endowed with active synaptic currents (Hennou et al. 2002). Toward the end of the first postnatal week, discontinuous oscillatory activity in theta-frequency band (4–12 Hz), defined as theta bursts, is present in the hippocampus (Brockmann et al. 2011; Hartung et al. 2016a). This activity evolves into continuous theta oscillations toward the end of the second postnatal week.

There are many important directions for future work: Different frequencies of oscillations are observed at different ages and determining their functional role is an important issue to address. The detailed cellular and molecular mechanisms underlying these transitions remain to be investigated. Changes in the number and identity of glial cells over development will impact neuronal signaling properties. The small number that are present up to P7 control GABA and other neurotransmitter levels that are very important for proliferation and migration. Another important issue needing substantial clarification is the mechanism governing the transition between discontinuous oscillatory activity during neonatal (rodent) or fetal (human) development and continuous rhythms at juvenile age.

Knowledge of the spatiotemporal patterns of waves of activity has clinical importance. Burst activity recorded with EEG in preterms predicts clinical outcome (Benders et al. 2015; Iyer et al. 2015). With so many factors involved, drugs can have disruptive effects in the pregnant mother by interfering with activity patterns and altering apoptosis (Nimmervoll et al. 2013), map formation, and perhaps other features.

What Are the Patterns of Activity and Their Roles in Developmental Plasticity?

What Do We Mean by Activity?

Electrical activity has many forms during the early stages of development. The conclusion from research to date is that there can be substantial variability in the forms of activity across animal models, although Ca entry seems to be a consistent feature. In rodents at early stages of development, GDPs occur spontaneously. They last several hundreds of milliseconds in duration, are triggered by subthreshold depolarization, and involve voltage-gated Ca channels. The GDP frequency is appropriate for activation of the PV gene and likely others as well. It is not clear whether it is permissive or instructive. GDPs are primarily generated in interneurons but seen also in pyramidal cells. The arrival of synaptic inputs is hypothesized to terminate this spontaneous activity.

In the developing kitten, the visual cortex exhibits distinct forms of plasticity in response to changes in activity. During the first postnatal week, thalamocortical and cortical activity driven by spontaneous waves of activity in the retina plays a crucial role in the formation of high-resolution topographical maps in the lateral geniculate nucleus, primary visual cortex (V1) and superior colliculus, as well as in connections between V1 and superior colliculus.

At the beginning of the "critical period" of V1 in the fourth week of life, visual receptive fields of single cortical cells are not similar when they are driven through the left and right eyes, and simultaneous binocular stimulation is required for them to match, as they do in all normally developing animals after this period. If one eye is deprived of vision and the other eye allowed to see normally during this period, responses to the deprived eye are rapidly reduced, and anatomical connections serving that eye are, more slowly, reduced. The reduction in cortical activity leads to a homeostatic synaptic scaling dependent on TNFα and increases the responses to the open eye, resulting in a change in the balance of responses to stimulation of the two eyes.

Juvenile plasticity can be reactivated in adult V1 by heterochronic transplantation of embryonic PV- or somatostatin-containing inhibitory neurons from the medial ganglionic eminence into postnatal V1 (Southwell et al. 2010). The reactivated plasticity appears in a second critical period, similar to that of the host animal, at the point in time when the transplanted donor cells would have reached the normal critical period had the donor animal survived. The factors which enable these cells to create a second critical period, time-locked to their age, are as yet unknown.

In *Xenopus* embryos, electrical activity—assessed by the ability to generate action potentials—begins at the closure of the neural tube. Neurons are already sensitive to neurotransmitters at this stage and are spontaneously active. Action potentials are 100 ms in duration and Ca dependent initially; they mature into brief 1 ms sodium-dependent spikes over the next few days. They do this even when isolated in culture in minimal medium, indicating that they follow an intrinsic program. Imaging intracellular Ca in neurons in culture, in the intact spinal cord, or in the brain reveals spontaneous transient elevations. This spontaneous activity is generated in response to activation of metabotropic GABA and glutamate receptors (Spitzer and Lamborghini 1976; Gu et al. 1994; Root et al. 2008).

What Are the Roles of Activity?

Altering the frequency of Ca transients in the developing *Xenopus* CNS can change the identity of the transmitter the neuron expresses, both in culture and *in vivo*, where it takes place in response to both artificial perturbations and sustained stimulation by natural sensory stimuli (Gu and Spitzer 1995; Borodinsky et al. 2004; Dulcis et al. 2013). These data suggest that it can be instructive. This transmitter switching involves the loss of one transmitter and the gain of another, with corresponding changes in postsynaptic receptors to maintain synaptic function. The most frequently observed switch is between excitatory and inhibitory transmitter or vice versa, switching the sign of the synapse. The transmitter switch causes changes in animal behavior (Spitzer 2017). Transmitter switching involves changes in the levels of transmitters, which is what seems to occur during psychiatric disorders, motivating studies of the role of transmitter switching in depression and schizophrenia. These findings raise the possibility that sustained differences in activity between sleep and wake could change transmitter identity (Levenstein et al. 2017). The relationship between transmitter switching and transmitter coexpression is, however, not yet clear. One can wonder whether GABA projection neurons are precursors to glutamatergic projection neurons.

An important test of the role of activity in development of the nervous system involved the study of the Munc-13 mutant mouse in which vesicular release of neurotransmitters is blocked (Verhage et al. 2000). Strikingly, normally the CNS assembles morphologically in the absence of activity, although it is not clear that all transmitter release is blocked and that all activity was suppressed (e.g., Ca waves).

Putting a cell in a new environment may change its properties in ways that allow it to develop abnormally. The medial ganglionic (MG) eminence makes many cell types, in addition to somatostatin and PV interneurons, but when neurons are removed from the MG, they no longer go through mitosis. While transplantation is often viewed as a challenge to progenitors to differentiate in the new environment, they often no longer divide, dramatically constraining their ability to adopt a new fate. Additionally, when cells are introduced into an environment that produces only a particular type of cells, those are the only cells that differentiate. Study of PV neurons illustrates the importance of intrinsic versus extrinsic cues. These cells are very sensitive to environmental context, which in turn is activity dependent. When animals are raised in the dark, brain-derived neurotrophic factor (BDNF) secretion is reduced, and overexpression of BDNF and other signaling molecules affects terminal maturation. The expression of clock genes may be the only case with completely intrinsic regulation (Kobayashi et al. 2015). It would be interesting to learn what happens if all cells expressed their critical period at the same time.

When we talk about two different critical periods, we may not be discussing the same thing. The visual cortex critical period, which occurs at four weeks postnatally, is sometimes compared to the somatosensory cortex critical period that occurs later. The effect of whisker plucking on the barrel cortex is compared to the effect of eye closure on ocular dominance and the visual cortex. However, whisker plucking constitutes an injury. If we want to compare apples to apples, it would be better to contrast the formation of the visual cortex map with the formation of the somatosensory cortex map. Whether the changes that occur in the visual system during the critical period are accompanied by changes elsewhere in the brain remains to be addressed.

All of the brain's activity, coupled with development and continuous remodeling, makes it an expensive organ. Although the human brain comprises only 2% of total body weight, the adult and infant brain require 20% and 40% of the total energy budget, respectively (Allman 2000; Herculano-Houzel 2011). The energy budget of the brain is divided between basal metabolism (which in the human brain accounts for ca. 25%), action potentials, synaptic signaling, and, in the developing brain, synaptogenesis. The cost of spontaneous versus evoked activity and the cost of excitation versus inhibition has been analyzed (Attwell and Laughlin 2001; Laughlin and Sejnowski 2003; Buzsáki et al. 2007). There are four times more excitatory cells than inhibitory neurons in the cortex, but the inhibitory cells fire at a faster rate and inhibitory synapses are more reliable, so on average excitatory and inhibitory synaptic drive are balanced. Action potentials vary in their duration, and the thin action potentials of fast-spiking basket cells in the cortex are optimized to reduce the number of sodium ions that need to be pumped out (Hasenstaub et al. 2010).

While connectivity is dominated by thalamic projections to the subplate at birth, this rapidly shifts postnatally. By P5 in mice, there is a large increase of thalamic afferents to somatostatin interneurons in deep cortical layers, which shifts to PV interneurons by P9. As such, rapid and dynamic changes occur in the wiring of the cortex during the early postnatal period. An important future goal will be to examine in close detail the dynamics of wiring shifts in the early cortex. Although we have a lot of descriptive data from development, we lack data related to mechanisms. Inferences from adult physiology and plasticity, while attractive, are dangerous.

Do Changes in Activity Prefigure Development of Pathology? How General Is This Pattern?

A central concept is that insults at early stages cause deviations from normal development. Good examples are afforded by migration disorders. If neurons do not migrate, they display immature properties, retaining the currents that they express at early stages (Ackman et al. 2009). Conditional knockout of *Sox6* leads PV neurons to wind up in layer I. The result is more severe epilepsy than a total knockout of PV neurons: miswiring is worse than no wiring. This is a kind of "neuroarcheology," in which the present arises out of the past.

If, however, extremely strong stimuli are applied later in the adult (e.g., seizures, autism, spinal cord injury), neurons are not able to cope and either commit suicide or activate genes that express immature properties. What is the benefit for neurons to respond by expressing immature properties? Perhaps reopening chromatin, which had been open at an earlier stage, leads to expression of the same genes that had been expressed at the previous stage. This neuronal response has important implications for the treatment of clinical disorders: simply reinstating normal genes and/or proteins in the adult is not

Repairing Broken Phenotypic Checkpoints

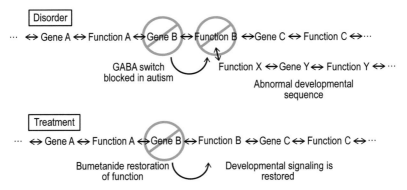

Figure 4.4   Circumvention of arrested phenotypic checkpoints is achievable. The normal developmental sequence is blocked when the GABA switch fails to occur in autism spectrum disorder (top), leading to abnormal development. Restoration of altered gene function with bumetanide bypasses the defect, providing treatment (bottom) that enables normal development.

likely to be effective in restoring function. Reinstatement of gene expression or function at the relevant site and at the relevant developmental stage would be a better approach. The example of bumetanide treatment for high Cl⁻ in autistic animal models (Tyzio et al. 2014) and patients (Lemonnier et al. 2017) is particularly interesting (Figure 4.4). Intrauterine insults appear to alter normal developmental sequences, disrupting them, with an outcome of misplaced neurons or inadequate connections and immature patterns of current expression. This underscores the difficulty of repairing such defects by inserting the appropriate gene (Manent et al. 2007; Ben-Ari 2008). In a similar case, a major study of vision in children was undertaken by the county government in Cambridgeshire to identify young persons with amblyopia. Appropriate treatments were determined and the burden of the disease was dramatically reduced (Atkinson et al. 2007). Unfortunately, the subsequent change in government resulted in the termination of plans to screen and treat throughout the rest of the United Kingdom.

The situation is even more complex and less well understood in neuropsychiatric disorders with a developmental time course. In schizophrenia, brain structure appears almost normal throughout all stages. Since psychotic episodes are devastating, psychiatrists ask if there is a way to identify vulnerability in advance. Is there any general feature of activity that allows prediction of the disease (Uhlhaas and Singer 2011; Andreou et al. 2014; Leicht et al. 2016)? Blind examination of six mouse multihit models of mental illness has shown differences in the hippocampus in the extent of coupling to the PFC already in neonates (Hartung et al. 2016b). These disorders are likely to be the result of multiple hits. For example, a single mutant gene may not be sufficient by itself; inflammation might allow the brain to compensate, but another insult could push the system over the threshold. Accordingly, we need to follow behavior over development to look for changes. A single biomarker may not be sufficient.

An alternative approach would be to look for a series of developmental checkpoints and correlate failure to pass them using a probabilistic model of origin of the disease. This is being done with human autism spectrum disorders: checkpoint analysis has shown that the GABA switch was not happening, but it appears possible to move through the checkpoint by applying bumetanide (Tyzio et al. 2014; Lemonnier et al. 2017; Figure 4.4). Neurons that do not complete their normal developmental sequence retain immature features, as evidenced in genetic disorders such as the double cortex syndrome, in which neurons that have not migrated have immature features (Ackman et al. 2009).

When addressing mental disorders, we need to decide on the strategies to prioritize. Here we can derive a cautionary tale from the history of Nixon's war on cancer. Initially, most money went into chemotherapy, but it was not until recombinant DNA technology was invented that cancer was shown to be a complicated genetic disease. Now immunotherapy is becoming successful (Kosik et al. 2016). Schizophrenia is similarly complicated, and the pathways generating it need to be investigated if we are to gain an understanding of its pathological background.

The use of mouse and other animal models in understanding and treating human disorders has both advantages and disadvantages. Mouse models may be valuable for understanding some aspects of disease etiology. However, they cannot capture the full range of human complexity, since the human brain is not a scaled-up version of the mouse brain (Herculano-Houzel 2009). Multiple animal models of Fragile X syndrome, for example, have responded to mGluR antagonists. However, proximal mouse models have not informed the human disease: clinical trials have either failed or are failing. Mouse models specify early treatment but that opportunity rarely exists with patients, who can only be treated once they search out medical help. Finally, there is the issue of doses and duration of drugs. Many unexpected checkpoints, possibly involved in metabolism, can be expected. In addition, the same mutation can lead to different disorders, implying that other factors must be involved (Guilmatre et al. 2014).

Curing disorders of developmental origin does not seem possible because the multiple insults are early and cannot be treated at the time they are generated using current technology. For instance, attempts to restore migration in a conditional knockout of doublecortin failed when the correction was applied later than a few days postnatally, underscoring the challenge of correcting deviations in the developmental sequence (Manent and LoTurco 2009). We should not, however, give up on animal models; it was necessary to screen 500 drugs before one was discovered for treating strokes. *Treating* disorders of developmental origin seems more feasible, is a more modest approach, and seems to work. Combining psychiatric/psychological treatment with biochemical/molecular treatment could result in even better outcomes.

It is useful to reevaluate what it means to be a model. Ideally, each patient should be his/her own control. At the start, as many biomarkers as possible should be obtained to compare against "normal" states; thereafter, machine learning could be used to predict which drugs should be used to treat symptoms. Included here should be the recognition that each disorder is really many disorders. The Adolescent Brain Cognitive Development study—a longitudinal study of 20,000 children, ages 9–19 years—will enable analysis of many aspects of development (Barch et al. 2017). It is now feasible to use iPSCs to generate a patient's neurons and study the genetic disorder. This may assist in developing treatments and is an important avenue to pursue, despite present challenges in implementation.

Another approach is to find a brain closer to ours, such as a nonhuman primate. Currently, the Allen Brain Institute is making physiological recordings from human tissue, and the membrane properties of the neurons are different from those of the mouse. Hideyuki Okano's group in Japan is using marmosets. Macaque monkeys are even closer, and the Institute of Neuroscience in Shanghai is mounting a big effort with macaques as part of the Chinese Brain Initiative.

While the mouse has proven a useful model for CNS development, the limits in the similarities between mouse and humans argue for the need of higher-order models to study human brains. The use of nonhuman primate models as well as *in vitro* models using organoid cultures of human iPSC-derived cells show promise for studying the specification of cell identity as well as the signaling in neural networks. Nonhuman primates provide the requisite neuronal types, but tools will need to be developed to allow for the targeting and manipulation of specific types in a manner akin to what is now possible with mouse genetics. Similarly, the advent of methods to use the neural equivalent of "Yamanaka" factors means that we are well advanced in our efforts to generate specific neuronal types from iPSCs. With these tools in hand, the ability to study brain function using *in vivo* (nonhuman primate) and *in vitro* (iPSC-derived organoids) models will be greatly enhanced, with relatively small advances in opto- and chemogenetic methods.

In conclusion, we do not know enough at this point to be able to determine whether changes in activity will be useful biomarkers. However searches in mice, nonhuman primates, and the use of human registries seem appropriate.

Future Directions

As we look to the future, it is clear that acquiring more data will be important and necessary to derive general principles by which dynamic brain coordination is achieved during development. Using the findings achieved to date to construct conceptual frameworks that focus the directions of experimental work should promote the most rapid progress, either by supporting, reorienting, or eliminating particular conceptual schemes. The formulation of computer models, both for data analysis and for testing our understanding of the operation of circuits at different scales, is now possible as never before and should be a component in all research programs. Among the many promising avenues to pursue, we wish to highlight the following:

- Determining the functional role of different frequencies of oscillations observed at different ages is essential.
- Understanding the basis of developmental changes in cortical wiring is crucial.
- Identifying the cellular and molecular mechanisms by which these patterns of activity exert their effects is a critical objective.
- Clinical diagnostics and treatment protocols will likely follow from this knowledge.
- Mice will remain an attractive model for many studies, but nonhuman primates and humans themselves, via registry databases, will be important.

Early Childhood

5

Factors that Initiate and Terminate Critical Periods

Takao K. Hensch

Abstract

During development, neural circuitry can be profoundly shaped by experience at well-defined periods of time. Using amblyopia as a model of postnatal synaptic plasticity, this chapter reviews the "triggers" and "brakes" that determine the onset and offset of these critical periods. Consideration is given to the molecular constraints that act on plasticity as well as to the physical and sensory environmental factors that impact function and cortical circuit plasticity. Reactivation of plasticity in primary visual cortex suggests that critical periods are not limited to early postnatal development. The extent to which the amblyopia model will generalize at a mechanistic level is discussed. Genetic diversity in mice and humans may provide insight into individual variability and the timing of critical periods and should be pursued. To permit comparison of developmental trajectories more readily across species and disease states, the call is made for better models of critical period plasticity and the identification of biochemical and electrophysiological correlates of these windows.

Introduction

It is well appreciated that defined windows in early life exist when neural circuitry can be robustly restructured in response to experience. These time-limited critical periods have been demonstrated for diverse brain functions across many brain regions and are thought to allow developing neural circuits to establish an individualized, optimal neural representation of a highly variable environment. The relative stability in cortical circuitry that follows the critical period may also allow for conservation of energy/resources. With age, however, enhanced stability also inhibits large-scale adaptations to changes in input during adulthood. Utilizing the power of molecular, genetic, and imaging tools, recent advances with mouse models are beginning to unravel the network, cellular, and molecular mechanisms controlling the onset and closure of critical periods of plasticity in primary sensory areas (Figure 5.1). A pivotal

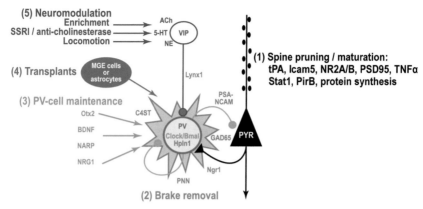

Figure 5.1   Five sites of critical period regulation: (1) Pruning and maturation of excitatory synapses onto dendritic spines involves proteases (tPA) and their cell-adhesion targets (Icam5), NMDA receptors (NR2A/B), and their postsynaptic partners (PSD95) which unsilence them, homeostatic factors (TNFα, Stat1), immune genes (PirB), and new protein synthesis. (2) Removal of brake-like factors on parvalbumin (PV)-positive basket cells enables plasticity outside the critical period, including GABA synthesis (GAD65) or cell-adhesion factors (PSA-NCAM) that control perisomatic inhibitory output, and perineuronal net (PNN) components (C4ST, Hpln1) or receptors (Ngr1) enwrapping their synaptic inputs. (3) Direct manipulation of cell-extrinsic PV-cell maintenance factors (Otx2, BDNF, NARP, NRG1) or intrinsic circadian gene regulation (Clock, Bmal) shifts plasticity onset. (4) Transplantation of immature inhibitory cell precursors (MGE cells) or astrocytes restores plasticity to the mature cortex. (5) Engaging upper layer inhibitory neurons (VIP cells) by neuromodulatory input (ACh, 5-HT, NE) through behavior or drugs disinhibits the core PV-pyramidal (PYR) circuit driving plasticity, and is actively counteracted by other brake-like factors (e.g., Lynx1 on ACh signaling) in adulthood.

player is the fast-spiking, parvalbumin (PV)-positive inhibitory neuron, which matures in register with these windows. Further evidence suggests that mechanisms enabling plasticity in juveniles are not simply lost with age, but rather that plasticity is actively constrained by the developmental upregulation of molecular "brakes." Lifting these brakes enhances plasticity in the adult visual cortex, which can be harnessed to promote recovery of function. Notably, most of the identified brake-like factors converge again upon the PV-positive interneuron, which is well-poised to generate rhythmic oscillations. Here, we discuss recent insights into the neurobiology of critical periods, and how our increasingly mechanistic understanding of these pathways can be leveraged toward improved clinical treatments.

The Amblyopia Model

The shift in ocular dominance of binocular neurons and blunted acuity (amblyopia) induced by discordant vision through the two eyes ("lazy eye") is the canonical model for synaptic plasticity confined to a postnatal critical period. The enhanced plasticity corresponds to peak phases of physical growth and may therefore allow for constant perception during expansion of the body surface. For example, visual receptive fields must repeatedly remap as the distance between the two eyes increases. Indeed, experience-dependent matching of stimulus selectivity of visual input from the two eyes occurs during the critical period (Wang et al. 2010). An asymmetry in the quality of visual input across the two eyes at this time leads to reduced visual acuity and visually evoked spiking response through the affected eye with no obvious pathology in the eye, thalamus, or cortex. The severity of amblyopia depends on the age at initiation and the type of asymmetry, which can be caused by unequal alignment (strabismus), unequal refractive error (anisometropia), or form deprivation (e.g., cataract). The critical period for developing amblyopia in children extends to eight years, and is relatively easy to correct until that age by improving the quality of visual input in the affected eye (reviewed by Daw 1998; Mitchell and MacKinnon 2002; Simons 2005) but becomes increasingly resistant to reversal with age. Developmental constraints on this plasticity lend stability to mature visual cortical circuitry but also impede the ability to recover from amblyopia beyond an early window.

In animal models, amblyopia is most often induced by monocular deprivation (MD)—eyelid suture, which significantly occludes the patterned visual input to one eye. Across various species, MD unleashes a sequence of functional and structural changes in V1 that shifts the ocular dominance of binocular neurons away from the deprived eye and toward the open eye, resulting in a reduction in deprived-eye acuity (Wiesel and Hubel 1963, 1970; Olson and Freeman 1975; Hubel et al. 1977; Movshon and Dürsteler 1977; Blakemore et al. 1978; LeVay et al. 1978; Shatz and Stryker 1978; Antonini and Stryker 1993; Fagiolini et al. 1994; Gordon and Stryker 1996; Hensch et al. 1998; Trachtenberg and Stryker 2001; Mataga et al. 2002; Taha and Stryker 2002; Prusky and Douglas 2003; Frenkel and Bear 2004; Sato and Stryker 2008).

While ocular dominance plasticity peaks during the postnatal critical period, it generally persists at some level in many species, including rodents and cats, beyond sexual maturity. For example, the adult cortex may retain the ability to express some forms of synaptic plasticity, which may be expressed differently from those utilized during the critical period. In this context, it is important to bear in mind that many measures are in current use to study *ocular dominance plasticity*. Originally defined as a change in the eye preference of spiking output of V1 neurons (Wiesel and Hubel 1963), it has grown to encompass visually evoked synaptic potentials, intrinsic hemodynamic signals, immediate early gene activation, thalamocortical axon or dendritic spine morphology and motility, and calcium responses in individual cell types. Each of these methods yields different resolution and may be variably sensitive to subthreshold inputs (Morishita and Hensch 2008), which are important considerations when informing therapies for recovery of visual function.

Pruning Connections

The initial response to MD during the critical period is a reduction in functional strength and selectivity of deprived eye visual responses (Gordon and Stryker 1996; Hensch et al. 1998; Trachtenberg et al. 2000; Frenkel and Bear 2004). Depression of deprived-eye responses may occur by synaptic depression at both thalamocortical and intracortical connections. Notably, the most rapid shifts in visual response are seen in PV-expressing inhibitory interneurons which may enable further functional changes within V1 (Yazaki-Sugiyama et al. 2009; Aton et al. 2013; Kuhlman et al. 2013). Depression is then followed by a relatively slower, homeostatic strengthening of open eye responses (Sawtell et al. 2003; Frenkel and Bear 2004; Kaneko et al. 2008).

Robust morphological plasticity is also induced by MD during the critical period. An initial degradation of the extracellular matrix by the upregulation of proteases occurs within the first 2 days after MD in the mouse, and may elevate spine motility (Mataga et al. 2004; Oray et al. 2004). Studies in cats, monkeys, and humans suggest that structural plasticity is facilitated by a reduction in the neurofilament-light protein within V1, and that this may destabilize the cytoskeleton and promote plasticity (Duffy and Livingstone 2005; Duffy et al. 2007; Duffy and Mitchell 2013). Brief MD during the critical period alters spine density on pyramidal neurons (Mataga et al. 2004; Tropea et al. 2010; Yu et al. 2011a; Djurisic et al. 2013) and induces a transient decrease in the density of synapses formed by thalamocortical axons originating from the lateral geniculate nucleus (Coleman et al. 2010). Long-term MD yields enduring alterations in the length and extent of thalamocortical arbors serving the two eyes (Hubel et al. 1977; Shatz and Stryker 1978; Antonini et al. 1999) as well as a significant reduction in dendritic spine density (Montey and Quinlan 2011).

Studies from humans and nonhuman primates suggest a protracted decline in visual plasticity that extends into adulthood rather than an abrupt closure of the critical period. The residual plasticity that persists in adult visual cortex, however, appears to differ from the plasticity during the critical period in several important ways:

- The shift in ocular dominance in adults is slower and smaller and may require a longer duration of deprivation to engage.
- It may not require depression of deprived eye responses for subsequent strengthening of responses to the nondeprived eye.

- It may be restricted to synapses in supragranular and infragranular lamina, as plasticity in layer IV has been shown to be constrained early in postnatal development.
- It may be restricted by saturated synapses, setting limits on the amount of recovery of visual function that can be accomplished using this pathway.

Additionally, MD in adults does not elicit the robust structural alterations that accompany ocular dominance plasticity during the critical period, such as increased spine motility and pruning (Mataga et al. 2004; Oray et al. 2004; Lee et al. 2006). Indeed, a general decline in structural plasticity is one of the hallmarks of the termination of the critical period. However, residual increases in the rate of formation and stability of dendritic spines may persist in adult layer I after MD (Hofer et al. 2009).

Inhibition and Critical Period Induction

Powerful new tools in neuroscience, especially those which enable molecular genetic control in mice, are beginning to elucidate the cellular and molecular mechanisms that may initiate and terminate critical periods. Ocular dominance plasticity peaks during the third postnatal week in rodents, demonstrating that elevated plasticity is not the initial state of immature circuits. Indeed, the maturation of specific inhibitory circuitry is necessary to initiate the critical period, which can be accelerated by activating inhibitory $GABA_A$ receptors with allosteric modulators such as benzodiazepines (Hensch et al. 1998; Fagiolini and Hensch 2000; Iwai et al. 2003; Fagiolini et al. 2004). Premature initiation of the critical period can be induced if early maturation of the specific class of inhibitory interneurons containing the calcium-binding protein PV is promoted by increasing levels of growth factors, BDNF (Hanover et al. 1999; Huang et al. 1999) and Otx2 (Sugiyama et al. 2008; Spatazza et al. 2013), or by removing cell-adhesion, PSA (Di Cristo et al. 2007), or DNA-binding proteins, MeCP2 (Durand et al. 2012; Krishnan et al. 2015).

The perisomatic inhibition mediated by these fast-spiking PV interneurons exerts powerful control over the excitability and plasticity of downstream pyramidal neurons, potentially sharpening the spike timing required for synaptic plasticity (Katagiri et al. 2007; Kuhlman et al. 2013; Toyoizumi et al. 2013). Several proteins that regulate synaptic strength and/or number are highly enriched at excitatory synapses onto PV interneurons and impact the timing of the critical period (NARP: Gu et al. 2013; NRG1: Gu et al. 2016; Sun et al. 2016). Accordingly, NARP-deficient mice fail to initiate a critical period unless rescued by enhancing the strength of inhibitory output or excitatory drive onto PV interneurons (Gu et al. 2013, 2016). The dynamic balance of excitation–inhibition within PV networks also drives oscillatory activity in the gamma-frequency (30–80 Hz) range (Bartos et al. 2007), which may shift dramatically over development and disease states.

A further increase in perisomatic inhibition is thought to terminate the critical period. Hence, the critical period can be reopened in adulthood by pharmacological reduction of inhibition (Harauzov et al. 2010) or the knockdown of Otx2 (Beurdeley et al. 2012; Spatazza et al. 2013). Treatment with an NRG1 peptide induces a precocious termination of the critical period, while inhibition of the activity of the NRG receptor (ErbB) reactivates the critical period in adults (Gu et al. 2016). Indeed, a developmental reduction of plasticity at excitatory synapses onto fast-spiking interneurons may explain the requirement for longer durations of MD with age (Kameyama et al. 2010). Together, these studies indicate that PV inhibitory cells exert bidirectional control over ocular dominance plasticity (van Versendaal and Levelt 2016).

Other classes of inhibitory neurons may influence the expression of plasticity, either independently or through the regulation of PV neurons. Interestingly, inhibitory neurons in layer I of the visual cortex and those expressing vasoactive intestinal peptide (VIP) are strongly activated during certain behavioral states and exert cortical effects by disinhibition of pyramidal neurons (Letzkus et al. 2011; Donato et al. 2013; Pfeffer et al. 2013; Pi et al. 2013; Fu et al. 2015). Locomotion activates VIP interneurons, which enhances neural activity in V1 (Niell and Stryker 2010) and promotes adult plasticity by increasing inhibition onto other interneuron subtypes that target pyramidal neurons (Fu et al. 2014, 2015). Similarly, reinforcement signals (reward and punishment) during the performance of an auditory discrimination task activate VIP neurons in auditory cortex, which increase the gain of a functional subpopulation of pyramidal neurons by disinhibition (Pi et al. 2013). Thus, disinhibitory circuits that transiently suppress other inhibitory interneurons may be a general mechanism for enabling plasticity in the adult cortex.

Molecular Constraints on Critical Period Plasticity

Increasing evidence demonstrates that removing molecular "brakes" in adulthood can enhance plasticity and promote recovery from amblyopia. For example, epigenetic mechanisms, such as histone deacetylase (HDAC) activity, may downregulate expression of genes that promote plasticity over development. HDAC inhibition then enhances plasticity in adult V1, allowing for recovery from amblyopia (Putignano et al. 2007; Silingardi et al. 2010). However, the downstream targets of histone acetylation at specific stages of development remain to be identified.

Alternatively, increased expression of specific genes over development can actively limit rewiring. The expression of *Lynx1*, an endogenous inhibitor of nicotinic acetylcholine receptors, emerges in V1 coincident with critical period closure, which would dampen neuromodulatory actions of acetylcholine (Miwa et al. 1999; Morishita et al. 2010). Both genetic deletion of *lynx1* and administration of acetylcholinesterase inhibitors enhance spine motility and the morphological plasticity induced by MD (Sajo et al. 2016) and enables recovery of visual acuity following MD throughout life (Morishita et al. 2010). The major histocompatibility complex class I (MHCI) receptor, PirB, is another molecular brake. Disruption of PirB signaling enhances ocular dominance plasticity throughout life and facilitates recovery from amblyopia in adults (Syken et al. 2006; Bochner et al. 2014). Another immune system molecule, Stat1, restricts the increase of open eye responses following MD, and its genetic deletion enhances this component of plasticity (Nagakura et al. 2014). The identification of specific molecules that actively suppress plasticity in the adult visual cortex may inform strategies for pharmacological interventions to reopen the critical period.

Molecular brakes can also present physical barriers to morphological plasticity. Perineuronal nets are highly enriched around PV neurons and reach maturity at the end of the critical period. Disrupting the molecular latticework of this extracellular matrix (Pizzorusso et al. 2002, 2006; Carulli et al. 2010) or the molecules which bind to it (Otx2: Beurdeley et al. 2012) enables ocular dominance plasticity and recovery from amblyopia in adults. Consistent with this, mice lacking (globally or only from PV cells) the Nogo receptor (Ngr1), a bimodal receptor for chondroitin sulfate proteoglycans and myelin-derived inhibitory factors (Dickendesher et al. 2012), also retain critical period plasticity into adulthood and spontaneously recover visual acuity following long-term MD (McGee et al. 2005; Stephany et al. 2014). Interestingly, PirB may act in concert with Ngr1 (Atwal et al. 2008) to dampen morphological plasticity of dendritic spines on layer V pyramidal neurons in adults (Bochner et al. 2014).

One recently identified molecular brake may lie within the dendritic spine itself. Postsynaptic density protein 95 (PSD-95), an intracellular scaffold highly enriched at excitatory synapses, is thought to accelerate maturation of excitatory synapses. PSD-95 promotes the incorporation of AMPA-type glutamate receptors into synapses containing only NMDA receptors, which are normally functionally "silent" at resting membrane potential. In contrast, the immediate early gene *Arc* promotes removal of AMPA receptors from cortical synapses and precludes visual plasticity when deleted (McCurry et al. 2010). Genetic reduction of PSD-95 in adulthood increases the number of silent synapses and reactivates the juvenile form of ocular dominance plasticity, characterized by a rapid and robust deprived-eye depression (Huang et al. 2015). Notably, no changes in GABAergic or NMDA receptor currents are observed, suggesting that the reactivation of plasticity by PSD-95 deletion lies downstream of the regulation of inhibitory circuitry. A conversion of "silent" to functional synapses has been proposed as a general mechanism to constrain plasticity across brain regions (Greifzu et al. 2013; Huang et al. 2015).

Environmental Reactivation of Critical Period in Adulthood

Characteristics of the physical or sensory environment strongly impact the function and plasticity of cortical circuits. Remarkably, adding social, sensory, or motor enrichment to the typically impoverished environment of the laboratory rodent influences the expression and time course of ocular dominance plasticity. Robust ocular dominance plasticity persists into adulthood when mice are raised in large complex cages with multisensory and motor enrichment (Sale et al. 2007; Greifzu et al. 2013). In fact, enriched rearing may better reflect the sensorimotor environment of primates, including humans. At a molecular level, exposure to enriched environments in adulthood increases H3 acetylation (Baroncelli et al. 2016), reduces the expression of PV and GAD67 within inhibitory neurons of the visual cortex, weakens GABA signaling, and fosters plasticity in both the cortex and hippocampus (Sale et al. 2007; Donato et al. 2013; Greifzu et al. 2013).

In this regard, it is intriguing that total visual deprivation also reactivates robust plasticity in adult V1 and promotes recovery from chronic MD (He et al. 2007; Montey and Quinlan 2011; Duffy and Mitchell 2013; Stodieck et al. 2014; Eaton et al. 2016; Mitchell et al. 2016). Several mechanisms, engaged by dark exposure, have been predicted to lower the threshold for synaptic plasticity in pyramidal neurons (Cooper and Bear 2012). For example, the composition of the NMDA type glutamate receptors is reset to a "juvenile" form (containing the NR2B subunit) which exhibits enhanced temporal summation (Yashiro et al. 2005; He et al. 2006). In addition, synaptic plasticity typically limited to juveniles is re-expressed (Huang et al. 2010; Montey et al. 2013), spines on pyramidal neurons are shifted toward immature structure and dynamics (Tropea et al. 2010), and immature excitatory synapses on pyramidal neurons are strengthened, thereby increasing excitability and expanding the integration window for spike timing-dependent plasticity (He et al. 2006; Goel and Lee 2007; Guo et al. 2012).

Dark exposure also decreases the excitability of PV interneurons, and the reactivated plasticity can then be reversed by increasing the strength of excitatory synaptic input onto them (Gu et al. 2016). A loss of specific neurofilament protein associated with cytoskeletal stability is observed in the lateral geniculate nucleus following dark exposure, which may further contribute to the reactivation of structural ocular dominance plasticity beyond the peak of the critical period (O'Leary et al. 2012; Duffy et al. 2016). Thus, the seemingly opposite interventions of environmental enrichment and dark exposure may ultimately enhance cellular plasticity through the removal of functional and structural constraints that normally accumulate over development to stabilize V1 circuitry.

It is important to note that dark exposure alone does not impact visual acuity or neuronal stimulus selectivity, which is regained only after repetitive visual experience (Montey et al. 2013; Eaton et al. 2016). Likewise, enrichment or locomotion alone does not strengthen visual performance (Kaneko and Stryker 2014; Greifzu et al. 2016). This suggests that environmental reopening of plasticity in adulthood is a two-stage process that requires (a) the reactivation of plasticity machinery (permissive step) and (b) focused sensory experience to stimulate perceptual learning (instructive step). One of the challenges, therefore, is to identify the optimal sensory stimulation to drive change. In addition, prolonged plasticity by environmental enrichment in mice raises the question whether complex environments better mimic those of primates including humans. At a minimum, it provides a valuable condition with which to better understand the biological basis of critical period closure.

Reactivating Plasticity to Enhance Recovery

The reactivation of plasticity in primary visual cortex has revised the idea that critical periods are strictly limited to early postnatal development (Bavelier et al. 2010; Takesian and Hensch 2013; Sengpiel 2014). As described above, early in the visual pathway, MD induces significant structural rearrangements in V1, including pruning of thalamocortical inputs that serve the deprived eye (Wiesel and Hubel 1963; Hubel et al. 1977; Shatz and Stryker 1978). Long-term MD yields a near complete loss of stimulus selectivity for input coming in through the chronically deprived eye (Montey and Quinlan 2011). Given these severe structural and functional deficits in V1, it is even more remarkable that full recovery of visual acuity has been demonstrated with some interventions.

Based on mechanistic studies (above), novel therapies with translational potential to reverse the developmental constraints have been identified. Several commonly prescribed drugs, such as cholinesterase inhibitors (Morishita et al. 2010), valproate (Gervain et al. 2013; Lennartsson et al. 2015), or selective serotonin reuptake inhibitors (SSRIs) (Maya Vetencourt et al. 2008), could be repurposed to rescue adult amblyopic patients. Interestingly, reduced PV interneuron function may be a mechanism common to several of these interventions. The SSRI antidepressant fluoxetine reduces basal levels of extracellular GABA (Maya Vetencourt et al. 2008) and the number of PV interneurons surrounded by dense perineuronal nets (Guirado et al. 2014). Similarly, dark exposure may rejuvenate intracortical inhibition by reducing the excitatory drive onto PV neurons (Gu et al. 2016). The use of action video games or vagal nerve stimulation, to recruit neuromodulatory pathways that engage attention and motivation (Mitchell and Duffy 2014; Hess and Thompson 2015; Levi et

al. 2015; Murphy et al. 2015), may also be effective in treating sensory abnormalities (Hess and Thompson 2015; Tsirlin et al. 2015).

Beyond Ocular Dominance

One now must wonder to what extent the amblyopia model will generalize at a
mechanistic level. Multiple critical periods are reported across a variety of modalities (for a review, see Hensch 2005). Armed with the molecular markers of
critical period initiation and termination from the visual cortex, we can expect
a "golden age" of critical period research across brain regions. For example,
novel windows of plasticity for higher cognitive functions such as multisensory integration in the insular cortex (Gogolla et al. 2014) or the acquisition of
preference behaviors in the medial prefrontal cortex (Yang et al. 2012) have
been shown to observe common principles of PV cell maturation and revers-
ibility by HDAC/Nogo receptor inhibitors, respectively. Of greatest interest is
the potential to monitor electrophysiological signatures of shifting excitatory–
inhibitory balance indicative of critical period timing which can be translated
noninvasively to the human.

Even in the visual domain, the primary aspects of visual system function assessed in animal studies of amblyopia are ocular dominance and spatial acuity.
Amblyopia, however, is associated with a range of visual deficits, including
loss of stereoscopic depth perception, crowding, impairments in shape discrimination, deficits in motion and direction perception, and object tracking
(reviewed in Daw 2013). Furthermore, separable neuronal response properties
of individual V1 neurons have distinct, overlapping critical periods (reviewed
in Kiorpes 2015). For example, in kittens, direction selectivity precedes ocular
dominance (Daw and Wyatt 1976), and in the primate visual system, critical
periods for basic spectral sensitivities end relatively early (6 months), whereas
those for complex representations, such as contrast sensitivity and binocularity, extend much later (25 months) (Harwerth et al. 1986). As critical periods
for different visual functions may depend on separate underlying mechanisms,
some manipulations may restore only selective features of V1 responses. For
example, a genetic deletion of PSD-95 disrupts the development of orientation
preference in mouse visual cortex without impacting the development or plasticity of ocular dominance in juveniles (Fagiolini et al. 2003).

Moreover, the magnitude of compromised vision observed in psychophysical experiments is often not mirrored by changes in the function of V1
neurons, suggesting that physiological changes may be propagated and amplified in higher cortical areas (Shooner et al. 2015). Indeed, psychophysical
and neural recording data indicate that amblyopia is also associated with abnormalities in extrastriate regions (reviewed in Kiorpes 2015). For example,
deficits in higher-order visual functions, such as motion perception have been
described in amblyopic monkeys (Kiorpes et al. 2006) partly explained by

aberrant development of extrastriate area MT/V5. Higher brain areas and neuromodulatory pathways are also potential targets to facilitate visual responses and plasticity within V1 of amblyopic adults (Masuda et al. 2008). Regions outside of the primary sensory cortices are thought to express late, prolonged windows of plasticity that extend well beyond that of V1. Thus, devising treatments to target these regions may be an effective strategy for recovery of visual function in adulthood that does not require the reactivation of plasticity in V1. Future primate studies, ideally with tools to monitor, activate or silence specific neural circuits, will also be essential to examine plasticity within higher-order visual regions.

Concluding Remarks and Outlook

During developmental "critical periods," neural circuitry can be potently shaped by experience. Although the brain retains the capacity to rewire beyond early life, adult forms of plasticity may utilize distinct underlying mechanisms. Understanding the differences between developmental and adult plasticity, including differences in how they are measured, will provide key insights into novel therapies for recovery of visual function from amblyopia in both children and adults.

Importantly, evolving tools in neuroscience have shed new light on the "triggers" and "brakes" that determine the onset and offset of critical periods. Strikingly, the brain's intrinsic potential for plasticity is *not* lost with age, but is instead actively constrained beyond the early critical periods. Indeed, lifting molecular "brakes" unmasks potent plasticity in adulthood. Ongoing work to determine how the various "brakes" act within common cellular and circuit networks will lead to targeted therapeutic strategies to promote plasticity; that is, biologically inspired clinical studies for functional recovery.

Future work should include the development of better models for critical period plasticity across animal species and humans. Several molecules implicated in regulating the timing of the critical period, including the constraints on adult plasticity, are known risk factors for neurodevelopmental disorders such as schizophrenia (e.g., redox imbalance, HDAC and NRG1) (Rico and Marín 2011; Penzes et al. 2013; Do et al. 2015). Curiously, male schizophrenics are two times less likely to have refractive errors (Caspi et al. 2009). Capitalizing on genetic diversity in mice and humans will provide insight into the individual variability that influences the timing of critical periods. Identifying biochemical and electrophysiological correlates of these windows will allow us to compare developmental trajectories more readily across species and disease states.

6

Epigenetic Mechanisms and the Developing Brain

Bridging the Nature–Nurture Divide

Sarah R. Moore and Michael S. Kobor

Abstract

Epigenetic mechanisms are critical to the developing brain. This chapter reviews epigenetic mechanisms, their involvement in the processes of brain development, and the literature suggesting that epigenetic mechanisms may account for the enduring effects of environmental factors on the brain and behavior in human development. Epigenetic factors guide the expression of the genome in response to the intrinsic signals inherent to the processes of embryogenesis, neurogenesis, cell migration, synaptic transmission, and the timing of developmental windows. Moreover, evidence suggests that epigenetic regulators may account for the embedding of early social experiences within neurobiology. These early modifications to the epigenetic code are hypothesized to have consequences for developing neural structures and function. Epigenetic changes might also channel or moderate the effects of genetic variation on emotional and cognitive processes, and psychiatric conditions. Thus, the study of the epigenetic consequences of early-life environments may shed light on the biological pathways of environmentally induced risk.

Introduction

Epigenetics refers to the processes that allow identical DNA sequences to give rise to a diversity of cells. Waddington first reasoned that there must be contextual factors acting "upon" (the Greek root meaning of "epi") the genome to guide developmental processes (Van Speybroeck 2002). Today, the elucidation of these mechanisms has been achieved in part through the study of epigenetics, more recently defined as the structural adaptation of chromatin in a manner that alters or regulates the activity states of genes without modifying the genetic code itself (Bird 2007; Meaney 2010). Hence, the expression of the genome (i.e., DNA and encoded nucleotide sequences) is better understood accompanied by knowledge of the epigenome, inclusive of the structures and molecules affecting the packaging of chromatin and activity states of DNA, which can be both the cause and the consequence of the transcription of genomic material (Jones et al. 2013).

Beyond unraveling the cellular mechanics behind the selective transcription of gene sequences, epigenetics has provided insight as to how developmental signals, ranging from intracellular to external stimulation, might impact the expression of the genome. Epigenetic processes are essential to the developing biology of the body and brain: in conjunction with programs initiated by transcription factors, epigenetic marks determine the stable histological fate of cells, yet also allow the organism's characteristics to develop and adapt appropriately to environmental context. Because the epigenome is plastic and modifiable, epigenetic marks not only alter gene expression, they may also carry the vestiges of developmental history. In this way, epigenetic patterns may serve as a link between the interplay of genotype, developmental context, and functional biology.

In the following, we begin with a brief review of epigenetic mechanisms and then describe how epigenetic modifications channel gene expression in the formation and strengthening of neural connections over the course of development. Specifically, we review detailed evidence that (a) epigenetic modifications are critical regulators in normative neurodevelopmental processes, including the organization of the nervous system in embryogenesis, neurogenesis, neuronal migration, and synaptic transmission and plasticity, and (b) epigenetic modulation might account for how early environmental experiences mold the developing brain in humans.

Epigenetic Mechanisms

Our focus here is on epigenetic mechanisms that are involved in the regulation of transcriptional potential. Some of these mechanisms act at the level of chromatin, the packaging of DNA within chromosomes which allows ~2 meters of DNA in each cell to be condensed within the cell nucleus. The basic unit of chromatin is the nucleosome, comprised of 147 base pairs of DNA wrapped around a histone protein octamer, with each cell containing 3×10^7 nucleosomes, connected together by linker DNA. Some epigenetic "marks" or chemical tags affect how loosely or tightly chromatin is wound around the nucleosomes, and consequently the degree of physical access of DNA to transcriptional machinery. Other epigenetic mechanisms act directly at the level of DNA structure, while others activate or inhibit transcription factor proteins, affecting their ability to enhance or inhibit expression. Finally, epigenetic mechanisms may involve noncoding RNA, which regulates gene expression at the transcriptional or posttranscriptional level.

DNA methylation is a well-characterized epigenetic mark, most popularly studied as the covalent, chemical modification to a cytosine base adjacent to a guanine base (i.e., a CpG dinucleotide). CpG islands are areas of the genome with high CpG content (Illingworth and Bird 2009). Rather than a random distribution across the genome, CpG islands are found proximal to 70% of gene promoters, the noncoding sequences preceding coding DNA regions where transcription elements bind to regulate gene activity (Saxonov et al. 2006; Illingworth and Bird 2009). Higher DNA methylation in promoter regions (which often means within CpG islands) is linked to lowered gene expression, whereas in the gene body or coding region, methylation is more often associated with enhanced gene expression (Jones 2012). This trend generally holds true when analyzing genes within individuals; however, in cross-individual comparisons of single genes, the relationship between DNA methylation and transcription is more complicated and may relate to a negative, positive, or null relationship with gene expression (Lam et al. 2012; Gutierrez-Arcelus et al. 2013; Klengel et al. 2014).

DNA methylation is not restricted to CpGs; non-CpG methylation (CpH; H = A, T, or C) patterns in the mammalian brain demonstrate conservation across species. CpH methylation, unlike the majority of CpGs, is established *de novo* during neuronal maturation, suggesting it could be a key regulatory mechanism for the neuronal genome. CpH methylation, like CpG methylation, can repress transcription *in vitro* and is bound by MECP2 (a protein essential to neuron function) *in vivo* (Guo et al. 2013). Finally, the abundance of CpH in the frontal cortex of the mammalian brain, with its levels inversely related to gene transcription, further supports functional relevance (Lister et al. 2013). It has also been shown that DNA can be actively demethylated via the hydroxymethylation of CpGs (Guibert and Weber 2013; Jones et al. 2013), which is critical for neuronal differentiation and function (Santiago et al. 2014). Thus, DNA methylation and demethylation, both within CpGs and at CpH sites, are avenues of epigenetic regulation.

Histone modifications describe posttranslational alterations to the histone proteins of nucleosomes: H2A, H2B, H3, and H4. The structure of nucleosomes and chromatin are affected by a number of modifications (e.g., acetylation, phosphorylation, methylation) that can occur at over 100 sites of protein N-terminal tails, as well as across histone core domains (Mersfelder and Parthun 2006; Bridi and Abel 2013). These numerous modifications can influence nucleosome stability and positioning, ultimately affecting the state of chromatin and accessibility of particular genes (Venkatesh and Workman 2015). It has been suggested that specific combinations of histone modifications, referred to as the "histone code," correspond to particular transcriptional states (Strahl and Allis 2000; for discussion of the debate surrounding the histone code hypothesis, see Rando 2012). The enzymes that modify histones occur together within regulatory complexes, guiding the co-occurrence of synergistic histone marks necessary for transcriptional outcomes (Day and Sweatt 2011). In addition to targeting specific genes, histone marks can localize within a gene to regulate specific locations within the exon–intron structure, providing selective gene readout (for further details, see Day and Sweatt 2011). Finally, the canonical histone proteins mentioned above can be replaced by histone variants independent of replication, and these variants can result in differentiation of chromatin with epigenetic consequences (Henikoff and Smith 2015). Taken together, modifications to histone proteins and histone variants offer multiple levels of complexity in terms of epigenetic regulation.

Noncoding RNA molecules (ncRNA), including long RNA, microRNA (miRNA), small interfering RNA, and small nuclear RNA, serve additionally as epigenetic marks with effects on activation, repression, and interference with expression. The majority of the mammalian genome is transcribed into ncRNAs (molecules that do not encode for proteins) and comprise an additional layer of internal cellular information (Mattick and Makunin 2006). Gene expression is controlled by ncRNAs at multiple levels (e.g., chromatin architecture, epigenetic memory, splicing, transcription and translation) for the normal processes of physiology and development. For instance, miRNAs bind to the 3′ untranslated messenger RNA regions or mRNA coding sequence, degrading mRNA or regulating its expression through translation (Day and Sweatt 2011). Composed of 20–25 nucleotides of noncoding RNA, miRNAs are particularly relevant, as they control the expression of the majority of genes in the genome. As will be reviewed in more detail below, miRNA regulation is a key mechanism for development and plasticity of the nervous system.

Two additional means of epigenetic regulation are noteworthy and relevant to this discussion. First, the regulation of the expression of specific isoforms of a protein arises through a process of alternative splicing of exons. Epigenetic marks, such as histone modifications (Bridi and Abel 2013) or DNA methylation at exon–intron junctions (Jones 2012), can regulate alternative splicing, thus leading to different splice variants of the same gene, which have different functions and affinities for effector proteins. For instance, modifications to histone proteins can affect the recruitment of splicing regulators, and thus the protein product outcome of splicing.

Second, genomic imprinting describes the acquirement of epigenetic modifications to DNA or histone proteins (discussed above) from one of the parental gametes in a manner that biases the expression toward only one gene copy (Perez et al. 2016). Genomic imprinting occurs in at least ~50 human genes (Ishida and Moore 2013), and these parental expression biases have important functional significance for both imprinted and nonimprinted genes within regulatory gene networks (Perez et al. 2016). Although imprinting was originally identified as the silencing of one parental allele and consequent monoallelic expression of the second parental allele, the prominence of parental allelic bias on a continuum from weak to monoallelic expression, rather than solely an all or none monoallelic effect, has since been identified (Perez et al. 2016). Whether monoalleic expression or some level of parental bias of an imprinted gene occurs is complex and often depends on tissue or developmental stage (Martinez et al. 2014).

In summary, epigenetic machinery can be thought of as an overlay to the genome, providing the flexible and specific gene readouts required for the complex patterns of expression that take place in the development of the organism.

Neurodevelopment

Given the connection between epigenetic modification and gene expression, it is intuitive that epigenetic mechanisms may play a role in bridging from the genetic code to complex neurodevelopmental processes. In neurodevelopment, cells must be able to express or repress sets of genes to ensure that cells differentiate and migrate to proper locations, and that synaptic connections form and adjust. Consistently, epigenetic shifts occur simultaneously with normative phases of brain development and plasticity in mammalian neurodevelopment. Below, we describe the role of epigenetic mechanisms in shaping embryogenesis, neurogenesis and migration, neuronal plasticity, and critical windows of development in which the fate of neurons and circuitries may be especially sensitive to the effects of external stimulation. The focus is primarily on DNA methylation and ncRNA epigenetic mechanisms, as these are extensively studied, but we also highlight other processes reviewed above when relevant.

Embryogenesis

In humans, embryogenesis occurs during the first eight weeks of development, in which a fertilized egg is transformed to a multilevel body plan. Although epigenetic mechanisms are highly involved in guiding the processes required for the differentiation of cells into the various types across the body and brain, here we focus on the differentiation of cells in the central nervous system.

Mammalian neurodevelopment involves a coordinated sequence of genomic methylation and demethylation in the creation of functionally distinct neuron and glia populations (Wu and Zhang 2014). In human embryogenesis, two waves of genome-wide DNA demethylation occur: the paternal genome is demethylated a few hours post fertilization, and the maternal genome is demethylated after the two-cell embryo stage (Haaf 2006). This global DNA demethylation is followed by epigenetic reprogramming, in which epigenetic factors intersect with transcription factors to assist the differentiation of cells into at least 200 different histological types through the calibration of ~20–25k protein-coding genes.

During this phase of epigenetic reprogramming, the patterns of DNA methylation which emerge during cell differentiation are reliably reproduced in daughter cells. The epigenetic patterns responsible for stable expression profiles specific to cell type, as well as random defects in epigenetic marks, are maintained across the human lifespan. The mitotic replications of differentiated cells are subject to stochastic errors, the rate of which is higher for epigenetic marks relative to DNA replication. In humans, it is very difficult to parse marks that stem from random developmental processes versus those that arise due to important cellular or environmental signals with potential functional consequences, though each of these sources may lead to stable epigenetic differences between individuals.

During embryogenesis, the differentiation of neural stem cells into neurons and glia requires the induction of multiple transcription factors that activate cell type-specific transcriptional programs. This process is highly regulated by imprinted genes. For instance, a maternally expressed miRNA cluster promotes the shift from neural stem cell proliferation to differentiation and migration (Rago et al. 2014) and neural stem cells express a paternally expressed zinc finger protein, PLAGL1 to express a maternally imprinted gene promoting the arrest of neural stem cell cycle, and subsequent differentiation (Hoffmann et al. 2014). Imprinted genes are required for higher-level specific structures as well, such as the differentiation of GABAergic interneurons and Golgi cells in the cerebellum (Chung et al. 2011) and midbrain dopaminergic neurons (Hoekstra et al. 2013).

In summary, the transition of a single cell to the high-level organization of the emerging mammalian brain is highly regulated by epigenetic processes. The creation and placement of new cells, discussed in the following sections, involves a similar set of epigenetic machinery.

Neurogenesis

Neurogenesis is the process by which neural stem or progenitor cells generate new neurons during embryonic and perinatal development. Neurogenesis also occurs in adulthood within the subventricular zone of the lateral ventricles as well as within the subgranular zone of the dendate gyrus (Ming and Song 2011). Epigenetic mechanisms guide neurogenesis during early development through coordinated responses to extracellular signals, which modulate the expression of transcription regulators controlling cell proliferation, cell-type specification, and the differentiation of neural progenitor cells. In adults, epigenetic modifications remain critical for maintaining neural progenitor cells and guiding their fate through spatial and temporal expression of transcription regulators (Yao et al. 2016).

Multiple epigenetic modulators are required for neurogenesis. Protein complexes that orchestrate histone methylation and demethylation are in tight control of gene expression during neurogenesis. Transcription factors that bind to DNA to regulate neurogenesis are guided to proper sequences by the presence and absence of DNA methylation marks (Wang et al. 2016). During embryogenesis, miRNA miR-19 is responsible for neuronal progenitor cell proliferation and radial glial cell expansion. Long RNAs (a type of noncoding RNA mentioned above possessing over 200 nucleotides) recruit transcription factors that bind to intergenic regions to modulate the expression of key homeobox genes (Yao et al. 2016). These long RNA molecules are also essential for the neurogenesis of GABAergic interneurons in the postnatal hippocampus (Yao et al. 2016). Hence, epigenetic mechanisms are critical regulators of neurogenesis across developmental time.

Neuronal Migration

Neuronal migration plays a critical role in establishing cell identity and functional connectivity in the developing brain, and involves key epigenetic modulators. For instance, through the regulation of transcriptional programs, histone methyltransferase Ezh2 controls the topographic neuronal guidance and connectivity of the pontine nuclei, which serve as the main relay point between neocortex and cerebellum (Di Meglio et al. 2013). Epigenetic mechanisms also inhibit neuronal migration once neurons have reached their destination. For example, cortical neuron migration is inhibited when DCX is silenced by maternally expressed miR134 (Gaughwin et al. 2011). Finally, much of the guidance of migration is dependent on neuronal activity, which shapes expression through epigenetic pathways. Maternally expressed KCNK9 controls resting potentials and excitability of neurons, and maternally transmitted mutations in this gene are responsible for impaired neuronal migration and maturation of dendrites in Birk-Barel syndrome (Bando et al. 2014). To conclude, epigenetic modulators guide the proper migration of neurons in the developing brain as well as, ultimately, the establishment of functional circuitries.

Synaptic Transmission and Plasticity

Consistent with a role for epigenetic regulation in plasticity processes, there is substantial evidence for their involvement in activity-dependent processes critical to the formation and plasticity of neural connections. Empirical investigations in animal models have exemplified how various markers, including histone modifications and DNA methylation, may orchestrate sets of changes to specific signals and behavioral experiences, and that these changes are necessary for plasticity. These changes are relevant from the level of individual cell adaptation to the multicellular plasticity events underlying the formation and consolidation of memory.

Brain-specific miRNAs, which are transiently and locally expressed in dendrites and responsive to neuronal activity, have been shown to mediate the regulation of gene functions that contribute to learning and memory (Bredy et al. 2011). A number of key miRNA proteins are expressed in dendrites and regulate spine formation and synaptic plasticity within hippocampal neurons via regulation of the expression of brain-derived neurotrophic factor (Ye et al. 2016). It has been suggested that miRNAs contribute to neural plasticity and memory by regulating dendrite morphogenesis in early development and by fine-tuning gene function via translation regulation within synapses (Bredy et al. 2011).

The proper wiring of neuronal circuits through regulation of synaptic transmission is highly dependent on genomic imprinting mechanisms, as imprinted genes are involved in neuronal transmission, and on activity-dependent alterations of neuronal excitability states. Maternally expressed KCNK9 regulates membrane-resting potential and changes in firing patterns (Musset et al. 2006; Brickley et al. 2007). Imprinting mechanisms also regulate presynaptic vesicles, important for the strength of postsynaptic signals, and long-term potentiation (LTP) of NMDA and AMPA glutamate receptors, which are mechanisms of excitatory synaptic plasticity (Fleming and England 2010). In addition, imprinted genes modulate the maintenance of excitatory–inhibitory balance within neuronal circuits (Wallace et al. 2012).

Histone modifications are responsible for the transcriptional flexibility observed in the cellular events of memory formation. Epigenetic networks regulate short- and long-term changes in the chromatin environment, modifications which are required for memory acquisition and synaptic plasticity in the cortex, hippocampus, striatum, and amygdala (Bridi and Abel 2013). Histone acetylation is important for reconsolidation and extinction; phosphorylation is involved in the transcriptional effects triggered by external stimulation; and histone methylation has been linked to the activation and repression of the protein complexes that regulate histone acetylation in plasticity processes (Bridi and Abel 2013; Ciccarelli and Giustetto 2014). For instance, interference with the molecular mechanisms regulating histone acetylation alters associative learning and the LTP cellular correlates of memory (for a review, see Bridi and Abel 2013).

It also appears that histone variants could be key players in guiding neuro-plasticity processes. A variant of histone H2A, H2A.Z, actively replaces H2A following fear conditioning in the hippocampus and cortex. H2A.Z appears to mediate gene expression in these brain areas in a manner that inhibits the formation of memory (Zovkic et al. 2014). Furthermore, a recent study in mice reported H2A.Z deposited at the promoter of activity-dependent genes is responsible for triggering their deactivation and interfering with dendritic pruning (Yang et al. 2016). Essentially, through its regulation of activity-dependent transcription, the presence of this histone variant has lasting implications for the patterning of dendrites and the coding of sensorimotor information in the brain. Histone variants, then, may be an important and to date understudied epigenetic mechanism relevant to sensorimotor and cognitive processes in key brain regions.

DNA methylation also plays a role in synaptic plasticity (Baker-Andresen et al. 2013). One study observed genome-wide CpG methylation before and after neuronal activation in the adult mammalian brain, reporting dynamic changes in DNA methylation (Guo et al. 2011). These alterations in methylation were prevalent in areas of lower CpG density and occurred within genes involved in brain development and neuronal plasticity. The authors concluded that there may be a key role of the DNA methylome in activity-dependent epigenetic regulation of neuroplasticity, which may concentrate around areas of low CpG density rather than CpG islands (Guo et al. 2011).

DNA methylation within various brain regions is also relevant to memory formation and consolidation. DNA methyltransferase (DNMT) enzyme activity regulates the induction of hippocampal LTP: DNMT expression is significantly enhanced within the hippocampus after contextual learning takes place, and blocking its activity in the hippocampus disrupts the formation of associative memory (Yu et al. 2011b). The brain-derived neurotrophic factor (BDNF) gene, which has been linked to the persistence of fear memories, shows dynamic changes in DNA methylation in the hippocampus in response to fear conditioning (Lubin et al. 2008). Moreover, changes in BDNF methylation are reversed with the application of a DNMT inhibitor and NMDA receptor blocker, both of which correspond with impaired memory formation. DNA methylation and histone acetylation within the amygdala have also been shown to support learning and memory processes. Finally, the cortex has been targeted in the study of epigenetic regulators of LTP, as the hippocampus and amygdala have been implicated in associative learning and memory, but are not essential for long-term memory. Studies focused on the cortex have shown that contextual fear conditioning has robust and enduring alterations in DNA methylation in the anterior cingulate cortex, at least 30 days following conditioning. Long-lasting memory can be reversed by inhibiting DNMT in the anterior cingulate cortex, suggesting the ongoing relevance of DNA methylation in the cortex for memory stabilization (Day and Sweatt 2011). Thus, DNA methylation appears to modulate learning and memory within multiple areas of the brain involved in associative and long-term memory formation.

In summary, epigenetic mechanisms have been implicated as part of the activity-dependent machinery responsible for the formation and molding of synaptic connections. Because neuronal activity is the critical mechanism by which external stimulation modulates neural circuits, this suggests that epigenetic modulators are indeed a critical factor in bridging signals from the external environment to functional neurobiology.

Epigenetic Regulation of Critical Periods of Plasticity

Early development marks a phase of heightened plasticity and malleability to contextual surroundings, encompassing critical periods of brain development. A critical period is a window of enhanced developmental plasticity in which experiences have accentuated, irreversible effects on neural circuits

(Fox et al. 2010). For instance, there is a critical period for exposure to linguistic experience in humans—including variation in sounds, language, and practice producing language—which is required during an early developmental window if humans are to develop the neural circuitries underlying the capacity for speech (Kuhl 2004). Critical periods of brain development also render infants and children more susceptible to the effects of the social environment. For example, in a study of Romanian orphans randomly assigned to foster homes at various ages, it was observed that children placed prior to the age of 2 years demonstrated substantial gains in cognitive and emotional outcomes relative to children who remained at the orphanages (an environment of extremely high social deprivation) until later ages (Zeanah et al. 2011).

These critical periods are initiated, guided, and terminated by epigenetic molecular events affecting the expression of neuroregulatory genes (Fagiolini et al. 2009). The epigenetic molecular substrates of these developmental windows, serving as triggers and breaks, can initiate and constrain brain plasticity (Takesian and Hensch 2013; Werker and Hensch 2015). Advances suggest that the brain's default state is plastic, with the solidification of cell functions and neural networks requiring a timed and synchronized suppression of this plasticity. For example, the end of the critical period for acquisition of ocular dominance involves a downregulation of vision-dependent acetylation and phosphorylation of histones (Putignano et al. 2007). In adults, the removal or inhibition of histone deacetylases (the enzymes that remove acetyl groups from histones) reactivates plasticity in the primary visual cortex via changes of chromatin organization that enhance the accessibility to transcription (Lennartsson et al. 2015). Thus, epigenetic molecular mechanisms not only drive ongoing plasticity processes, they open and close critical windows in which brain development is particularly sensitive to incoming signals.

Summary

The architecture of the brain is established and modified by a continuous series of dynamic interactions between the genome and the developmental signals which modulate its expression via epigenetic machinery. Chromatin structure is dynamic and incorporates hundreds of signals from the cell surface to achieve the coordinated transcriptional outcomes that guide each of these neurodevelopmental steps. Epigenetic marks upon chromatin and DNA integrate these signals, creating the enduring signatures that determine cell fate, guide neurogenesis and migration, synaptic plasticity, and even the opening and closing of developmental windows. In addition, epigenetic mechanisms bridge external environmental signals to neurodevelopment within critical periods of plasticity, as will next be explored.

Neurobiological Embedding of the External
Environment during Critical Periods

Beyond intrinsic cellular signals, epigenetic machinery appears to be responsive to external environmental cues and possibly responsible for encoding these signals within developing neural circuitries. Since the brain is highly plastic during early critical windows, and associations between stressful early experiences and later developmental outcomes in humans have been observed, researchers have sought to explore the possibility that epigenetic mechanisms account for the impact of early experiences on the developing brain and subsequent psychological outcomes.

Here we briefly discuss the animal work that spearheaded investigations of epigenetic modulators in relation to human early social experiences. We then explore the existing human literature that links early environments to epigenetic markers and neural structure and function.

Animal Research: Potential Link between Early Social
Environments and Epigenetic Machinery

The potential role of epigenetic marks as a biological consequence of early forms of social environmental adversity was first studied by leveraging natural variation in maternal behavior in rats (Weaver et al. 2004). Epigenetic markers in stress-related areas of the brain were compared between rat pups that experienced low versus high maternal care (measured by frequency and duration of licking and grooming). Low levels of care corresponded with upregulated hypopituitary adrenal axis (HPA) reactivity through an epigenetic mechanism: pups reared in low-care early environments demonstrated increased DNA methylation and decreased histone acetylation of the glucocorticoid receptor (GR) gene *NR3C1*, which ultimately was associated with diminished expression of GR and upregulation of CRH secretion and HPA activity. Although this seminal study sparked a large body of research exploring methylation of the GR promotor, it is important to note that efforts to replicate the original finding in terms of direction and effect size are still underway (Pan et al. 2014; for a discussion, see Boyce and Kobor 2015).

Further investigations in mice and rats reported links between maternal care, maternal separations, and communal rearing and altered methylation and histone acetylation in the hippocampus, as well as links between maternal care and epigenetic markers within the hypothalamus, amygdala, pituitary gland, and prefrontal cortex (for a review, see Kundakovic and Champagne 2015). There is additional evidence in rhesus macaques that rearing conditions and social rank associate with subsequent epigenetic variation within blood and prefrontal cortex (PFC) cells, as well as gene regulation (Kinnally et al. 2011; Provençal et al. 2012; Tung et al. 2012). These primarily experimental findings suggest a causal link between early social experiences and subsequent epigenetic profiles within critical regions of the brain. Next, we discuss the evidence for this connection in human studies utilizing both peripheral and central tissues.

Human Research on Association between Early Environments and Epigenetic Markers

Although limited in the ability to distinguish cause from correlation, human work on early environmental factors and DNA methylation has demonstrated potential biological signatures of early adverse experiences. Links to methylation patterns may suggest that early social experiences had effects on neurodevelopmental processes, although this cannot be determined. Beyond the correlational nature of human studies, another noteworthy issue is the typical use of peripheral tissues (including blood, saliva and buccal epithelial cells) rather than brain tissue to obtain epigenetic profiles. Brain tissue is obtained postmortem or during a required surgical procedure, and thus is not a viable option for large-scale studies. Because different tissues show distinctive epigenetic patterns, and variation in methylation is most largely driven by cell-type composition (Jaffe and Irizarry 2014; Farré et al. 2015), the biological variation associated with the cell type of the collected tissue must be taken into consideration to identify any meaningful interindividual differences in methylation profiles. Finally, it is worth noting that although the focus on DNA methylation in this literature is typically tied to an interest in the expression of genes in the developing brain, the relationship between epigenetic marks and gene expression is complex. DNA methylation might be the cause or the consequence of gene expression, which can depend on the direction of a DNA methylation change, location relative to the gene, and specific function of CpGs. With these caveats in mind, let us look at the existing literature linking prenatal and postnatal environments to patterns of DNA methylation in humans.

Prenatal Environment

Substantial correlational work in humans has linked prenatal experiences to differential methylation patterns at later developmental periods. This suggests that epigenetic marks are involved in potential prenatal programming of subsequent phenotypes. Most of this work has targeted candidate genes. Here we briefly highlight interesting findings, but refer the reader to Cao-Lei et al. (2016) for a more detailed review.

Work in human prenatal exposures is generally limited to maternal reports of mood and exposures encountered during pregnancy. Much of the candidate gene work has focused on *NR3C1*, the gene encoding the gluco-corticoid receptor, which was initially implicated in experience-dependent epigenetic regulation of the stress response in animal work (for a review, see Turecki et al. 2016). In a number of studies, prenatal stress in the form of maternal depression, exposure to intimate parent violence, psychological well-being, and extreme trauma was associated with methylation of *NR3C1* in offspring at later developmental periods (Oberlander et al. 2008; Radtke et al. 2011; Hompes et al. 2013; Perroud et al. 2014). Moreover, the link between intimate partner violence and *NR3C1* methylation status in children was specific to the prenatal period, as no association was found for mater-nal stress prior or subsequent to pregnancy (Radtke et al. 2011). Notably, this popularly targeted promoter region is largely invariable, with very low levels of DNA methylation across individuals. Consistently, the effect sizes reported are extremely small. Perhaps more problematic, the majority of can-didate gene methylation studies do not adjust for cell-type composition of samples, the largest contributor to variability in DNA methylation. These pitfalls warrant interpreting these findings, as well as additional candidate findings targeting invariable promoters discussed below, with a healthy dose of caution.

DNA methylation of several additional candidate genes has been studied in relation to prenatal stressors. Methylation status of *HSD11B2* (involved in glucocorticoid responses) and *SLC6A4*, but not BDNF, in neonatal cord blood were associated with prenatal socioeconomic deprivation (Appleton et al. 2013). In a well-powered study involving 500 pregnant women, the dif-ferentially methylated regions of several candidate imprinted genes were in-vestigated and found to associate with severe maternal depression, increased low birth weight (Shapero et al. 2014), and heightened DNA methylation levels (Liu et al. 2012b). Finally, periconceptual exposure to famine during the Dutch Hunger Winter was found to associate with differential methyl-ation of developmental and immunological genes in late adulthood (Tobi et al. 2009).

Several studies have tackled associations between prenatal stress and infant methylation at the genome-wide level. In one study, differential DNA meth-ylation at *CYP2E1* (initially discovered via array findings and followed up by pyrosequencing the gene for greater coverage), a gene involved in metabo-lism, was predicted by maternal mood, selective serotonin reuptake inhibitor (SSRI) exposure, and their interaction. In turn, methylation status related to infant birth weight (Gurnot et al. 2015). In a prospective study, women were recruited who had experienced the 1998 Quebec Ice Storm while pregnant. Reported objective hardships experienced due to the disaster and subjective distress were related to methylation at numerous sites, with sets of CpGs reported that were both specific to and overlapping between objective and subjective measures. The functions of identified genes were predominantly related to immune function, and DNA methylation was found to mediate a link found between subjective stress and child immune and metabolic status (Cao-Lei et al. 2014).

In another study, minimal differences were found in DNA methylation in cord blood between infants born to mothers who had or had not experienced depression, with very weak differences reported for two significant CpG sites; however, the control group may not have sufficiently differed from the depressed group, as women in the control group had previously been diagnosed with mood disorders (Frey et al. 1990). In another study, nonmedicated depression or anxiety during pregnancy related to differential methylation at 42 CpG sites, with significant clusters related to the regulation of transcription, translation, and cell division. No differences were found between groups exposed to SSRIs *in utero* relative to controls (Non et al. 2014). A recent study compared the methylation profiles within buccal epithelial cells of infants with fetal alcohol spectrum disorder (FASD) and healthy controls, reporting 658 differentially methylated sites (Portales-Casamar et al. 2016). The majority of differentially methylated genes, when tested in cortical tissue from the Allen Brain Atlas, demonstrated high mRNA expression as well as high correlations with methylation patterns in corresponding buccal cells, supporting the potential functional significance of the sites identified in the FASD sample. Finally, one study assessed the link between prenatal environment and neonatal methylation, focusing on variably methylated regions, and tested whether several prenatal factors, genotype, or their interaction best explained methylation outcomes in independent models. Interestingly, the majority of variably methylated regions were best explained by an interaction between genotype and prenatal environment (Teh et al. 2014). It is possible that future investigations could more effectively detect prenatal environmental effects on DNA methylation if genotype is taken into account, as Teh and colleagues found that prenatal environment on its own was not the best predictor of DNA methylation in variable regions in any tested models.

Postnatal Environment

Similar to research on prenatal exposures, postnatal environments have demonstrated associations with methylation patterns in offspring at later developmental time points. The majority of these studies have targeted the candidate gene *NR3C1* (which raises the same concerns as mentioned for the prenatal literature above). In a recent systematic review of this candidate literature, it was reported that child adversity and parental stress in early life related to increased methylation (Turecki et al. 2016). One very small cohort study reported that childhood abuse was related to methylation of the GR promoter within hippocampal tissue of suicide victims, consistent with patterns reported for methylation in blood (McGowan et al. 2009).

In another study the entire region of the *NR3C1* gene was investigated in the hippocampi of suicide victims who had or had not experienced abuse as children and compared to profiles obtained from rats. Numerous DNA methylation differences were identified that were conserved between human and mouse, and appeared to target regulatory sites such as gene promoters (Suderman et al. 2012).

Another popular candidate of interest is the serotonin transporter gene (5-HTT). Increased methylation of this gene was observed among monozygotic twins who were bullied relative to nonbullied siblings (Ouellet-Morin et al. 2013). In another study, a number of childhood adversities were significantly associated with 5-HTT promoter methylation status, as well as increased depressive symptoms (Kang et al. 2013). Finally, 5-HTT methylation has been linked to childhood abuse and sexual abuse (Beach et al. 2011).

Growing numbers of whole epigenome studies have tested associations between critical postnatal environments and DNA methylation patterns. In one study, maternal stressors during infancy and paternal stressors during preschool years were related to DNA methylation patterns in adolescents (Essex et al. 2013). Two studies have reported associations between methylation and early-life socioeconomic status: one suggested that a cluster of variably methylated CpG sites was correlated with socioeconomic status (Borghol et al. 2012); the other found socioeconomic status to associate with DNA methylation, perceived stress, cortisol, and inflammatory responses within peripheral blood mononuclear cells (Lam et al. 2012). These studies suggest that a broad and complex prenatal exposure, like socioeconomic status, may have implications for subsequent methylation profiles.

Several studies have investigated the effects of severe postnatal stressors on DNA methylation. In an investigation of methylation from peripheral blood mononuclear cells of adopted and nonadopted youth, adopted youth demonstrated substantial differences in white blood cell-type composition, as well as differences in methylation of genes functionally enriched for neural and developmental processes. Moreover, differences in methylation were only observed in relation to early and not later experiences of trauma (Esposito et al. 2016). In another study, orphanage rearing related to genome-wide increases in DNA methylation in the blood of children 7–10 years of age relative to parent-reared controls (Naumova et al. 2012). Finally, a study assessing differential methylation of gene promoters reported that DNA methylation differences in individuals who were abused as children were related to cell signaling pathways relevant to transcription regulation and development, including 39 miRNAs (Suderman et al. 2014).

In all, prenatal and postnatal environments appear to link to methylation signatures in neonates and across developmental time. The next body of evidence to be reviewed, linking epigenetic marks to neural outcomes, suggests that these environmental effects may indeed have implications for neurobiological development.

Human Research on Link between Epigenetic
Markers and Neural Outcomes

A nascent area of research has begun to target the structural and functional neural correlates of epigenetic patterns in humans. Given the empirically supported possibility that prenatal and postnatal environments modify epigenetic patterns, and that these critical environmental factors are linked longitudinally to physical and mental health outcomes, it is predicted that experience-dependent epigenetic marks induce functional consequences on developing neurobiology, or at least serve as biomarkers of experience-dependent neural development. The vast majority of this work has focused on DNA methylation, in both peripheral and central tissues.

Human studies relating methylation to brain function are largely limited to adult samples and single genes, but they suggest that the methylation of candidate genes relate to structure and function in regions responsible for emotion and stress regulation, such as the amygdala, hippocampus and prefrontal cortex (for a review, see Nikolova and Hariri 2015). (These candidate studies, however, suffer the same pitfalls mentioned above, in terms of focus on invariable promoter regions and failing to account for cell-type composition of samples.) Interestingly, some links between candidate gene methylation and functional activity have been further substantiated by molecular brain measures using positron emission tomography (PET). For instance, methylation of the 5-HTT promoter has been linked to limbic functional activity in response to emotional tasks in three studies (Nikolova et al. 2014; Frodl et al. 2015; Swartz et al. 2016). These functional MRI findings are consistent with a PET study which demonstrates that 5-HTT methylation status predicts serotonin synthesis within the orbitofrontal cortex, a critical region for higher-level emotional processing (Wang et al. 2012).

A recent study investigated the expression of hundreds of miRNAs following a social stress task in conjunction with functional imaging (Vaisvaser et al. 2016). The authors reported that the miRNA miR-29c expression in blood was related to perceived stress following the task, and differences in ventromedial PFC functional connectivity within regulatory regions. Vaisvaser and colleagues suggest that miR-29c may serve as a blood biomarker for neural responsiveness to stress.

To date, only three studies correlating DNA methylation to neural outcomes have additionally incorporated genotype. One well-executed study demonstrated that the association between childhood exposure to abuse and DNA methylation (from blood samples drawn during adulthood) of intron 7 of *FKBP5*, a functional regulator of the glucocorticoid receptor complex, was related to hippocampal volume and dependent on *FKBP5* genotype (Klengel et al. 2012). In a second study, genotype (Val66Met) and methylation of the BDNF gene, known to underlie synaptic plasticity was investigated (Chen et al. 2015). Chen and colleagues found that the degree to which antenatal maternal anxiety was associated with neonatal DNA methylation depended on BDNF genotype. Specifically, greater effects were observed for met/met genotype relative to met/val or val/val genotypes. A greater number of CpG sites were identified in which methylation level was related to right amygdala volume for met/met genotype; whereas the opposite pattern was observed for the left hippocampus, in which more CpG sites were correlated with volume for infants with the val/val genotype. This was the first study to demonstrate that an interaction between antenatal environment and genotype on the epigenome is reflected in substructures of the brain that are important for stress and regulation.

Finally, consistent with the notion that DNA methylation signatures may have implications for neurological function, studies on psychiatric disorders have reported links to epigenetic markers in peripheral and brain tissue. A few studies have compared methylation profiles in the peripheral tissues of monozygotic twins discordant for psychiatric disorders and found differences in the methylation of candidate genes related to neurotransmitter function (Petronis et al. 2003; Mill et al. 2006). In a genome-wide study, monozygotic twins diagnosed with major depressive disorder were reported to have greater variance overall in methylation relative to unaffected siblings (Byrne et al. 2013). A few human studies of major depressive disorder have validated epigenetic findings discovered in animal models of chronic stress (for a review, see Nestler et al. 2016). For example, a repressive histone mark, H3K27me3, implicated in the link between chronic stress and suppression of BDNF expression in the hippocampus (Tsankova et al. 2006) was found at elevated levels within the synapsin gene family of PFC tissue in individuals with depression and bipolar disorder (Cruceanu et al. 2013). An epigenome wide study of CpG-rich regions in PFC tissues of individuals with schizophrenia, bipolar disorder, and controls implicated a number of epigenetic modifications in these disorders related to glutamatergic and GABAergic signaling, neuronal development, and metabolism (Mill et al. 2008). Finally, at least two loci have been implicated in psychiatric disorders in multiple cohorts and across tissues: *HLA9* has been implicated in psychiatric disorders in postmortem brain tissue, blood, and sperm (Nestler et al. 2016) and *GAD1*, which originally demonstrated dysregulated expression and epigenetic differences in multiple brain tissues in schizophrenic individuals (Akbarian and Huang 2006), similarly demonstrated differential methylation in hippocampal tissue from individuals with bipolar disorder (Kaminsky et al. 2012). These studies provide promising evidence that epigenetic patterns found across tissues are relevant to the manifestation of psychiatric disorders.

A recent article highlights a promising possibility; namely, that epigenetic mechanisms may explain associations between genetic variability and psychiatric conditions (Bavamian et al. 2015). Specifically, the miRNA miR-34a, shown to be reduced by pharmacological treatment for bipolar disorder, was found to regulate two bipolar disorder risk genes during neuronal differentiation. Elevation of the expression of miR-34a affected mRNA and protein expression of these two genes and resulted in defects in neuronal differentiation, whereas suppression of miR-34a enhanced dendritic growth. Moreover, 25 genes found to be targeted by miR-34a overlapped with genes containing single nucleotide polymorphisms associated with bipolar disorder in genome-wide association studies. Thus, epigenetic mechanisms may ultimately help to elucidate the functional links between genetic variation and psychiatric conditions (Bavamian et al. 2015).

Synthesis of Human Research

Collectively, a growing body of human research has linked early environments to epigenetic patterns, and epigenetic patterns to neural and mental health outcomes. The following limitations associated with this literature must be emphasized (Jones et al. 2018). First, human research primarily relies on peripheral tissues, and methylation patterns may not be consistent between peripheral and central tissues. It is critical for studies that report DNA methylation findings, interpreted as potentially relevant to the brain, to utilize biobanks (e.g., Allen Brain Atlas) with peripheral and central tissues from the same individuals, and to report the correlations of methylation between tissues at discovered sites. Second, studies focused on candidate gene methylation should strategically select sites which are more likely to vary across individuals. As DNA methylation research continues to progress, it has become clear that methylation at gene promoters may not be as telling for individual differences or early exposures as originally anticipated, and that more variable DNA methylation sites with functional relevance may be found elsewhere (e.g., at enhancers in intergenic region). Third, a good deal of the early DNA methylation literature did not control for cell-type proportions in samples. Moving forward, it is highly recommended that studies utilizing array data use deconvolution methods to estimate proportions of cell types, and to control for these proportions or to sort cell types for the analysis of DNA methylation. Finally, a large proportion of the variability in DNA methylation is due to genetic influences (Henikoff and Greally 2016). Given the findings reviewed above (i.e., Teh et al. 2014), studies concerned with environmental effects on DNA methylation will be better positioned to detect these effects with accompanying genotype data so that those exposures with genotype-dependent effects can be identified.

Despite these limitations, consistent findings that DNA methylation correlates with early social environment and neural phenotypes (especially in studies that were well designed to account for the above issues) suggest that epigenetic marks capture meaningful variation in early environments as well as concurrent neurological measures and mental conditions. This evidence offers an intriguing possibility: that epigenetic marks in human central and peripheral tissues reflect an important biological substrate of experience-dependent plasticity relevant to current mental health status. As more studies begin to incorporate full genome and epigenome markers alongside neural measures, there will certainly be much to learn about the potential for epigenetic marks as (a) biomarkers of the developmental history of gene–environment interplay in the developing brain and (b) biological signatures relevant to functional neurobiology and associated mental conditions.

Conclusions

Epigenetic studies in humans offer an exciting new avenue for understanding how critical environmental factors induce enduring effects on the brain and behavior in human development. In addition to potentially guiding the expression of the genome in response to the intrinsic signals inherent to the processes of embryogenesis, neurogenesis, cell migration, synaptic transmission, and the timing of developmental windows, epigenetic regulators are also implicated in the embedding of early social experiences within neurobiology in an enduring fashion. These early modifications to the epigenetic code are hypothesized to have consequences for developing neural structures and function, channeling genetically driven effects on emotional and cognitive processes in a manner consistent with the emotional and social realities of an individual's outer experience.

7

Early Childhood

Matthias Kaschube, Charles A. Nelson III, April A. Benasich,
Gyorgy Buzsáki, Pierre Gressens, Takao K. Hensch,
Mark Hübener, Michael S. Kobor,
Wolf Singer, and Mriganka Sur

Abstract

During the first years after birth, infants face the enormous task of building a comprehensive and predictive internal model of the external world, allowing them to navigate
and interact successfully with their environment. This chapter explores the frontiers involved in understanding the neural bases of this process and how such knowledge could
be leveraged to treat and prevent neurodevelopmental disorders. It begins by describing
how developing brains form dynamical networks that integrate genetic, epigenetic, and
sensory information, emphasizing the interplay between molecules and neural activity.
Strategies are highlighted that the brain uses to tightly control the impact of sensory
input onto its developing networks, which are manifest at the molecular, neural activity, and behavioral levels, and which appear pivotal as the brain strives to maintain a
fine balance of flexible yet stable configuration. While suitable animal models have
greatly contributed to our basic understanding of neural development, revealing the
neural basis of cognitive development in humans remains a challenge. To overcome this
barrier, new directions are discussed that combine animal and human studies. Finally,
this chapter discusses implications of the complexity of the human brain and highlights
the potential of data-driven formal models of neurodevelopmental trajectories to enable
early detection and individualized treatment of developmental disorders.

Introduction

Early childhood (0–3 years) is a period in life associated with enormous potential as well as significant functional narrowing. A fascinating example can

Group photos (top left to bottom right) Matthias Kaschube, Chuck Nelson, April
Benasich, Michael Kobor, Takao Hensch, Matthias Kaschube and Mark Hübener,
Mriganka Sur, Wolf Singer, Chuck Nelson, Gyorgy Buzsáki, April Benasich, Pierre
Gressens, Mark Hübener, Michael Kobor, Matthias Kaschube, Wolf Singer, Mriganka
Sur, Pierre Gressens, Gyorgy Buzsáki, Takao Hensch, Yehezkel Ben-Ari and Wolf
Singer

be found in the acquisition of language, arguably one of the major tasks of infancy. While at birth, a human infant can learn any of ~5000 languages known to humankind, by the age of three or four years, a child is able to speak with native fluency only those few languages to which s/he has been exposed.

This remarkable *flexibility* of the human brain to be susceptible—at least for a limited amount of time—to the particular statistics of the environment may come at a cost; namely, the risk of reducing *stability*, which could be a potential precursor of psychiatric diseases. One of the overarching goals of this chapter is to address the fundamental dichotomy between building a flexible yet stable brain, focusing on the period of postnatal development. At birth, the brain is exposed— for the first time—to the full spectrum of sensory input with which it has to cope for the rest of its life. Thus, the *external* world becomes more effective in driving neural activity, allowing the brain to learn relevant statistics of the sensory environment. The brain, at this point, is already equipped with an elaborate *internal* structure, both in terms of anatomical connections and endogenous neural activity patterns. In the first years of life, a shift from more internally to more externally generated neural activity occurs, during which an internal model of the external world is established. Strategies of the brain appear pivotal to control its susceptibility to sensory input, allowing it to navigate through this complicated process between the poles of flexibility and stability.

Our discussion here is guided by five main questions:

1. How does neural activity evolve during early childhood? How does the continuum from internally to externally driven activity shape the brain?
2. What is the role of critical periods? What are the factors that initiate and terminate these periods? How does critical period plasticity differ from adult plasticity?
3. What is the interrelation between activity and epigenetics?
4. What measurements and interventions do we have at our disposal?
5. What are the signatures of typical and atypical development?

We begin by reviewing the neural underpinnings of postnatal brain development in infants, emphasizing the shift from internally to externally generated neural activity and the interrelation with different molecular processes. We identify different strategies that may help to increase the overall level of robustness at this stage in development. We then focus our discussion on the neural basis of critical period plasticity and its relation to adult plasticity. In addition, we highlight novel directions in identifying the role of epigenetics in neural development. Using the important example of language acquisition in humans, we review current methods for measuring cognitive abilities in typically developing infants as well as those displaying atypical development. Here, we address both behavioral assays and noninvasive methods for measuring brain activity, and discuss current and future opportunities for combining animal and human studies to understand and develop novel treatments for developmental disorders. Moreover, we discuss recent conceptual frameworks for thinking about typical and atypical cortical development to leverage these concepts toward devising novel strategies for individualized treatments of developmental disorders.

Neural Activity in Early Childhood

Building a Cortical Scaffold and Using It to Establish a Model of the World

As discussed by Singer (this volume), the basic layout of the brain develops according to the same principles of cellular recognition and is based on the same molecular signaling cascades as any other organ. However, once nerve cells become electrically excitable, neuronal activity assumes a self-organizing role, which gains in importance with increasing differentiation of neuronal networks. Initially, activity serves a trophic function as it promotes secretion and uptake of molecules that promote growth and differentiation. As soon as primitive networks are formed, this spontaneous activity undergoes specific spatiotemporal patterning (e.g., traveling waves, periodic bursting, synchronous oscillations), and the emergent correlation structure is exploited for the definition of neighborhood relations. These data are used to refine projection patterns, to establish precise correspondence of maps, and to set up and fine-tune circuits acting as central pattern generators required for the later coordination of movement. The circuit changes follow the Hebbian principle: "neurons that fire together wire together." Once sensory signals become available (beginning in the late prenatal period), the self-generated activity is structured by further taking into account the influence of signals from the environment; thus, signals resulting from active exploration of the environment come to impact development (Held and Hein 1963).

During this postnatal developmental stage, activity is used to translate the statistical contingencies of events and features of the surrounding world into neuronal architectures, to complement the genetically prespecified model with information from the actually experienced environment (see Hensch, this volume). This additional information is used to refine connections to a degree not attainable with genetic instructions alone. Well-examined examples include the establishment of precise binocular correspondence (Wang et al. 2010), the matching of visual and auditory maps in the barn owl tectum (Knudsen and Knudsen 1989; Knudsen 2002), and generation of visual response features in rewired auditory cortex (Sharma et al. 2000; Sur and Rubenstein 2005). Others are the imprinting of species-specific song patterns or the templates of kinship (Brainard and Doupe 2002; Bolhuis and Gar 2006). The rules and molecular mechanisms supporting these experience-dependent circuit modifications are essentially the same as those mediating the effects of self-generated activity and closely resemble those supporting learning in adulthood. As in the latter case, many of these experience-dependent modifications are supervised by gating systems to minimize the danger that inappropriate or spurious correlations induce dysfunctional connectivity changes. This is why passive stimulation alone is rather inefficient in causing circuit changes, as first documented by the seminal experiment of Held and Hein (1963). A host of subsequent studies indicate that the developing brain seeks, via active exploration, the signals required for optimization of its functional architecture and is capable of evaluating the consistency of activity patterns before allowing them to modify circuitry. For instance, normal language development depends on normal sociolinguistic interactions; children whose primary exposure to language comes from TV or who are socially isolated, for example, have abnormal language (Kuhl et al. 2003).

When deprivation prevents exposure to signals required for circuit adaptation to the typical environment, the genetically prespecified scaffold of connections is not simply frozen in a state prior to sensory signal onset, but rather deteriorates to a level of specificity below that reached at the beginning of the critical period (Crair et al. 1998; White et al. 2001). In other words, the scaffold becomes less precise.

Different Modes of Activity Are Matched to Different Genetic and Molecular Processes during Development

While there is now little doubt about a general central role for electrical activity in successive maturational stages, how this relates to genetic and molecular events is less clear at present; nature and nurture are inseparable. Very early in development, neurogenesis and axon guidance require calcium currents in migrating neurons and axonal growth cones to decode diffusible and membrane-bound molecular recognition signals for path finding and contact formation. These activity-based cues are likely sufficient to establish a coarse scaffold, making activity permissive for growth. Once axons reach their target, they organize topographically into maps, guided by gradients of recognition molecules and broadly structured activity, such as waves of glutamate-driven activity in the early retina or calcium waves in the early hippocampus and cortex. Gene and molecular expression analyses have revealed the crucial role of molecular markers for axon guidance and map formation, and the complementary role of structured electrical activity in regulating gene expression, pathfinding, and contact formation (Sur and Rubenstein 2005; Assali et al. 2014). These processes are fundamentally self-organizing. We still need to resolve how activity intersects with molecules, what mechanisms are engaged, and to what extent the pattern of activity may also have an instructive role. Another open question is how waves of activity across space, which convert space into time, influence the timing and pattern of gene expression that may also underlie map formation.

Later, the fine structure of activity has an instructive role in shaping the specificity of connections. Spike timing-dependent plasticity (e.g., narrow windows when individual presynaptic and postsynaptic spikes can bidirectionally adjust synaptic strength) is a physiologically relevant mechanism that translates causal relations into modifications of synaptic connections, and is based on molecular processes involving coincidence-detecting mechanisms such as NMDA receptors. This process can be either unsupervised or controlled by top-down modulation, implementing a mechanism for supervised learning. In this mode, development and learning share many commonalities and the boundary between them becomes fuzzy. Exactly how supervised learning influences development, and its instantiation, remains an important open question.

Plasticity, Stability, and Development

Adaptive processes are deeply embedded in living systems, including the mammalian nervous system, and are present throughout life to a greater or lesser extent. They are observed following manipulations, such as monocular deprivation, which reveal the range of possibilities or potential for plasticity early in life. The development of stereopsis by matching inputs from both eyes is a good example of how activity-dependent circuit selection can build on external information to optimize correspondence between maps (Wang et al. 2010). Another example is the tuning of auditory maps underlying perception (Sanes and Woolley 2011).

Plasticity and stability are fundamentally related processes, and indeed processes of stability are essential elements of plasticity. Hebbian increases in synaptic weight need to be countered by homeostatic processes to constrain (normalize) levels of total drive and to permit neurons to operate within an optimal dynamic range. This process necessarily reduces the weights of unstimulated synapses, thereby providing greater "contrast" for stimulated synapses to influence the target cell.

Developmental plasticity involves strengthening and growth as well as weakening and elimination of synapses (see Figure 7.1). Synapses that undergo activity-dependent weakening may eventually be removed (pruning). Synapse elimination may exist as a process at early stages of development, sometime even prior to birth, as revealed for instance by the elimination of multiple climbing fiber inputs to Purkinje cells (Hashimoto and Kano 2005; Uesaka et al. 2015). In cortex, synapse elimination and synapse formation coexist as dual mechanisms of plasticity (Mataga et al. 2004), though many questions remain about how widespread the phenomenon is and how it might be implemented. In addition, synapse elimination may involve nonneuronal phagocytic cells such as microglia. Some of the molecular bases of this neuron–microglia cross talk have been identified (Paolicelli et al. 2011; Schafer et al. 2012), but several questions remain unanswered, such as the role of neural activity in this interaction.

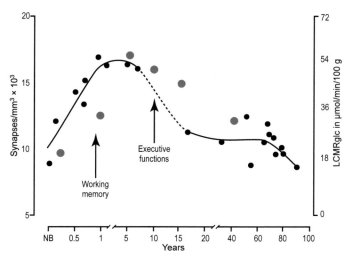

Figure 7.1 Variation in the human prefrontal cortex during development: Density of synapses in layer III of the medial frontal gyrus (black circles) and resting glucose uptake (LCMRglc/PET) in the frontal cortex (gray circles). Arrows point to the approximate periods of emergence of various prefrontal cortex functions. NB: newborn. Adapted with permission from Huttenlocher (1990) and Chugani (1993).

The Log-Normal Scaffold Hypothesis

Interestingly, evidence stemming largely from the hippocampus and neocortex suggests that brain circuits are poised to maintain a robust balance between flexibility and stability, owing to the observed diverse distribution of synaptic weights and firing rates. Diversity is the main "drive" of biology. The brain not only contains numerous neuron types, but a large diversity of neurons within the same type is the norm. The spontaneous firing rates of cortical neurons, and probably all neuron types, in the brain and spinal cord, are distributed over three orders of magnitude in a highly skewed, log-normal distribution. In addition to firing rates, the distribution of the fraction of neurons firing together in a given time window, the strengths of synapses on a given neuron, the volume of spines, and the macroscopic connectivity also show a skewed, typically log-normal form. The multiple levels of such systematic overarching organization may support the competing requirements of sensitivity, robustness, resilience, and plasticity in neuronal circuits (Buzsáki and Mizuseki 2014).

The dynamic range of firing rates increases during awake states and decreases during sleep. This implies that faster neurons decrease whereas slow neurons increase their rates during non-REM sleep, resulting in a narrowing and widening of rate distribution during sleep and waking, respectively. Faster firing neurons are more strongly connected to each other, forming a "rich men's club" organization, and have longer axon arbors. Recent findings indicate that high firing rate cells are "rigid" and form the backbone of local circuit dynamics. Their largely preserved firing patterns provide "good enough" and fast answers in most situations, since they generalize across situations and carry critical features of the brain's existing knowledge base. In contrast, slow firing neurons act as a large pool of "plastic" cells whose patterns change gradually during behavioral experience (Grosmark and Buzsáki 2016). The developmental origin of log-normal distributions is not known. An intriguing hypothesis is that the fast-firing, rigid end of the distribution represents early-born neurons with high resilience, whereas later-born cells comprise a large reservoir of slow firing, plastic, and perhaps more vulnerable populations.

Critical Periods

Critical periods of brain development are observed at distinct times for different modalities. As such, they constitute windows of malleability during which neural activity can robustly reorganize the circuitry that underlies specific functions. In the most thoroughly studied example, acuity loss (amblyopia) by discordant input to the primary visual cortex (V1), the duration of the critical period scales with the average life span of the species, from weeks (mice) to years (humans). This suggests a biological process that transiently optimizes circuits to best fit the individual's environment early in life followed by enduring stability, a learning rate strategy found to be efficient in artificial neural networks (Takesian and Hensch 2013).

The transition to critical period plasticity arises as inhibition emerges to preferentially suppress responses to spontaneous activity, relative to visually driven input activity, switching learning cues from internal to external sources (Toyoizumi et al. 2013). Thus, the critical period can open without changes in plasticity mechanisms when activity patterns become more sensitive to sensory experience. More broadly, hierarchical organization of sensorimotor pathways may develop through a cascade of critical periods induced as inhibitory circuit maturation progresses from "lower" to "higher" cortical areas (Condé et al. 1996; Werker and Hensch 2015).

Detailed mechanistic exploration of factors which initiate (inhibitory parvalbumin, PV+ cells) or terminate (brakes) a plastic window in V1 offers deeper insight into potential roles for critical periods (see Hensch, this volume). First, as the maturation of inhibition may preferentially suppress responses to spontaneous activity relative to visually driven input (Toyoizumi et al. 2013), critical period onset reflects a switch in learning cues from internal to external sources. Second, rapid plasticity of PV+ cells per se may generate changes in associated (gamma) oscillations relevant for synapse pruning. Third, critical period cascades can be coordinated through a sequential interaction of PV-cell intrinsic clocks with extrinsic, locally released activity-dependent cues. Finally, plasticity is eventually suppressed by molecular brakes. Glycosaminoglycans in the perineuronal nets, which gradually enwrap mature PV+ cells, buffer reactive oxygen species generated by these fast-spiking neurons and, in the human brain, are least evident in higher associational areas which have evolved to be plastic throughout life (Brückner et al. 1999). This, in turn, may render those areas most vulnerable to neural degeneration (i.e., Alzheimer disease). Likewise in mice, loss of the Lynx1 "brake" molecule prolongs critical period plasticity and neurodegeneration with age (Morishita et al. 2010; Kobayashi et al. 2014). Thus, closure of a critical period may ultimately be neuroprotective. Notably, all four aspects can be impacted in mental illness, mistiming the typical trajectories of development and contributing to differential rates of emergent psychopathology (Lee et al. 2014).

Normal and Enhanced Ocular Dominance Plasticity in the Adult Visual Cortex

While ocular dominance plasticity in V1 has long been thought to be strictly limited to a critical period, a number of recent studies have shown that monocular deprivation (MD) can also induce changes in adulthood, at least in mice (Sawtell et al. 2003; Tagawa et al. 2005; Hofer et al. 2006). It has also become clear, however, that there are differences between critical period and adult ocular dominance plasticity: the magnitude of MD-induced shifts in adult mice is smaller (Hofer et al. 2006), longer deprivation episodes are needed to induce them (Hofer et al. 2006), it fails to impact visual acuity (Morishita and Hensch 2008), and there is an age-dependent decline in the degree of adult ocular dominance plasticity (Lehmann and Löwel 2008). In addition, the underlying eye-specific changes are different: whereas during the critical period a rapid drop in closed-eye driven responses is followed by a delayed increase in open eye-evoked activity (Frenkel and Bear 2004; Mrsic-Flogel et al. 2007), the latter component alone seems to be the prevalent change in adult mice (Hofer et al. 2006; Sato and Stryker 2008). However, the exact contribution of both components to adult ocular dominance plasticity depends crucially on the conditions during deprivation: when deprived mice are exposed to prolonged stimulation with high contrast moving gratings, clear closed-eye depression is observed (Matthies et al. 2013; Rose et al. 2016), mimicking phenomenologically the situation during the critical period.

A number of behavioral interventions have been shown to promote critical period-type plasticity in the adult visual cortex. Among these are environmental enrichment (Baroncelli et al. 2010) or enhanced social interaction (Balog et al. 2014), as well as high contrast visual stimulation (Matthies et al. 2013; see above) and motor activity coupled to visual feedback (Kaneko and Stryker 2014). Also, an earlier MD episode enhances the effect of a second episode later in life (Hofer et al. 2006). This is likely mediated by the formation of new dendritic spines that outlast the first MD episode and might become reactivated later to promote the second ocular dominance shift (Hofer et al. 2009). Finally, dark-rearing prolongs developmental plasticity (Tropea et al. 2010), and a period of dark exposure before MD causes prominent ocular dominance shifts in adult rats (He et al. 2006), a species that seems to show little ocular dominance plasticity in adulthood. This latter observation bears particular relevance, as a clinical study is currently underway[1] to test whether complete light deprivation might mitigate amblyopia in adult humans, on the premise that unlocking plasticity in V1 allows the amblyopic eye's inputs to reestablish connections there. A common theme to several of the above-mentioned interventions is that they induce reduced levels of inhibition and thus rejuvenate cortical excitatory–inhibitory balance, which is known to play a crucial role in the induction of critical period plasticity. More direct molecular manipulations have also provided new insights (see Hensch, this volume).

Interrelation between Neural Activity and Epigenetics

One intriguing pharmacological intervention capable of reopening critical period plasticity in adulthood is epigenetic modification. Both amblyopia in rodents (Silingardi et al. 2010; Lennartsson et al. 2015) and the acquisition of absolute pitch in humans (Gervain et al. 2013) is enabled by histone deacetylase (HDAC) inhibitors. Major epigenetic mechanisms include DNA methylation and demethylation, histone modifications, genomic imprinting, alternate splicing, and noncoding RNAs (micro-RNAs, long noncoding RNAs, small interfering RNAs, and small nuclear RNAs) (see Moore and Kobor, this volume). Significant changes in the epigenome have been associated with all steps of brain development, including neurogenesis, neuronal migration, synaptic transmission, and plasticity, including critical period plasticity in visual cortex (Mellios et al. 2011).

Links between neuronal activation and impact on epigenome are emerging but represent a new avenue of research. As examples, in adult brains, neuronal activation has been shown to induce changes in DNA methylation, especially in genes involved in brain development and plasticity. This suggests a role of DNA methylome in activity-dependent epigenetic regulation of plasticity (Baker-Andresen et al. 2013). Similarly, studies in young adult rodents have shown that natural behavior, exploration of a novel environment, and raising neuronal activity through sensory or optogenetic stimulation increased neuronal DNA double-strand breaks (DSBs) in postmitotic neurons of relevant, but not of irrelevant, networks (Suberbielle et al. 2015). These DSBs were repaired within 24 h through a nuclear phosphoprotein that plays a role in maintaining genomic stability, BRCA1. DSBs have been associated with epigenetic remodeling and, in particular, with removal of DNA methylation marks.

---

[1] https://clinicaltrials.gov/ct2/show/NCT02685423 (accessed Oct. 21, 2017).

To achieve a more complete picture of child development, epigenetics provides numerous opportunities that include, but are not limited to,

- creating an "epigenetic growth chart" in the pediatric age range,
- careful intersection of allelic variation and epigenetic marks and their respective contributions to developmental programs,
- mapping concordance and discordance of epigenetics marks between accessible peripheral tissues and the brain during development, and
- determining the utility of epigenetics in measuring the impact of interventions.

At present, minimal information exists about the formation of epigenetic marks in early life in rodents and humans. Longitudinal assessments would be ideal to close this important gap, although for mapping these dynamics in the brain, the constraint of having to use postmortem tissues would necessitate the careful integration of a series of cross-sectional measures. Regardless, such maps of early-life epigenome "evolution," ideally across different salient tissue and involving samples from multiple subjects, offers the opportunity to make progress on several important aspects. It would provide a blueprint for normative variation of the epigenome that could be contrasted with patterns obtained from children with atypical development. As such, the high-dimensional assessment of the epigenome would become an integral part of the multimodal assessment of child development that could lead to personalized assessment of a given atypical child (see below). In addition, the existence of such a map in combination with careful environmental measures would also enable a rigorous assessment to which extent epigenetic patterns during development are shaped by intrinsic (i.e., genetic) factors versus extrinsic ones (i.e., environment), or a combination thereof.

For humans, epigenetic measures will be limited to accessible peripheral tissues (e.g., blood, buccal epithelial cells, saliva). Evaluating the information content of epigenetic patterns derived from these sources for epigenetic patterns in the brain is challenging as it requires parallel analysis of postmortem brain samples. In part, creating cross-tissue blueprints of epigenetic patterns in model organisms could close this knowledge gap. At the same time, a stronger focus on model organisms would allow a much deeper interrogation of the epigenome of individual cell types in the brain, and how the epigenome might change in relationship to their physiology and to pathological conditions. This would then serve as the basis for mechanistic research on the role of epigenetics in brain development and function, in typical and atypical developmental trajectories. Indeed, innovative modeling of early-life environments relevant for human development in model organisms will be crucial to tease apart complex variables (e.g., socioeconomic status, parental neglect, exposure to potentially toxic conditions such as alcohol, heavy metals, pollutants, and inflammatory stimuli associated with the human epigenome) as well as to move from correlations toward causality.

Measurements and Interventions

For a number of selected animal models, tools are rapidly emerging that allow us to directly measure and manipulate different aspects of the developing nervous system. However, in humans, the tools we currently have at our disposal are much more limited and indirect. A major issue in the future will be to bridge this gap so that the insights gained from animal studies can be leveraged to interpret measurements in the human brain, with the end goal of developing adequate treatments of developmental disorders. To illustrate some of the current challenges, we turn to language acquisition in infants—one of the most important developmental milestones in early childhood.

Acquisition of Language in Infants

Although there is individual variation in the rate of development, mastery of language is a universal phenomenon that occurs across cultures. Research into early infant abilities clearly shows that, as a group, even very young infants preferentially pay attention to and discriminate the sounds of language. As early as in the first few weeks after birth, infants can discriminate phoneme contrasts such as /pa/ and /ba/, not only for their own language, but for those in other languages as well. However, by 6–12 months, infants (like adults) are only able to discriminate these contrasts present in their own language. Over the course of the first two years, children produce their first words and begin to combine these words into short sentences, incorporating many aspects of the syntactic structures present in adult grammar (Bates et al. 1987; Benasich and Tallal 2002; Ortiz-Mantilla et al. 2013).

Thus, during the first year of life, perceptual abilities in an infant develop from a wide-ranging capacity to discriminate general sensory information to a finely tuned capability that favors the processing of selective, more relevant input from their environment. Ontogenic specialization (perceptual narrowing) promotes neural representation as language-specific phonemic maps are established and essential information is processed efficiently, and is particularly important as infants assemble the foundations of their native language. As early as the 30th gestational week, the cortical organization of premature neonates allows discrimination of phonemic variations between syllables (Mahmoudzadeh et al. 2013; Maitre et al. 2013). Subsequently, exposure to native language in the natural environment fosters construction of language-specific phonemic maps and commitment to their native language (Kuhl et al. 2006). As infants become language experts, they preferentially process characteristic features of their native language (Werker et al. 2012; Ortiz-Mantilla et al. 2013). Newborns and young infants begin favoring distinctive suprasegmental elements (rhythm, intonation, stress) of their own language (e.g., Mehler et al. 1988). Shortly after birth, they show enhanced electrophysiological responses (Cheour et al. 1998) and better behavioral categorization of familiar versus nonfamiliar vowels (Kuhl et al. 1992; Moon et al. 2013). At 6–8 months of age, infants still discriminate most native and nonnative consonant contrasts, but by 10–12 months, the ability to discriminate foreign contrasts attenuates while discrimination of native language phonemes strengthens (Werker and Tees 1984; Rivera-Gaxiola et al. 2005; Tsao et al. 2006). Despite strong evidence of this transitional time line from universal to native language phoneme specialization, the neural mechanisms underlying this transition remain unclear. Some progress has been made, however, in tracking this maturational process.

Specifically, there is accumulating evidence that dynamic coordination and oscillatory mechanisms underlie the establishment of language across development. Event-related oscillations in the mature system have been characterized by two phenomena: nested phase-locking and asymmetry of temporal processing. In nested phase-locking, evoked oscillations in the lower-frequency bands of delta and theta synchronize to the slower temporal dynamics of sound, such as the speech envelope (Abrams et al. 2009; Luo and Poeppel 2012), whereas fast oscillations in the gamma-frequency range are associated with the encoding of rapid feature analysis, temporal binding of stimulus events, and attention control (Tallon-Baudry and Bertrand 1999; Fries et al. 2007) as well as resolution of segmental/phonemic information (Poeppel et al. 2008; Giraud and Poeppel 2012). Similar to adults, syllable processing in infants is resolved in a multi-time fashion through synchronized activity in low- and high-frequency ranges, and it has been shown to track the evolving time-frequency dynamics supporting phonemic perceptual narrowing as well as processing of temporally modulated nonspeech (Musacchia et al. 2013; Ortiz-Mantilla et al. 2013, 2016). High gamma power in auditory cortex has been shown to index mapping of segmental/phonemic information (Steinschneider et al. 2011) in adults; similarly, high gamma has been shown to index the evolution of perceptual narrowing in infants (Ortiz-Mantilla et al. 2016). Robust associations have also been shown between gamma power in resting EEG at 16, 24, and 36 months of age and later language, specifically phonological memory (nonword repetition) and syntactical skills (Gou et al. 2011). Adding to this picture is research which demonstrates that early, targeted interactive auditory experience in infancy can enhance and accelerate the maturation of acoustic temporal processing in typically developing infants (Benasich et al. 2014). This enhancement of acoustic processing extends to both nonlinguistic and linguistic input, has been shown to be associated with changes in oscillatory encoding and acoustic cortical mapping, and may index neuronal maturation of auditory cortex (Musacchia et al. 2017).

These insights raise a set of fundamental questions:

- How is brain coordination accomplished at multiple levels across age?
- Can we identify particular "oscillatory signatures" (such as those generated by PV+ networks) that will index evolving dynamic coordination as the brain matures?

- What central mechanisms are critical to maturation of the developing brain, and what role do critical/sensitive periods play as well as the many extrinsic and intrinsic factors that impact developmental trajectories?

To address such questions, we need to define normative/typical developmental trajectories, identify potential biomarkers, measure evolution (or variations) in brain dynamics across maturation, and critically assess how these might reflect the underlying functional local and large-scale circuitry and dynamics. Unfortunately, it is difficult to examine these mechanisms and brain dynamics directly in humans. Thus, we need to use animal models as well as indirect means of measuring brain activity and link these different approaches to establish a more comprehensive understanding of the neural underpinning of acquisition of language and other skills during childhood.

Perceptual-Cognitive Paradigms: Assessing the Speed and Efficiency of Information Processing

A number of perceptual-cognitive paradigms can be used to assess the speed and efficiency of information processing in global and rapid temporal domains, short- and long-term memory and learning of contingencies: habituation, recognition memory, preferential looking, auditory-visual integration, cross-modal transfer (e.g., tactual-visual; auditory-visual), and operant conditioning in infants (e.g., nonnutritive sucking, conditioned head-turn, eye movements, two-alternative forced-choice, and anticipatory looking) (for reviews see Bornstein and Sigman 1986; McCall and Carriger 1993; Rose et al. 2004a, b).

It has been suggested that the lack of correspondence seen between the commonly used infancy tests and later assessments of later child cognitive competence may be a function of the nature and content of the standardized tests themselves (Rose et al. 2008). The items that comprise global tests of infant intelligence are strongly dependent on sensorimotor capabilities (reaching, grasping, hand-eye coordination), skills which apparently do not relate to differences in cognitive ability later in life. As tests in later childhood add items that utilize discrimination, memory, categorization, and abstraction, the correlations rise substantially with cognitive competence later in life. Therefore, the key to demonstrating continuity and thus predictability of cognitive ability is to utilize conceptual processes similar to those exhibited in psychometric tests of IQ.

Recent literature suggests that promising candidates for an infant analogue for later information processing may be habituation and recognition memory. In addition, operantly conditioned nonnutritive sucking, head-turn or eye movements, two-alternative forced-choice paradigms as well as delayed match (or mismatch) to sample tasks can be useful in assessing various emerging abilities, including language acquisition. Another strength of these types of paradigms is that they echo those used in animal models, thus providing continuity between animal and human studies. Relatively few studies, however, have examined the neural substrates of such developmental tasks in human infants, although more sophisticated techniques—including dense array EEG, magnetoencephalography (MEG) and functional near infrared spectroscopy (fNIRS)—are increasingly being employed, thus opening a developmental window on the neural correlates and underlying brain dynamics of human cognition (e.g., Mash et al. 2013; Nordt et al. 2016; Emberson et al. 2017; Lee et al. 2017a).

Understanding the Neural Processes That Underlie Behavioral Observations in Humans

Brain dynamics change over the course of development as well as across different states of consciousness and cognition (including sleep). The question arises as to what we already know about these processes, and how we might capture information about the underlying neural processes. Even more importantly, can we determine whether the observed behavioral/imaging developmental data collected in human studies corresponds to information from animal models regarding the underlying dynamics and brain function?

Given the difficulty in observing brain dynamics in humans, indirect means of measuring brain response must be used. These include data from behavioral paradigms (psychophysics, cognitive and linguistic testing), identifying associations between candidate genes and behavioral phenotypes, and various forms of imaging (e.g., fMRI, DTI, resting-state fMRI, EEG and MEG). All can be employed in pursuit of these goals. However, how do we bridge the gap between invasive animal studies and noninvasive human studies? Do EEG or MEG scalp measurements of frequency and power differences map onto oscillatory processes and local field potentials taken from *in vivo* recordings in rodent or nonhuman primate studies?

Many of the hypotheses, models, and paradigms we use arise from animal models, allowing us to examine critical genetic and neural–behavioral links and associations. At the very least, one can derive localization information. However, it is clear that data from invasive and noninvasive recordings in both animals and humans support the view that brain development and maturation are critically dependent on synchronized neuronal activity and activity-dependent plasticity (see Singer, this volume). Moreover, such data further highlight the specific relationship between brain maturation and changes in the frequency, amplitude, and synchronization of neural oscillations.

More often, a particular study using noninvasive methods in humans is motivated by a series of initial findings derived from animal studies. In translating these findings to the human, studies often begin with adults, use similar paradigms, and follow as closely as possible the animal protocol. For example, when examining gamma in dense EEG, gamma oscillations have been shown to represent synchronized activity of local neuronal populations during sensory and cognitive processes (Ward 2003; Herrmann et al. 2004; Ribary 2005; Buzsáki 2006; Fan et al. 2007), but they may also play a role in coupling of remote cortical areas (Buzsáki and Schomburg 2015). Low-frequency gamma oscillations appear to support object representations, temporal binding, arousal, attentional selection, and working memory (for reviews see Engel and Singer 2001; Buzsáki 2006; Fries et al. 2007; Uhlhaas et al. 2011).

Cortical activity in the gamma-frequency range has also been linked in humans and animals to a wide variety of higher cognitive processes and language. In addition, it has been hypothesized that correlations between the occurrence of higher amplitude activity centered in the high gamma range (over 70 Hz) and cognitive performance (as observed in human adult subjects using EEG and MEG recordings) may reflect increased synchronization of neural ensembles important for cognitive processing (Ribary et al. 1991; Rodriguez et al. 1999; Singer 1999). Specifically, synchronization of neuronal firing, often associated with gamma-frequency oscillations (Engel and Singer 2001; Varela et al. 2001), appears to be critically involved in the organization of cortical networks. Moreover, increasing evidence suggests that oscillatory mechanisms may support the coordination of distributed neural responses that underlie cognitive and perceptual function, and may thus be critical to normative development of cortical circuits. The emergence of specific patterns of oscillatory activity in relation to cognitive developmental milestones as well as the correlation between the appearance of certain brain disorders at different developmental periods and electrocortical signs of abnormal temporal coordination support the view that "neural synchrony is not epiphenomenal but plays a role in the functions of cortical networks" (Uhlhaas et al. 2010:79).

Further, gamma oscillations seem to be developmentally regulated. EEG studies that examined frontopolar, central, and occipital scalp locations in 3- to 12-year-old children showed that gamma power increased significantly across age, most strikingly over frontal regions (Takano and Ogawa 1998), which is parallel with the slow maturation of inhibitory PV+ networks in these higher-order areas (Condé et al. 1996). As children mature, there is a gradual shift in peak power as activity in the lower-frequency bands decreases and activity increases in the higher-frequency bands (Clarke et al. 2001; Uhlhaas et al. 2010). Importantly, EEG power indices of children that significantly deviate from this pattern seem to reflect distinct differences in the maturational time course of brain development (John et al. 1980).

In many instances of EEG and MEG measurements, specific features of evoked potentials can, at least partly, be related to ongoing activity within specific cortical and even subcortical structures. Certain spontaneous oscillations, such as awake resting-state alpha oscillations, have well-known origins within sensory cortical and thalamocortical circuits. The evoked potentials can in part be accounted for by stimulus-induced phase reset (synchronization) of ongoing oscillations. Source localization analyses, of either evoked potentials or

spontaneous oscillations, can provide additional insights by identifying putative cortical regions or networks that may be generating the scalp signals.

Bridging Human and Animal Studies to Understand Developmental Disorders

Despite these advancements and the substantial gains made over the last two
decades in our ability to image the developing human brain, the vast majority of
imaging studies (whether EEG, NIRS, MRI or MEG) tend not to be grounded
in neurobiology. This is particularly problematic when it comes to the study of
neurodevelopmental disorders, where one hopes to gain insight into the neurobiological mechanisms that underpin a given disorder (e.g., autism). Accordingly,
the ability to conduct translational research—where one moves from, say,
mouse to human and back again—is very limited. One approach (LeBlanc et al.
2015) adopted by Takao Hensch, Charles Nelson III, and Michela Fagiolini is to
use EEG, the visual-evoked potential (VEP), and the auditory-evoked potential
(AEP) in both species, thus testing mice and human children under very similar
conditions. Resting EEG is recorded from the scalp surface, followed by visual-
and evoked-potential testing: in the case of VEP, phase reversing checkerboards
are presented whereas in AEP, tones of different frequencies are presented. In
so doing, one can characterize the same "spontaneous" (EEG) versus "evoked"
(VEP, AEP) brain state in both species.

To extend their investigations to the study of neurodevelopmental disorders,
Fagiolini and Nelson (LeBlanc et al. 2015) recorded the VEP from girls with
Rett syndrome and the equivalent mouse mutant. Their studies show remarkable similarity across both species; specifically, the spatial frequency of the
visual stimuli and the underlying specific mutation affects the amplitude and
latency of the VEP in the same way in both mouse and human.

By using identical bridging tools across species and testing both species
under near-identical conditions, it becomes possible to ground the human brain
response on more solid neurobiological footing. This, in turn, offers promise for developing new therapeutics for rare genetic disorders that can first be
tested in the mouse and then extended to the human, based on shared mechanisms of dysfunction in synapses and circuits (Banerjee et al. 2016).

Signatures of Typical and Atypical Development

Capturing High-Dimensional Neurodevelopmental Trajectories

Typical development is a progressive unfolding of developmental programs:
while functional development of the brain and the expression of behavior
and cognition grows steadily and even monotonically through the first few
years of life (and is very pronounced in years 0–3), the actual dynamics of the

underlying processes is far from monotonic. Different sets of genes, together with available modes of activity and environmental influences, have different patterns of expression during this period (Sur and Rubenstein 2005; Majdan and Shatz 2006). Genes related to neurogenesis and migration are expressed early and then decline; molecules of axon growth and targeting have developmental programs with specific onsets and offsets; synapses require a large number of molecules to be constructed; some of these may decline after an early construction phase, others may persist into adulthood, and still others may be expressed pleiotropically to affect multiple processes necessary for synapse maintenance or plasticity throughout life (Sur et al. 2013).

The development of behavior and cognition also has different components with different onsets and rates. The measurement of cognition and hence tracking its development is presently challenging in infants. Thus, a major question for the field is if, and how, more refined and even rapid measurements are possible.

Such multidimensional measurements in humans—spanning cognition, genomes, epigenomes, molecules, brain structure, and brain function at different points of time—would, in principle, provide a rich description of the trajectory of development (Figure 7.2). Such data would allow us to "model" development in a principled way, potentially within a hierarchical Bayes framework with priors being updated with new information as development unfolds. If such rich descriptions could be standardized and made available for large numbers of children, it would enable the construction of a population model of developmental trajectories. Of course, such measurements need not be made all at once; they could be done progressively as the project evolves. Indeed many different research and clinical groups are already involved in making subsets of measurements or have subsets of data that could be explored for concatenation within this framework. Such a model, however, will require large numbers of children explored at multiple time points. In addition, behavioral-cognitive data will need, in terms of items, to match, up to some level, the rich data provided by different omics, genetics, and epigenetics of sophisticated MRI. To combine, store, and process such heterogeneous, cross-domain data, advanced technologies appear necessary to solve this important challenge.

Such a model is a formal description of a complex multidimensional process with many latent variables, similar to atmospheric (climate, weather) or geophysical (seismic phenomena, earthquakes) models. These types of models rely on large quantities of data, which are drastically reduced in dimension to features that capture large amounts of variance, and which drive increasingly sophisticated statistical and dynamical models of the phenomenon with enormous explanatory and predictive power. In developmental and clinical neuroscience, this predictive power of formal models would be at the heart of their utility and importance for neurodevelopmental disorders and atypical development (Sahin and Sur 2015).

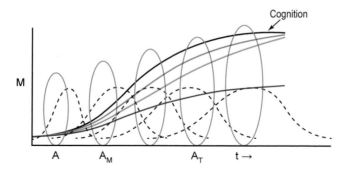

Figure 7.2 Schematic of developmental events underlying brain maturation and cognitive development. Brain development unfolds in time (x-axis) as a series of events involving gene expression, molecular signaling events, and cell–cell interactions (depicted schematically as dashed Gaussian curves). The effect or magnitude (M) of these events is depicted schematically on the y-axis. At each time point, development can be described quantitatively and computationally as the net vector average of these events (elongated gray ovals at multiple time points). Five are shown in the figure, though in principle as many multidimensional vectors as possible could be obtained. The evolution of these vectors in time describes the systems biology of brain development, which is manifest as growth of brain regions, connections between regions, and synaptic architectures (depicted as sigmoidal curves). Together, these events and circuits underlie the development of cognition (black sigmoidal curve). The role of electrical activity is different at different points in time as development proceeds. At the earliest stages of development, the presence of activity, A, may be sufficient for gene and molecular expression. Later, the magnitude of activity, A_M, may be permissive for neuronal and axon growth. Still later, the timing of activity, A_T, may play an instructive role in development of synaptic connections and circuits.

Our current understanding of atypical development is intuitive: it is based on a few measurements within each investigator's field of expertise, building on the understanding ("model") of typical development that the investigator brings or has derived from previous data. Variations that characterize atypical development are weighed against the range of normal variation. A large-scale, multidimensional set of measurements, coupled with machine learning and formal statistical and dynamical models, would help formalize and enrich this understanding.

Such an understanding is critical for weighing or evaluating atypical development. The "observables" of behavior and cognition, particularly in neurodevelopmental disorders, are the expression of latent variables in the developing brain, which span interactions between brain regions, circuits within and across structures, synaptic and neuronal processing, the molecules that underlie function and structure, and the genes that express these molecules. These variables change and evolve as development proceeds. Changes build upon previous deviations and may rectify or exacerbate deviations from typical profiles.

Detecting Atypical Development: Early Identification of Autism and the Issue of Specificity

There are a variety of childhood disorders and conditions that greatly increase the risk of developing autism, which is currently estimated at 1:68 in the United States, with comparable prevalence figures reported in many other countries. For example, children with a number of different monogenic and polygenic disorders (e.g., Down syndrome, Fragile X syndrome) have a greatly elevated risk of autism, as do children who are born prematurely (especially in the context of chorioaminionitis, intrauterine growth restriction, or ventriculomegaly) or endure their early years of life in deprived settings (e.g., orphanages). Infants who have at least one older sibling with the disorder are, however, at highest risk for developing autism: from 1:68 to 1:5. If more than one older sibling is affected, the risk for males can increase to 1:3.

Using a variety of neuroimaging tools (EEG, VEP, NIRS), Nelson and his collaborators have studied such at-risk infants for more than a decade. They consistently find that by three months of age, EEG measures can detect a difference between high-risk infants who subsequently go on to receive a diagnosis of autism at 3 years of age and those who do not. Thus, it would appear that EEG (in this case beta and gamma power) is able to predict who does and does not develop autism. This approach has important implications for the early identification of autism and other disorders, which is important given that children who receive early intervention have better outcomes than those who receive later intervention.

As promising as this work is, there are several conundrums that must be addressed. First, how specific to autism is this pattern of EEG findings? Is it possible that all the EEG is doing is telling us which brains have gone off track generally, rather than which brains will develop autism specifically? Second, the findings to date apply solely to the group level; at least thus far, such tools could not be applied at the level of the individual. Finally, until comparable studies are performed in the animal, we are unable to determine the neurobiological mechanisms that underpin these EEG patterns. For example, is there an imbalance between excitation and inhibition? Do these patterns reflect differences in long- versus short-range connectivity?

A deep understanding of atypical development requires in-depth description of typical development that integrates as many measurements as possible, spanning both observable and latent variables, to construct as rich models as possible.

Assessing an Individual Child's Trajectory and Individualized Treatments

Developing a population model of developmental trajectories would set the stage for evaluating an individual child's trajectory on the basis of his/her profile. Significant deviations from the population in a cognitive variable, for example, can be weighed alongside other measurements. This would allow the assessment of risk at an individual level for a particular neurodevelopmental outcome. It might also enable the application of noninvasive therapies intended to return the child to typical profiles.

The vision of individualized assessments may be the future for neurodevelopmental and neuropsychiatric disorders, where symptoms are highly variable and the outcome of events that span individual genetic variations, genetic backgrounds, and experiences—most of which influence the latent variables of the brain. The effects of these latent variables are expressed as the observables of brain activity, behavior, and cognition, and are evaluated as the "symptoms" of neuropsychiatric disorders.

In the far-off future, this approach may be used to develop combination therapies utilizing multiple pharmacological and behavioral approaches (Sahin and Sur 2015). In effect, our present one disorder–one treatment approach also invokes a model, but a very simple one. We need larger quantities of data, richer ways to capture their essence, sophisticated models to tell us what they mean, and a range of approaches to utilize this information meaningfully for treating the very complex signs that mark deviations in development. Without this, we believe there is no other way to understand, treat, and prevent disorders of neurocognitive development!

These approaches, however, could pave the way for developing and testing novel intervention techniques and provide us with unprecedented opportunities to assess the empirical evidence for their efficiency. Potential candidates currently include gene therapy in some cases, specific cognitive training, targeted psychophysical training, transcranial magnetic stimulation, transcranial direct-current stimulation as well as pharmacological interventions and selective deprivation. For example, Benasich and colleagues have used an infant-driven, interactive auditory training intervention to induce more robust and precise acoustic mapping in 4- to 7-month-old human infants: dense array EEG was used to obtain measurements of altering patterns of theta- and gamma-band expression and compared against passively exposed or nonintervened naïve infants (Benasich et al. 2014; Musacchia et al. 2017). The observed dynamic changes resulted in improved automatic sound processing and more accurate detection of rapid frequency change, and thus may influence the course of developmental language learning disorders.

Observation of changes in human brain dynamics as a function of a targeted intervention and assessment of outcomes using cutting-edge advances in imaging techniques provides the means to begin to investigate complex causality patterns previously only explored in animal models. More formal models that better capture the complexity of neural development and are calibrated to explore differences in individual trajectories will allow us to make more accurate predictions about which combination of interventions has the largest likelihood of success.

Flexibility and Stability Revisited

Given the considerable heterogeneity of genetic and epigenetic risk factors presumed to underlie the development of psychiatric diseases such as autism spectrum disorder, schizophrenia, and bipolar psychosis, one might ask whether there is perhaps a common final pathway and, if so, whether it would not be more efficient to attempt to identify the traits of the final outcome and then work backward. One example is epilepsy, where the causes are numerous and quite heterogeneous but the final outcome is stereotypical. Once the dynamics of seizures were well understood, it was relatively easy to backtrack the various causes with a targeted search.

Could psychiatric disorders be the evolutionary consequence of an increasingly complex brain? If so, it might be worthwhile to examine which specific problems are faced by highly evolved brains. It could be that the processes in these brains are relatively less constrained by sensory input. Compared to intrinsic interactions, input from the sensory periphery plays an increasingly smaller role in controlling cortical activity. Only a small fraction of the synapses in layer four of primary sensory cortical areas, the recipient layer of subcortical sensory projections, actually comes from the thalamus; the rest is of intracortical (intrinsic) origin. The phylogenetically recent cortical areas, in particular in the prefrontal cortex, only communicate with cerebral structures. In a sense, highly evolved brains are "autistic" and concerned mainly with generating internal models and working constructively on stored information. They are coupled only loosely to the environment and are thus relatively little constrained by the embedding environment. This could be one reason for the disturbance of high-level cognitive processes in neuropsychiatric disorders. Others could be related to the high degree of complexity and distributed nature of "big" brains (e.g., problems with cross-modal integration, the inability to distinguish between self-generated and externally induced activity patterns, difficulties with integration of widely distributed processes over large distances). Indeed, epigenetic mechanisms seem to contribute disproportionately to the etiology of neuropsychiatric disorders (Mellios and Sur 2012).

Could it be that complexity is the consequence of a selection process that favors resilience? Complex systems may be more resilient than simple systems because they are capable of self-organization. It could thus be that lesions (genetic, environmental) which would be lethal in simple systems (e.g., the loss of a particular receptor subunit) can be managed by compensatory strategies of complex systems, the outcome of which may be a psychiatric disease. The disease state could thus be the manifestation of an escape strategy. This raises the question as to whether all of the symptoms should actually be treated—if some of them fulfill a protective function. If not, which are the protective and which the deleterious symptoms (e.g., the negative and positive symptoms in schizophrenia)?

These considerations raise the important question: To what extent can valid animal models be generated to model specific traits of complex psychiatric diseases? Currently, mice serve as the major model system for diseases with a genetic component, even though their brains are orders of magnitude less complex than human brains, because techniques to generate transgenic animals have been developed for this species. Much of our knowledge about the neuronal underpinnings of higher cognitive functions has been obtained through studies on awake, behaviorally trained monkeys. It would be a great step forward if we had valid nonhuman primate models of disease. This approach has been extremely fruitful in the investigation of Parkinson disease, but such models are not yet available for psychiatric diseases. With the advent of recent genetic engineering methods such as the CRISPR-Cas technique (Doudna and Charpentier 2014; Heidenreich and Zhang 2016), this option is now within reach, and large programs investigating the possibility of inducing mutations identified as risk factors for psychiatric conditions in nonhuman primates, in particular the marmoset, have now been initiated.

Conclusion

During the first three years after birth, the human brain faces the enormous task of building a comprehensive internal model of the external world that will allow it to perceive, interpret, and predict the vast amount of parallel streams of input entering the sensory periphery. On the basis of an elaborate scaffold, set up prenatally, this task is achieved through self-organizing dynamical networks which integrate genetic, epigenetic, and sensory information. We are only beginning to understand the mechanisms that orchestrate this process across molecular, neural activity, and behavioral levels. The brain appears to use several strategies for tightly regulating the impact of sensory input on its developing networks, ranging from the molecular control of neural plasticity (as evident in the closing of critical period plasticity), to the diversification into stable "scaffold" and more volatile "plastic" elements, to attentional and behavioral control of sensory input. Such mechanisms appear necessary to safeguard the brain in its difficult journey during the first years of life: on one hand, brain circuits are required to learn how to process specific sets of features imposed by a particular environment (an illustrative example of which is language acquisition), yet on the other, this process must be protected against forgetting and insufficient or detrimental sensory input. This is a nontrivial task, requiring a fine balance between flexible and stable design on various levels. Several neurodevelopmental disorders, including certain forms of autism, appear associated with a malfunction intended to keep just this balance, the details of which remain to be elucidated in the future.

Tracing the neural basis of cognitive development in humans remains challenging, as behavioral assays and noninvasive imaging techniques provide only an indirect account of neural activity. In addition, our grasp of the developing epigenome of the brain is still very limited. The vast majority of insights gained on the neural basis of brain function stems from studies using suitable animal models. Thus, one of the most important tasks in the years to come will be to link animal and human studies.

Here we have presented examples of promising directions that will result in observables that are comparable between human and nonhuman primates (e.g., properties of oscillatory neural activity). Such methods can then be used to examine developmental disorders caused by mutations in single genes, which can be studied in equivalent mouse mutants. We have identified opportunities that modern data acquisition, analysis, and computing methods provide, which could be leveraged to develop high-dimensional formal statistical and dynamical models of normal and abnormal neurodevelopmental trajectories by integrating vast amounts of chronic data from a large number of individuals. Such approaches could be exploited to achieve early detection of a developing disease and may enable individualized treatment, both of which could greatly improve the rate of successfully treated or even prevention of disease outbreaks. Progress will rely on computational efforts to develop better dynamical and statistical neural circuit models as well as to establish or adapt machine-learning tools to cope with the vast amounts of molecular, neural, and behavioral data. Finally, the size and complexity of the human brain appear to be important factors in understanding typical and atypical development. To represent these crucial features in humans, appropriate animal models, beyond rodents, must be established. This remains an important task to be addressed in the future.

Early Adolescence

8

How Do Pubertal Hormones Impact Brain Dynamics and Maturation?

Cheryl L. Sisk

Abstract

Adolescence is characterized by maturation of reproductive and other social behaviors and social cognition. Although gonadal steroid hormones are well-known mediators of these behaviors in adulthood, the role these hormones play in shaping the adolescent brain and behavioral development has only come to light in recent years. This chapter reviews the organizational effects of pubertal hormones on sex-specific behaviors that mature during adolescence and the neurobiological mechanisms of structural organization of the adolescent brain by pubertal hormones. Important questions are identified to direct further study of the relationship between pubertal hormones, the adolescent brain, and experience.

Introduction

The adolescent transition from childhood to adulthood requires a metamorphosis of brain and behavior as individuals acquire the ability to function independently in adulthood. This gain of function involves the reorganization of neural circuits, especially those regulating sex-typical reproductive function and social behaviors. Recent work in both animals and humans reveals that reorganization of the adolescent brain involves many of the same developmental processes used during initial construction of the nervous system, including neurogenesis, programmed cell death, elaboration and pruning of dendritic arborizations and synapses, and sexual differentiation (reviewed in Juraska et al. 2013; Schulz and Sisk 2016; Herting and Sowell 2017). Given the extent of neural plasticity during this time, the adolescent brain is particularly sensitive to experience and nervous system insult, which likely contributes to the

adolescent emergence of a number of psychiatric illnesses that disproportionately affect either females or males (Merikangas et al. 2010).

The onset of puberty marks the beginning of adolescence, and a growing
body of evidence supports the view that gonadal steroid hormones, which become elevated during puberty, play a major role in shaping the adolescent brain
and behavior. Indeed, gonadal steroid hormones influence virtually all of the
early developmental processes noted above, yet research on how hormones
influence these processes during adolescence, and the consequences for behavioral maturation, is just beginning to shed light on how pubertal hormones
influence brain dynamics during adolescence. Not all structural and behavioral
changes that occur during adolescence are driven or modulated by hormones.
This review focuses on those that are.

Organizational and Activational Effects of Gonadal Hormones

Gonadal steroid hormone action in the nervous system can be dichotomized
as activational or organizational. Activational effects refer to the ability of steroids to modify the activity of target cells in ways that facilitate expression
of particular behaviors in specific contexts. Activational effects are transient;
they come and go with the presence and absence of hormones and are typically
associated with steroid action in the adult brain. In contrast, organizational
effects refer to the ability of steroids to sculpt nervous system structure and
function during development. Organizational effects are long-lasting, persist
beyond the period of developmental exposure to hormones, and program activational responses to hormones in adulthood.

Conceptualization of the relationship between organizational and activational effects of steroid hormones has evolved over the past fifty years. To
explain sex differences in behavioral responses to hormones in adulthood,
Phoenix and colleagues first proposed that sex-typical adult behavioral (activational) responses to steroid hormones are programmed (organized) by steroid
hormones acting on the nervous system during early development (Phoenix
et al. 1959). Subsequently, scores of experiments led to the identification of a
sensitive period for hormone-dependent sexual differentiation (organization)
of the brain during prenatal and early neonatal development in nonhuman
primates and rodents (reviewed in Baum 1979; Wallen 2005). Research over
the past twenty years has revealed that in addition to the perinatal period of
hormone-dependent organization of behavioral neural circuits, adolescence is
another period of development during which gonadal hormones organize the
nervous system (reviewed in Schulz and Sisk 2016).

The current conceptual framework of organizational and activational effects
of gonadal steroid hormones is a two-stage model of development in which the
perinatal period of hormone-dependent organization is followed by a second
wave of organization during puberty and adolescence (Figure 8.1). During this

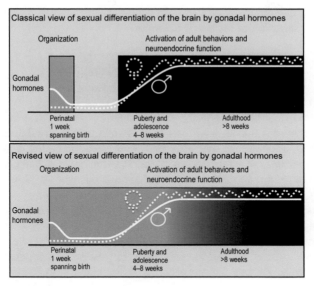

Figure 8.1 Schematic representation of the classical and revised views of sexual differentiation of the rodent brain and behavior. In the classical view, brain architecture is permanently masculinized by exposure of the male brain to testicular hormones during a brief perinatal period. In the absence of testicular hormones during this period, the default developmental trajectory is a feminine brain. When gonadal hormones become elevated at the onset of puberty, they activate sex-specific behaviors. In the revised view, the period of organization/structural differentiation is extended, continuing well through puberty and adolescence, during which both testicular and ovarian hormones organize the male and female brain, respectively. There is some evidence to support the idea that the perinatal and peripubertal periods comprise an extended postnatal window of decreasing sensitivity to organizational effects of gonadal hormones, and that this window of sensitivity closes by the end of adolescence. The two periods of brain organization and sexual differentiation are driven by the two naturally occurring times of elevation in gonadal hormone levels. From Juraska et al. (2013).

second wave, pubertal hormones organize neural circuits in the developing adolescent brain by inducing long-lasting structural changes in the nervous system that program adult activational responses to hormones and socially relevant sensory stimuli. In this model, hormone-driven adolescent organization reinforces and refines the sexual differentiation that occurred during perinatal neural development (i.e., what occurs during perinatal brain organization determines the substrate upon which pubertal hormones act during adolescent organization). One important distinction between the perinatal and pubertal periods of organization is the contribution of ovarian hormones. Perinatal organization is accomplished primarily through the masculinizing and defeminizing effects of testicular hormones; ovarian hormones, which are not elevated perinatally, do not play a major role, and the developing brain is not actively feminized.

In contrast, both testicular and ovarian hormones are actively involved in the pubertal organization of brain and behavior.

Hormone-Dependent Organization of Behavior during Adolescence

The general experimental strategy used to determine whether behavioral circuits are organized during adolescence is to manipulate circulating levels of gonadal hormones during that time and then assess behavior in adulthood. Typically, animals are gonadectomized prior to the onset of puberty to allow adolescent development in the absence of endogenous gonadal hormones, and then the hormone is replaced in adulthood prior to behavioral tests. The behavior of animals treated in this way is compared with that of those similarly treated, except that gonadectomy, washout period, and hormone replacement all occur in adulthood. With this experimental design, the observed deficits in adult behavior of animals that did not experience gonadal hormones during adolescence can be attributed to the absence of organizational effects if hormone replacement in adulthood does not reverse these deficits. Studies employing this general paradigm in a number of different species provide a growing body of evidence that gonadal hormones organize a variety of social and nonsocial behaviors in both males and females.

Males

When testosterone is absent during adolescence, a wide range of male-typical adult social behaviors is compromised. For example, prepubertally gonadectomized male Syrian hamsters display lower levels of sexual behavior compared with male hamsters that are gonadectomized in adulthood. The deficits resulting from prepubertal gonadectomy are not reversed, either by prolonged testosterone replacement therapy or sexual experience in adulthood (Schulz et al. 2004). Other male-typical adult behaviors organized by pubertal testosterone include aggression, scent marking, play fighting, and nonaggressive social interactions (reviewed in Schulz et al. 2009a; Schulz and Sisk 2016). Thus, the absence of testicular hormones during adolescence results in long-lasting impairments in sociosexual behaviors. Conversely, the presence of testicular hormones during adolescence masculinizes neural circuits underlying sociosexual behaviors and enhances activational responses to testosterone in adulthood.

Which features of sociosexual behaviors are organized by pubertal testosterone? It does not seem to be the performance or motor execution of the behaviors per se, because males deprived of testosterone during adolescence do display the consummatory components of sexual behavior, aggression, and scent marking, albeit at lower levels compared with males that did experience testosterone during adolescence. Instead, research suggests that pubertal testosterone organizes social proficiency: the ability to make behavioral adaptations as a function of social experience (De Lorme et al. 2013; De Lorme and Sisk 2013, 2016). For example, male hamsters gain social proficiency over the course of repeated encounters with another male in a neutral arena. During the first social encounter between two unfamiliar males, an aggressive interaction occurs initially and a dominant-subordinate relationship is established within a few minutes. In subsequent encounters, there is little aggression, but the dominant-subordinate relationship is maintained through flank marking by both males. This experience-dependent pattern of behavior is disrupted in males deprived of testosterone during adolescence: these males display low overall levels of flank marking, even if they are dominant, and the dominant-subordinate relationship is not maintained by flank marking, but is instead reestablished via aggression in subsequent encounters (De Lorme and Sisk 2013). Thus, during adolescence, pubertal testosterone organizes neural circuits that govern social cognition, the mental processes by which an individual encodes, interprets, and responds to sensory information from a conspecific. To identify neurobiological correlates of behavioral organization, future research should thus focus on brain regions that evaluate social stimuli and govern behavioral flexibility, such as the amygdala and components of the mesocorticolimbic reward circuit.

On average, men outperform women in tests of spatial cognition, and this sex difference in humans may be organized by pubertal hormones. Evidence for this comes from a study of men with either prepubertal- or adult-onset idiopathic hypogonadotropic hypogonadism (IHH), a condition in which a deficiency in gonadotropin-releasing hormone (GnRH) levels or pituitary insensitivity to GnRH results in reduced levels of gonadotropins, gonadal steroid hormones, and fertility (Hier and Crowley 1982). Men with prepubertal-onset IHH had low or undetectable levels of circulating gonadal steroids during the normal time of puberty and adolescence, whereas men with adult-onset IHH experienced normal levels of pubertal gonadal hormones during adolescence. Spatial cognition is impaired in men with prepubertal-onset IHH, both in comparison to healthy control subjects as well as to men with adult-onset IHH, suggesting that the presence of testicular hormones during puberty organizes circuits underlying spatial cognition. In a separate study, women with a variation of congenital adrenal hyperplasia, which leads to slightly but chronically elevated levels of adrenal androgens during childhood and early puberty, performed better in a virtual Morris water maze (a test of spatial memory) compared with healthy subjects. These data suggest that exposure to adrenal androgens during adolescence organizes (masculinizes) spatial ability in females (Mueller et al. 2008). Rodent work demonstrates that spatial memory is hippocampus-dependent, and synaptic plasticity in the hippocampus appears to be organized by pubertal androgens. Specifically, activation of androgen receptor during puberty results in long-term depression in CA1 in response to a tetanizing stimulus in adulthood, whereas if androgen receptor activation is blocked during puberty, long-term potentiation occurs in response to a tetanizing stimulus in adulthood (Hebbard et al. 2003). These findings in rodents provide a potential mechanism by which pubertal testosterone could organize hippocampus-dependent learning and memory in humans, including spatial cognition.

Females

Behavioral receptivity to males is feminized during adolescence by ovarian hormones, specifically estradiol, as shown in studies using an aromatase knock-out mouse model in which estrogen is not synthesized, but estrogen receptors are fully functional. Female knockout mice display significantly less lordosis behavior compared to wildtype or heterozygous mice following adult ovariectomy and hormone treatment, suggesting that endogenous estrogen normally feminizes reproductive responses to estradiol and progesterone in adulthood (Bakker et al. 2002). In another study, estradiol was systematically administered during development either prior to the onset of normative ovarian secretions of gonadal steroid hormones (postnatal days 5–15), or the earliest time frame for ovarian secretions of gonadal steroid hormones (P15–P25). Whereas administration of estradiol between P5–P15 had no effect on lordosis behavior in wildtype or knockout animals, administration between P15–P25 significantly increased lordosis behavior in the aromatase knockout animals (Brock et al. 2011). These data provide compelling evidence for the feminization of female reproductive behavior by estradiol during early adolescent development in female mice. Other social behaviors that are organized by ovarian hormones during puberty include the female pattern of rough and tumble play as well as maternal behavior (reviewed in Schulz and Sisk 2016).

For females, reproductive success depends on being fertile and finding a mate as well as on being physiologically prepared to sustain a pregnancy and provide nutrition for her young. It appears that elevated levels of ovarian hormones during puberty organize behaviors related to fertility, which is contingent on metabolic signals that predict sufficient energy availability to sustain pregnancy, lactation, and maternal care. In rats, defense of food is a sexually dimorphic behavior, with males and females displaying different postural strategies for guarding their food source. Prepubertal ovariectomy alters the defense strategy to be more phenotypically male, whereas adult ovariectomy has no effect on this behavior; this indicates that ovarian hormones during adolescence actively feminize postural strategies for food defense (Field et al. 2004). Pubertal estradiol also feminizes ingestive responses to metabolic signals in rats. Treatment with mercaptoacetate, a drug that interferes with fatty acid oxidation, increases food intake in male but not female rats. Prepubertally ovariectomized (OVX) females display a male-like response to mercaptoacetate and increase their food intake in adulthood, whereas adult OVX females do not increase food intake in response to mercaptoacetate. Furthermore, the effect of prepubertal ovariectomy is prevented by estradiol replacement during puberty, indicating a role for estradiol in organizing (feminizing) the response to metabolic challenge (Swithers et al. 2008).

Neurobiological Mechanisms Underlying Hormone-Dependent Organization of the Adolescent Brain

It is presumed that adolescent organization of brain structure has something to do with adolescent organization of behavior. Currently, however, we can only point to correlational relationships between hormone-dependent organization of structure and behavior during adolescence. Hormone-dependent organization of brain structure during adolescence involves many of the same developmental processes that are in play during the perinatal organizational period: cell proliferation and differentiation, cell death and survival, synapse proliferation and selective elimination, and myelination. In this section I review what is known so far about how gonadal hormones influence these developmental processes during puberty and adolescence, with a focus on brain regions known to be involved in behaviors that are organized during this same time.

Anteroventral Periventricular Nucleus and Posterodorsal Medial Amygdala

Hormonal regulation of adolescent development of the anteroventral periventricular nucleus (AVPV) and posterodorsal medial amygdala (MePD) has been the focus of research because both cell groups are sexually dimorphic, undergo structural (organizational) change during puberty, are rich in steroid hormone receptors, and are involved in reproductive function and social behaviors that mature during puberty and adolescence. Gonadal hormones contribute to the adolescent organization of the AVPV and MePD by influencing cell proliferation and survival as well as synaptic and glial cell complexity.

The rat AVPV is one of the few examples of a female-biased sexual dimorphism: it is larger and more cell-dense in females than in males. The AVPV integrates a hormonal signal from the ovaries (elevated estradiol levels) with a circadian signal from the suprachiasmatic nucleus to provide the neural trigger for generation of the preovulatory surge of luteinizing hormone (LH) that results in ovulation (reviewed in Simerly 2002). The AVPV has also recently been linked to aggressive behavior in males and maternal behavior in females (Scott et al. 2015). The sex difference in AVPV volume emerges during pubertal development (Davis et al. 1996), along with the capacity for females to generate an LH surge. The neuroendocrine positive feedback trigger for ovulation is sexually differentiated: male rats are incapable of generating an LH surge at any age.

The male rat MePD is larger and contains more neurons and glial cells than the female MePD; it evaluates chemosensory information from conspecifics and integrates these external cues with internal hormonal signals to regulate a variety of social behaviors that mature during adolescence. Although the MePD is sexually dimorphic prior to puberty, the dimorphism becomes significantly greater across adolescent development due, at least in part, to increases in astrocyte number as well as astrocyte branching in males. Furthermore, males carrying the testicular feminization mutation (tfm) of the androgen receptor and females do not display MePD increases in astrocyte number and branching during adolescence. In contrast, wildtype males display normative increases in MePD astrocyte number and branching, indicating that normative androgen receptor function is necessary for the adolescent development of this sexual dimorphism (Johnson et al. 2013).

The increase in volume of the female AVPV and the male MePD during puberty prompted us to ask whether addition of new cells may be another mechanism of structural change in these brain regions. In an initial study, a daily injection of the cell birthdate marker BrdU (200 mg/kg, intraperitoneal injection) was given to male and female rats on P30–P32, which is right around the onset of puberty (the initial rise in gonadal hormone secretion) in rats (Ahmed et al. 2008). Brains were collected 21 days later and immunohistochemistry was performed to visualize BrdU-immunoreactive (BrdU-ir) cells that were born on P30–P32. We found sex differences in the number of pubertally born cells, with more cells added to the female AVPV than to the male and more cells added to the male MePD than to the female. Gonadal hormones drive these sex differences in the number of pubertally born AVPV and MePD cells. In females, prepubertal ovariectomy reduces the number of pubertally born AVPV cells to a number indistinguishable from that of males but does not affect the number of pubertally born MePD cells. Conversely, in males, prepubertal castration reduces the number of pubertally born MePD cells but does not affect the number of pubertally born AVPV cells (Figure 8.2). Cell addition during puberty may be an active mechanism either for *preserving* structural and functional sexual dimorphisms in the face of remodeling of the adolescent brain or for *creating* new sex differences that emerge during adolescence.

Phenotyping studies using double-label immunohistochemistry for co-localization of BrdU and cellular markers for mature neurons, astrocytes, or microglia show that ~50% of pubertally born AVPV cells differentiate into one of these cell types within a month of proliferation (see Figure 8.3; Mohr et al. 2016). In the female AVPV, ~10% of pubertally born cells express estrogen receptor alpha. We do not know whether an even larger proportion of the cells would be activated or steroid receptor-expressing if these newly born cells were asked to be functionally active during differentiation and maturation.

Some pubertally born MePD and AVPV cells appear to be functionally integrated within the cell group. In male hamsters, some pubertally born MePD cells are activated (express fos) after a sexual encounter with a receptive female (Mohr and Sisk 2013). In female rats, some pubertally born AVPV cells express fos after estrogen and progesterone priming to induce an LH surge (see

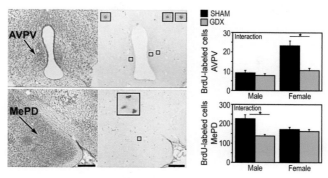

Figure 8.2 New cells are added to the anteroventral periventricular nucleus (AVPV) and posterodorsal medial amygdala (MePD) during puberty in rats. The cell birthdate marker bromodeoxyuridine (BrdU) was given to male and female rats at the onset of puberty (postnatal days 30–32, 200 mg/kg/day, intraperitoneal injection) and brain tissue was collected 21 days later. More cells are added to the female AVPV than to the male AVPV, and more cells are added to the male MePD than to the female MeDP. Prepubertal gonadectomy abolishes these sex differences. *significant effect of prepubertal gonadectomy, $p < 0.05$; GDX, gonadectomized rats; SHAM, control rats. From Ahmed et al. (2008).

Figure 8.4; Mohr et al. 2017). We used a mitotic inhibitor, cytosine arabinoside (AraC), to block cell proliferation during puberty to ask whether this blockade affects the ability to generate a LH surge following hormone priming. Female rats received intracerebroventricular infusions of AraC or vehicle (which also contained BrdU) for 28 days during puberty, then were ovariectomized and

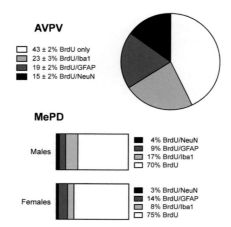

Figure 8.3 Proportion of pubertally born cells that co-localize markers for mature neurons (NeuN), astrocytes (GFAP), or microglia (Iba1). Over half of pubertally born AVPV cells in the female rat differentiate into neurons, astrocytes, or microglia within 21 days of proliferation. Longer survival times may allow additional cells to mature. From Mohr et al. (2016).

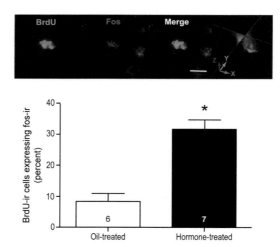

Figure 8.4   Pubertally born AVPV cells are activated by estradiol and progesterone. Hormone induction of the luteinizing hormone surge caused a significant increase in the proportion of immunoreactive bromodeoxyuridine (BrdU-ir) cells that express fos-ir (*p < 0.001); x, y, and z indicate planes of section in the 3-dimensional orthogonal view (right-most image). Adapted after Mohr et al. (2017).

hormone-primed to generate an LH surge. AraC significantly reduced the number of cells added to the AVPV during puberty by 50–60%. This inhibition of cell proliferation during puberty blunted and delayed the hormone-induced LH surge (Figure 8.5; Mohr et al. 2017). These findings support the idea that pubertally born cells become functionally incorporated into the circuitry of the AVPV.

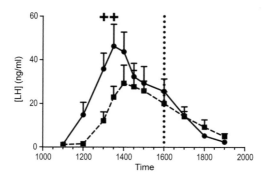

Figure 8.5   Knock-down of cell addition with the mitotic inhibitor cytosine arabinoside (AraC) during puberty dampens and delays the hormone-induced luteinizing hormone (LH) surge. Rats were on a 14:10 light:dark cycle; dotted vertical line denotes lights off. Female rats received intracerebroventricular infusions of AraC or vehicle (which also contained bromodeoxyuridine, BrdU) for 28 days during puberty; they then were ovariectomized and hormone-primed to generate an LH surge. AraC reduced the number of cells added to the AVPV during puberty by 50–60%. + significant difference between control and AraC-treated rats, p < 0.01. Adapted after Mohr et al. (2017).

Cerebral Cortex

Neuroimaging studies demonstrate that gray matter volume decreases during human adolescence (reviewed in Herting and Sowell 2017). Earlier studies reported a curvilinear pattern of development, in which volumes peak at approximately 11 years of age in girls and 12.5 years of age in boys, followed by a significant decline in volume that tapers off in the 20s in both sexes (Giedd et al. 1999). More recent studies indicate that this curvilinear pattern is not present in all cortical areas, with some showing a more linear decrease in volume across development (Ducharme et al. 2016). The timing of adolescent-related decreases in volume is not uniform across cortices. The dorsal parietal regions decrease the earliest whereas the dorsolateral prefrontal cortex is one of the latest areas to show gray matter decline (Gogtay et al. 2004). The temporal cortex reaches peak volume at approximately 16.5 years of age in both girls and boys, followed by less volume decline than one sees in the frontal and parietal regions (Giedd et al. 1999). Some of these volumetric changes in cortical structures can be predicted by the Tanner stage, and recent studies indicate that estradiol and testosterone predict volume and rate of change in cortical volume (Peper et al. 2011; Herting and Sowell 2017).

Rodent studies provide insight into the mechanisms by which prefrontal cortical gray matter volume changes occur during adolescence. Sex-specific cortical volume decreases also occur across adolescence in rats, with the adult volume of the medial prefrontal cortex (mPFC) being greater in males than in females. While no sex difference is present in the number of ventral mPFC neurons early in adolescence at 35 days of age, males display significantly more neurons by 90 days of age (Markham et al. 2007). Furthermore, prepubertal gonadectomy (GDX) prevents the decline in neuron number in females but not in males, suggesting that ovarian hormones drive the emergent sexual dimorphism during adolescence (Koss et al. 2015). In addition to decreases in neuron number observed in females, dendritic spines significantly decrease in both sexes between days 35 and 90, but only females show a loss of mPFC basilar dendrites (Koss et al. 2014). Since the majority of excitatory synapses are found on dendritic spines, these changes suggest a sexually dimorphic adolescent remodeling of synapses and, in particular, excitatory synapses.

White Matter

While gray matter volume decreases during adolescence, white matter increases linearly across adolescence, with similar growth curves and trajectories in frontal, parietal, and temporal cortices (Giedd et al. 1999; Paus et al. 1999). Boys, however, show a steeper age-dependent slope of increase in white matter volume than girls, resulting in larger white matter volumes in boys (Lenroot and Giedd 2006). Just as gonadal hormones shape adolescent gray matter changes, gonadal hormones have likewise been implicated in pubertal-related increases in white matter volumes (reviewed in Herting and Sowell 2017). For example, pubertal maturation is associated with increases in white matter density in frontal, parietal, and occipital lobes in boys (Perrin et al. 2009). The extent of the increase in white matter volume in adolescent boys is positively correlated with androgen receptor (AR) activity, as assessed by the number of CAG repeats in the AR gene (Perrin et al. 2008). Furthermore, pubertal levels of testosterone and estradiol are associated with cortical microstructural development in boys and girls, respectively. Pubertal testosterone predicts white matter increases in boys, whereas pubertal estradiol is associated with white matter decreases. Thus, these relationships between estradiol and testosterone and cortical white matter may partly explain the emergent sexual dimorphism during adolescence.

Rodent studies provide further insight into the mechanisms by which pubertal hormones may influence cortical white matter development. Like humans, white matter volume increases across adolescent development in rodents, resulting in a male-biased sexual dimorphism (Willing and Juraska 2015). Testosterone may drive increases in white matter in males by increasing axon diameter via an androgen-receptor mediated mechanism (Pesaresi et al. 2015). Estradiol in females may also contribute to the sexual dimorphism in white matter. Prepubertal ovariectomy in females significantly increases white matter volume in females, whereas prepubertal castration does not impact white matter volume in males (Koss et al. 2015). Thus, under normative developmental conditions, the pubertal onset of ovarian secretions may slow the development of white matter volume in females resulting in the male-biased dimorphism in adulthood.

Remaining Questions

What Is the Role of Social Experience in Survival, Differentiation, and Functional Incorporation of Pubertally Born Cells?

Physical exercise promotes neurogenesis, whereas social isolation and stress impair neurogenesis. We found that an enriched environment (running wheel in the home cage) increased the number of pubertally born MePD cells and the proportion that are activated by social interaction (Mohr and Sisk 2013). This indicates that experience can promote the functional incorporation of new cells into existing neural circuits. An important question for further research is whether and how specific types of social experience influence not only cell proliferation and survival during adolescence, but also cell fate and function. Our research shows that about half of pubertally born AVPV and MePD cells have not differentiated into mature neurons or glia when examined within one month after they were born. However, in our experiments, rats were singly housed and thus not given opportunities for social interaction. Would social experience facilitate the differentiation and integration of pubertally born cells into neural circuits that govern social behaviors, perhaps in sex-dependent and brain region-specific ways? What are the cellular and molecular mechanisms by which experience determines pubertally born cell fate? Answers to these questions will advance understanding of brain dynamic coordination and plasticity during adolescence.

Is Adolescence a Sensitive Period Distinct from the Perinatal Period, or Is It Part of an Extended Period of Postnatal Sensitivity for Hormone-Dependent Organization?

From a developmental perspective, it is of interest to know whether the perinatal and pubertal periods are distinct windows of sensitivity to organizational effects of gonadal hormones, or whether the two periods of hormone-dependent organization are driven by naturally occurring elevations in gonadal hormones. We tested the hypothesis that adolescence marks the opening of a second sensitive period for the organizing actions of testosterone on adult male reproductive behavior (Schulz et al. 2009b). This hypothesis predicts that exposure to testosterone during adolescence, but not before or after adolescence, will result in full activational responses to testosterone in adulthood. Male hamsters were gonadectomized at 10 days of age (after the perinatal period of sexual differentiation), and then treated with testosterone for 19 days either before (10–29 days of age), during (29–48 days of age), or after (64–83 days of age) the normal time of puberty. In adulthood, males were again treated with testosterone for one week prior to a behavioral test with a sexually receptive female. Testosterone treatment before and during puberty, but not after puberty, activated male sexual behavior in adulthood, demonstrating that (a) adolescence is not a discrete sensitive period for the organizing actions of testosterone on adult reproductive behavior and (b) the window of sensitivity to organizational effects closes at the end of adolescence. Furthermore, prepubertal testosterone treatment had the greatest impact on adult reproductive function, suggesting that the potential for testosterone to organize reproductive behavior decreases across postnatal development.

I propose that the classical view of organizational and activational mechanisms of steroid action be revised to incorporate an extended window of decreasing postnatal sensitivity to the organization of adult behavior by gonadal hormones. This proposed framework is based on our study of male reproductive behavior, and it will be important for future research to determine whether the proposed window of decreasing sensitivity to hormone-dependent organization generalizes to females and other behaviors. If it does prove to generalize, then it has implications for how the timing of puberty affects hormone-dependent organization of brain and behavior. For example, would precocious puberty result in a greater degree of organization or limit capacity for experience-dependent plasticity later in development?

A recent study in humans lends support to the possibility of an extended postnatal window of decreasing sensitivity to gonadal steroid hormones. Beltz and Berenbaum (2013) hypothesized that if sensitivity to organizational effects of gonadal steroid hormones decreases across adolescence, then the age at which adolescents undergo puberty should be inversely associated with the effectiveness of gonadal steroid hormones in organizing spatial (men) or verbal (women) ability. Participants reported whether they experienced specific pubertal events much earlier, somewhat earlier, the same, somewhat later, or much later than their peers to determine a pubertal timing score; their verbal and spatial abilities were also assessed. Among men, an effect of pubertal timing on three-dimensional mental rotation test scores was found, with early maturers outperforming late maturers. In contrast, no effects of pubertal timing on verbal or spatial ability were detected in women. Beltz and Berenbaum conclude that their findings are consistent with the hypothesis of declining sensitivity to the organizing actions of testosterone throughout adolescent development.

Is Adolescent Brain Development Experience Expectant or Experience Dependent?

The distinction of experience-expectant and experience-dependent development is based on two categories of environmental information or experience that influence nervous system development (Greenough et al. 1987; see also Kolb, this volume). The first category is experience that is ubiquitous for all individuals of a given species throughout most of its evolution. Experience-expectant development of neural circuits involves a critical period during which this experience *must* occur, otherwise the underlying function is severely compromised. Two examples of experience-expectant development in humans are (a) the requirement for visual sensory experience for normal wiring of visual cortex and binocular vision and (b) exposure to spoken words for normal acquisition of oral language. The second category is experience that is unique to a particular individual and sculpts neural circuits in a more refined way. Examples of experience-dependent development include exposure to one's native language, and growing up in an enriched or impoverished environment. Experience-dependent development does not involve a well-defined critical period, although there may well be certain times during development that a particular experience exerts more profound influences than at others; such times are more accurately described as sensitive periods.

Can we neatly categorize adolescent brain development as being either experience expectant or experience dependent? At first glance, it might seem obvious that adolescence is experience dependent, because it is hard to name an experience that *must* occur during that time to create a functional adult brain; conversely, it is easy to cite experiences that shape the trajectory of adolescent brain development. However, there is one adolescent experience that has been ubiquitous for all humans throughout our evolution: the appearance of gonadal hormones during puberty. By that definition, adolescence would be an experience-expectant developmental period during which the absence of gonadal hormonal influences would result in seriously compromised maturation of neural circuits underlying social behaviors. Nevertheless, existing data suggest that adolescence is *not* an experience-expectant critical period of development during which the absence of exposure to gonadal hormones would totally incapacitate an individual; instead it is an experience-dependent sensitive period for influences of gonadal hormones on brain and behavioral development.

As reviewed above, when male rodents are gonadectomized prior to the onset of puberty they are capable of expressing social behaviors, such as sex and aggression in adulthood, but show impairments in interpreting social cues received from conspecifics. Thus in rodents, it appears that gonadal hormones program aspects of social cognition and behavioral flexibility, and not social behavior per se. Experiments of nature in humans point to the same idea. For example, men with congenital hypogonadotropic hypogonadism do not undergo a natural puberty and typically do not begin testosterone replacement therapy until 17 years of age or older, effectively resulting in much of their adolescent development occurring in the absence of testicular hormones. Once on testosterone replacement therapy, these men are able to have sexual relationships but report long-lasting psychosexual problems, such as difficulty with intimate relationships and body image concerns (Dwyer et al. 2015). Thus, the trajectory of adolescent maturation of social cognition depends on (is influenced by) gonadal hormones, but the adolescent social brain does not expect (does not require) gonadal hormones for adult social behaviors to be expressed.

A related concept is metaplasticity: the plasticity of plasticity. At the cellular level, metaplasticity entails a change in the physiological or biochemical state of neurons or synapses that alters their ability to generate synaptic plasticity at a later time. As applied to adolescent development, metaplasticity means that the history of activity of neural circuits during adolescence (i.e., the experiences that are encountered) can make these same circuits either more or less plastic in adulthood. Metaplasticity during adolescence results in some paradoxical outcomes which, at first glance, can be hard to digest. Take, for example, the organizational effects of pubertal testosterone on male social cognition. When a neural circuit is organized by testosterone, by definition the circuit becomes less plastic than it was before the organizing action. Yet one of the long-lasting organizational effects of testosterone on behavior is to render the male rodent more flexible during social encounters, better able to adapt his behavior as a result of social experience. How does testosterone jell a circuit during adolescence yet, in doing so, make that circuit seemingly more plastic in social situations later in adulthood? Answering this question is a major challenge for future research.

9

Brain Plasticity in the Adolescent Brain

Bryan Kolb

Abstract

Adolescence is a time of enhanced neural plasticity, including both experience-expectant plasticity and experience-dependent plasticity. Experience-expectant changes are likely related to socioaffective behaviors, including play, sex, and social interactions, all of which come to dominate the life of adolescents. The most likely candidates driving plasticity in adolescence include the generation of new neurons and/or glia, the formation of connections either by axon extension or synapse formation, pruning or growth of dendrites and thus synapses, thinning of the cortex, epigenetic changes, and changes in the excitatory–inhibitory balance. A range of factors influence plasticity in the adolescent brain (e.g., play, drugs, sensorimotor experiences, stress, diet, cerebral injury, and the immune system). The onset of the sensitive period is around the onset of pubertal gonadal hormone production, but may or may not be triggered by the hormone release. The offset of the sensitive period may be related to myelination, which reduces plasticity, and the timing of the offset likely varies in different cortical regions.

Introduction

The general term "sensitive period" refers to times when the brain is unusually responsive to experiences during development. The term "critical period" is a special type of sensitive period in which specific experiences result in irreversible changes in brain organization or function (Knudsen 2004). In this chapter, I will use the concept of sensitive period, in part because although much is known about irreversible experience-dependent changes in the infant brain, far less is known about such changes in adolescence.

As Knudsen (2004) pointed out, the effect of experiences during sensitive periods can be seen in behavior, but behavioral changes are really a property of changes in neural circuits. Although behavioral changes can be obvious and even dramatic, they can also be subtle. Measurement is dependent upon behavioral tests, which often underestimate long-term behavioral changes. As a result, a clear definition of the adolescent sensitive period requires other measures of brain/behavioral correlates, ranging from noninvasive imaging (structural, connectional, and functional MRI, electrophysiological measures) to measurements of structural changes of neural networks, and eventually to molecular measures including changes in the expression of genes (epigenetics) and their products.

What is, however, adolescence? This is hardly a simple question. Some psychologists have argued that adolescence is a social construct that functions to allow adults to control adolescents (for further details, see Steinberg 2016). Given that nonhuman animals clearly show behavioral and brain changes during adolescence, it is hard to argue that it is cultural or social in nature. The real challenge is to identify when this period of development begins and ends. One common onset marker is the onset of puberty, which in humans is around 12 years of age and in rats around 28–30 days (Spear 2000). The more difficult question is: When does it end? The end of obvious behavioral changes often has been used, but when we consider changes in the brain, this becomes less certain. There are multiple potential markers—e.g., reduced neural pruning, myelination, gliogenesis (aside from oligodendrocytes), mature connectivity—but all of these changes take place according to different timetables and manifest differently in different parts of the brain. I am persuaded by the general argument that adolescence is a period of heightened neural plasticity relative to the juvenile and adult brain (e.g., Knudsen 2004; Fuhrmann et al. 2015; Steinberg 2016). The onset of enhanced plasticity likely coincides with the release of gonadal hormones, but the timing of the reduction in plasticity at the end of adolescence has not been well studied. I will argue, however, that at least for some cortical regions, it is likely later than it is often considered to be.

A sensitive period in adolescence is characterized by greatly increased plasticity; it is thus likely adaptive as there is considerable learning about the environment, and especially the social environment (Blakemore 2008). But enhanced plasticity is a double-edged sword: it renders the brain vulnerable to a wide range of experiences, such as stress, psychoactive drugs, brain trauma (e.g., concussion), and variations in the patterns of play that influence brain organization and function differently than in adults. It is no accident that many forms of mental illness become apparent in adolescence (e.g., Tottenham and Galván 2016).

In this chapter, I explore the nature of sensitive period plasticity, especially in adolescence. After a discussion of the factors likely to influence plastic changes in adolescence, I examine the role of preadolescent experiences on trajectories in adolescence. Specific examples of brain plasticity during adolescence are then examined, and I conclude with a set of questions intended to prompt future enquiry.

The Nature of Plasticity in the Adolescent Brain

To understand that nature of plasticity, we must confront two distinctly different issues: the types of plasticity and the mechanisms by which plasticity plays out in the adolescent brain.

Types of Plasticity

Three types of plasticity can be distinguished in the early developing brain: experience-independent, experience-expectant, and experience-dependent.

Experience-independent plasticity results from the fact that it is impractical for the genome to specify all connections in the brain; instead, it generates a rough approximation of connectivity that is modified by both internal and external events, both prenatally and in the early postnatal period. For example, projections of the retinal ganglion cells to the lateral geniculate nucleus arrive from both eyes and eventually segregate into layers having projections from just one eye. However, the initial projections overlap: to separate the layers, the retinal ganglion cells are spontaneously active, allowing them to correlate their activity with nearby cells but independent of the other eye, which has a different pattern of spontaneous activity. Neurons that are active together increase their connections whereas those that are not coincidentally active weaken their connections.

Experience-expectant plasticity occurs mostly during early postnatal development. A good example is found in the development of sensory processing, when the brain "expects" to be stimulated by a range of sounds, visual and tactile inputs, etc., that will vary depending upon the environment. The brain becomes expert in discriminating inputs that it receives and loses the ability to make fine discriminations when an input is not experienced. For instance, a human infant may be born into a family that speaks one of hundreds of languages, each of which has a distinct phonetic structure. A child raised hearing Korean will thus be exposed to different speech sounds than a child raised in an English-speaking environment. Early in life, infants are able to discriminate the speech sounds of all languages, but over the first year of life, the auditory system begins to change such that the infant becomes expert in discriminating sounds in the language of its environment but loses the ability to discriminate sounds not experienced.

Experience-dependent plasticity, a process whereby the connections of ensembles of neurons are modified by experience, begins in early postnatal life and continues over the entire course of life. One common example is seen in the effects of so-called "enriched experiences." When laboratory animals are housed in highly stimulating environments, the patterns of neural connections are modified, resulting in animals that have enhanced cognitive and motor functions. Although such changes are often thought to reflect the generation of new synapses, there is also a loss of synapses as the networks

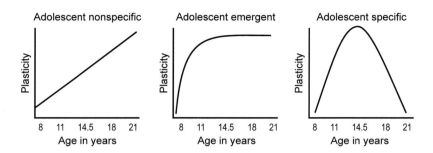

Figure 9.1   Patterns of plasticity in the developing brain. Adapted after Casey (2013).

are remodeled. This is especially evident when animals are given psychoactive drugs, which results in regionally specific increases and decreases in synaptic density.

What is the nature of plasticity in the sensitive adolescent period? It seems likely that experience-dependent plasticity plays a major role. However, in principle there could be some form of experience-expectant plasticity, likely related to the production of gonadal hormones in puberty and changes in social behavior. Casey (2013) has suggested that there are three distinct forms of adolescent developmental changes (Figure 9.1). Changes in brain and behavior may be *adolescent nonspecific*, which reflect a continuation of processes already in progress; *adolescent emergent*, which increase rapidly up to adolescence and then are relatively constant for some period of time; or *adolescent specific*, which emerge for the adolescent period and then decline. The latter type would best reflect the contention that adolescence is a period of enhanced plasticity.

An alternate version of the Casey model was proposed by Fuhrmann et al. (2015). Like Casey, they imagine an adolescent specific and nonspecific period, but also suggest that plasticity may decline continuously from childhood through adolescence and into adulthood. The problem, of course, is that different scenarios can only be discriminated if we know how the plastic changes appeared before and after adolescence. This is a significant area of ignorance.

Mechanisms of Plasticity

In the search for understanding plasticity in the adolescent brain, several potential mechanisms deserve consideration:

- neurogenesis,
- gliogenesis,
- changing neural networks (e.g., fcMRI),
- cortical thinning,
- changes in neuronal or glial morphology,

- changes in synaptic structure and number,
- epigenetic changes, and
- changes in the excitatory–inhibitory balance.

Although none of these have been found to be specific to adolescence, they are driven by factors that may be relevant to adolescence. The most likely candidates, considered in turn below, include the generation of new neurons and/or glia; the formation of connections, either by axon extension or synapse formation, pruning, or growth of dendrites and thus synapses; epigenetic changes; as well as changes in the excitatory–inhibitory balance.

Although neurogenesis in the cortex is complete by birth in placental mammals, it can be stimulated postnatally under special circumstances. We have shown that it can occur spontaneously in rats or mice after injury to the olfactory bulb or midline cortex in the second, but not the first, week of life (e.g., Kolb et al. 1998). It can also be stimulated by growth factors, such as FGF-2, after injury in the second week of life (e.g., Monfils et al. 2006). In both cases, the new neurons form functional connections that support at least partial behavioral restitution. There is no evidence from our studies that spontaneous neurogenesis occurs after injury in adolescence, and I am unaware of any studies administering growth factors (e.g., EGF, FGF-2) after cortical injury in adolescence. It seems likely that neurogenesis could occur, however, because administration of growth factors in adult rats following cortical stroke does stimulate the production of neural precursors, which can be driven to differentiate with erythropoietin. Although the neurons do not mature normally or form functional connections with the adjacent brain, they do enhance behavioral recovery (Kolb et al. 2007).

Not only can neurons potentially be generated, glia can as well. Glial cells constitute about 50% of the cells in the human brain, with astrocytes constituting the largest population. Although astrocytes arise from radial glial cells in the subventricular zone in early development, the major source of astrocytes in postnatal rodents is the proliferation of undifferentiated astrocytes already located in the white matter and cortical layers V–VI. This proliferation derives from symmetric division in which the progeny integrate into the existing glial network (Ge et al. 2012). It is known that postnatal experiences, such as enriched housing, can increase the production of astrocytes, but little is known about any special factors that might influence astrocytosis in the adolescent brain. This important gap in knowledge, however, needs to be addressed, given the critical role that astrocytes play in synaptic connectivity.

In addition to astrocyte proliferation in adolescence, myelin formation increases, partly to increase conduction speed along axons. Functional MRI studies have shown that increasing myelin formation in adolescents increases the efficiency of communication across brain regions compared to younger children. Myelin formation in adolescence, however, may serve other functions. First, there are changes in myelin related to learning. For example,

Sampaio-Baptista et al. (2013) showed that rats trained to reach through a slot to grasp food exhibit changes in white matter tracts in somatosensory cortex. Similar changes can be seen in humans as well. In a study where subjects were trained for two hours on a car-racing video game, diffusion tensor MRI revealed changes in white matter in hippocampus and parahippocampal gyrus (Hofstetter et al. 2013). More recently, the micro changes were found in white matter during a language task (Hofstetter et al. 2017). In the latter study, Hofstetter et al. introduced lexical items (flower names) that were new to participants for about an hour and found rapid changes in white matter tracts underlying the cortex. The extent of change correlated with behavioral measures of the lexical learning rate.

Second, as myelin continues to form, it may act to close the adolescent sensitive period by inhibiting axon sprouting and the creation of new synapses (Fields 2008).

MRI studies have also used structural MRIs to calculate changes in cortical volume, cortical thickness, and surface area. Although there are inconsistencies in the literature, a recent study of four independent longitudinal data sets have demonstrated decreasing cortical thickness and cortical volume that increases with age during late childhood (as of 7 years of age) and across adolescence before leveling off at around 20 years (Tamnes et al. 2017). In addition, there is a small decrease in cortical area during adolescence. Thus, the major change in the cortex during adolescence is cortical thinning.

Considerable evidence indicates that adolescence is an active time of changes in connectivity. One powerful way to examine connectivity changes is by using resting-state functional interactions and networks with rs-fcMRI. This technique allows us to analyze how cerebral activity changes over age within regions, as well as how the interactions between ages also change with age. Vogel et al. (2010) reviewed such studies and identified two general properties: (a) regional interactions, primarily local in children, change during development to become interactions that span longer cortical distances; (b) these developmental changes segregate local regions and integrate them into disparate subnetworks (see also Khundrakpam et al. 2013). However, it is not just neocortical networks that change during adolescence. Cortical connections with the amygdala, striatum, and hippocampus are also changing (see Figure 9.2). For example, Casey et al. (2015) provided an oversimplified illustration of the types of changes in prefrontal-subcortical circuitry: the interconnections and their relative strength change with development, providing insight into the emotional, social, and other nonemotional behaviors of adolescents. An important principle is that changes in connectivity must be precise enough for an altered circuit to process information differently and carry out the altered (or new) function.

Connectivity can also be inferred from measurements of synaptic changes. The most accurate, and labor intensive, method is to use electron micrographs to count synapse numbers throughout columns across different cortical regions.

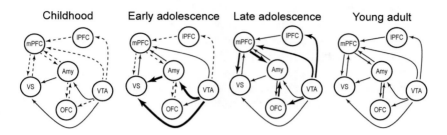

Figure 9.2 Simplistic illustration of hierarchical changes in connectivity from sub-cortico-subcortical to cortico-subcortical circuits with age. mPFC: medial prefrontal cortex; lPFC: lateral prefrontal cortex; OFC: orbitofrontal cortex; VS: ventral striatum; Amy: amygdala; VTA: ventral tegmental area. Modified after Casey et al. (2015).

Rakic et al. (1994) summarize such heroic studies in five different cortical regions in the rhesus monkey from the E50 through 20 years of postnatal age. Across all areas a rapid increase in synapses occurred postnatally, followed by a stable period for several years with a sharp decline beginning about age 3, which is roughly the onset of puberty, depending upon the area studied. The end of the rapid pruning varies widely, ending at about 4 years in visual cortex but continuing for many more years in prefrontal cortex (PFC). Once again, the end of the sensitive plastic period appears regionally specific, likely reflecting an extended plastic period in association regions.

A simpler method for estimating synapse number is to use a Golgi impregnation stain, which allows the analysis of dendritic complexity as well as length and spine density. The plasticity of dendritic morphology is high early in development but becomes stable by adolescence. Dendritic spines (the location of most excitatory synapses) remain, however, highly labile across adolescence (e.g., Koleske 2013). The relative stability of dendrites coupled with the synaptic plasticity allows the brain to fine-tune spine-based synaptic connections. Although there have been many studies of spine density at specific ages in development, I am unaware of a study parallel to the Rakic et al. electron microscopy (EM) studies in any laboratory species. In our own studies with rats, we have compared spine density in early adolescence (P30) to late adolescence (P55) to adulthood (P120), finding little change in parietal and occipital cortex from P30 to P55 but a sharp drop into adulthood. This result would seem to be at odds with the EM studies, yet there is one fundamental difference: the Golgi studies focused on the distal tips of the dendrites of pyramidal neurons, which are usually the most sensitive region to experiences, whereas the EM studies did not have this bias. The advantage of the Golgi studies is that a much larger range of cortical areas can be sampled fairly quickly.

If the brain is perturbed at P10 or P35, large changes occur during adolescence in dendritic complexity and spine density relative to controls not seen after similar perturbations after P90. This suggests that there is more plasticity than is apparent in the "normal" brain. Finally, one unexpected finding is that changes in two prefrontal regions are starkly different: there is little change in spine density in medial PFC from P60–P90, but a significant increase in spine density in orbital PFC over the same time period. In view of the importance of social behavior in adolescence, it is likely that the longer plasticity in the orbitofrontal cortex (OFC), which is central to social interaction and especially social context, should continue longer than other prefrontal regions.

Epigenetics can be viewed as a second genetic code, the first one being the genome, which is an organism's complete set of DNA (see also Moore and Kobor, this volume). Epigenetics refers to the changes in gene expression that do not involve alteration of the DNA sequence but rather the processes by which enzymes read the genes within the cells. Thus, epigenetics describes how a single genome can code for many phenotypes, depending on the internal and external environments. Epigenetic mechanisms influence the brain throughout the lifespan and are integrated with environmental changes characteristic of developmental milestones (e.g., Kanherkar et al. 2014).

Although genome-wide association studies have identified many genes associated with the regulation of the time at puberty, these genetic variants account for only a small fraction of the variation in the timing of puberty in humans. Epigenetic differences thus appear to play a significant role in the timing of puberty and adolescence as well as in the integration of hormonal, social, environmental, and genetic information. This complex interaction includes exposure to adverse experiences such as chronic stress and drug exposure (alcohol, cannabis) that significantly alter gene expression, often in a sexually dimorphic manner (e.g., Morrison et al. 2014). One fundamental difference between childhood and adolescence is that environments tend to be less structured for adolescents, at least in some cultures, leading to greater individual differences in experiences and thus in gene expression.

Considerable work has shown that the critical period in infancy results from an appropriate balance of excitatory and inhibitory (E–I) inputs (see Figure 9.3). The maturation of inhibitory GABA circuits underlies the timing of onset of the critical periods, which vary across brain regions. Premature onset of the critical period is prevented by various factors; for example, polysialic acid acts on neural cell adhesion molecules, which act on parvalbumin (PV) in GABA interneurons. When other factors promote PV cell maturation, the critical period begins. The critical period closes as molecular brakes emerge to dampen plasticity, and thus limit adult plasticity (for a review, see Takesian and Hensch 2013). A key point is that it is possible to reopen the critical period by manipulating the E–I balance chemically, such as by using valproate. It is thus possible that the critical period could be reopened in adolescence by some type of endogenous process, possibly gonadal hormones. As a general rule

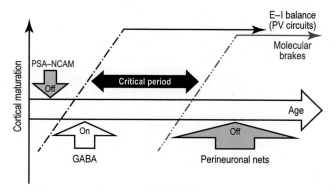

Figure 9.3 Proposed mechanism for turning on and off critical periods. Plasticity is blocked initially by factors such as polysialic acid (PSA) on neural cell adhesion molecule (NCAM), thus limiting the parvalbumin system (PV). Critical period onset is triggered as various factors enhance the GABA PV system. The critical period closes as molecular brakes (e.g., perineuronal nets) emerge to dampen plasticity. Adapted after Takesian and Hensch (2013).

of thumb, when there are changes in gonadal hormones, there are changes in neural plasticity.

The question here is whether the onset of adolescence acts to release the brakes in place in early development, or if there is a *de novo* set of molecular experiences that might be in play. One other possibility is that experience-expectant plasticity renders the brain more sensitive to certain kinds of experiences (e.g., social), which in turn drives circuits to self-organize. This could account for the opening of an adolescent sensitive period but provides no clear mechanism for closing it.

Perinatal Programming and Plasticity in the Adolescent Brain

There is considerable evidence that adult disease may have developmental origins (e.g., Barker 2004). Expanding on this idea, it can be argued that there is a sensitive period during which early experiences alter epigenetic programming that initially result in no obvious behavioral effects in infancy, but later surface in life to alter the brain and behavior. For instance, pre- or postnatal stress, nutrition, or drug exposure can alter the epigenome and induce later changes in adolescent brain and behavior (see review by Mychasiuk and Metz 2016). Exposure to high doses of alcohol on P4–9 triggers the production of astrocytes in adolescent rats (Helfer et al. 2009). In humans, the effects of early-life chronic stress is often first observed when periadolescent children (11–12 years of age) start to diverge in their developmental trajectories (Andersen and Teicher 2008). Thus, depressive symptoms are rarely observed in children when they are initially exposed to early stress; they are delayed about 9 years after the abuse (for further examples, see Andersen 2016). Early experiences may not only lead to behavioral pathology but may result in reduced plasticity in adolescence. For example, gestational exposure to nicotine blocked the effects of enriched housing in adolescence (Mychasiuk et al. 2014).

The effects of early-life experience may not be apparent and remain hidden until the adolescent brain is exposed to special experiences (e.g., alcohol, nicotine, enriched housing). Similarly, experiences in adolescence may change the manner in which the adult brain changes to experience in adulthood. Exposure to psychoactive drugs alters the neuronal structure in the adult PFC (for a review, see Robinson and Kolb 2004). Similar effects on the immature PFC have been proposed to interfere with the development of prefrontal circuits and increase risk for psychiatric problems in adulthood, including substance abuse (e.g., Jordan and Andersen 2017).

Special Examples of Plasticity in the Adolescent Brain

Before special examples in the adolescent brain are addressed, it is worthwhile considering what types of factors influence the brain in early development (for a more extensive review, see Kolb et al. 2013).

Factors Influencing Plasticity in the Early Developing Brain

Since the 1950s there has been considerable interest in the effects of early experiences on brain development. Although it was initially assumed that large changes in experience (e.g., being raised in darkness) would be required to influence brain development, it has become clear that an unexpectedly large range of experiences can alter brain development:

- age,
- sensory and motor experience,
- pre- and postnatal stress,
- psychoactive drugs,
- parent-child relationships,
- peer relationships,
- diet,
- gut bacteria, and
- immune system.

In addition, even fairly innocuous-appearing experiences can profoundly affect brain development. Furthermore, the magnitude of the changes is much larger than expected. A brief discussion of these factors follows, with an emphasis on studies of laboratory animals. Although other factors most certainly alter early development, our focus here is on those that are also likely to influence the adolescent brain.

Age

Precise embryological age is a critical factor in understanding plasticity in early development. Although there is a tendency to think of early postnatal development as a time of enhanced neural plasticity, this is not always the case. For example, if the cerebral cortex of rats is damaged during the first few days of life, the functional outcome is poor and the brain does not show successful neural compensation. However, similar damage during the last few days of gestation or in the second week of life stimulates remarkable plasticity, and behavioral outcomes are surprisingly good (e.g., Kolb 1995). A difference of just a few days in early development makes an enormous difference to brain plasticity.

Sensory and Motor Experiences

Sensory and motor experiences can easily be manipulated through extreme deprivation (e.g., rearing animals in total darkness) or by placing animals in quasi-enriched environments. Raising animals in severely deprived conditions interferes with development and often leads to permanent loss of function, often associated with reduced dendritic branching in the neocortex. By contrast, rearing animals in enriched environments stimulates synaptic changes in most cerebral regions, with increases in the number and density of blood vessels, neuron soma size, dendritic elements, synapses, gene expression, and glia. Animals do not actually need to be housed in these environments because if either of their parents lived in them prior to conception, the offspring's brains will be similar to the animals who actually lived there.

Fairly minor modifications of sensory inputs can also modify both brain and behavior. For example, when we tactilely stimulated newborn rats with a light brushing for 15 min three times daily until weaning, there was a significant increase in brain weight, dendritic length in cortical pyramidal neurons, and skilled motor behavior (e.g., Kolb and Gibb 2010). We have also shown that the tactile stimulation increases the production of FGF-2 in the skin, and FGF-2 crosses the developing blood–brain barrier to stimulate receptors in the cortex.

Pre- and Postnatal Stress

The general findings of studies of gestational stress is that behaviorally, offspring exhibit increased anxiety, altered play behavior, impaired skilled reaching, and slower spatial learning (reviewed by Kolb et al. 2017). Postmortem analyses revealed decreases in overall brain (but not body) weight and decreased spine density in OFC, but increased spine density in mPFC and changes in gene expression in these regions.

There is a large literature on the effects of maternal separation in rodents with consistent evidence that the duration is critical. Short duration (3–15 min daily) is beneficial to the offspring as it alters the hypothalamic-pituitary-adrenal axis, making it more efficient allowing better recovery from stress. In contrast, longer periods of maternal separation (e.g., 3 hr per day) is associated with increased anxiety in adult male offspring, disrupted play behavior in both sexes, decreased brain weight in males (but not females), increased dendritic complexity, and spine density in both mPFC and OFC.

Psychoactive Drugs

All psychoactive drugs, including prescription drugs, appear to change the structure of neurons, especially in PFC and nucleus accumbens (e.g., Robinson and Kolb 2004). Less is known about the effects of drug exposure during development, although it has long been known that early exposure to alcohol is deleterious for brain development. There is growing evidence that prenatal exposure to a wide range of drugs—including nicotine, amphetamine, fluoxetine, valproic acid, morphine, and marijuana—alter behavior, neuronal structure, and epigenetics in the adult offspring (e.g., Vassoler et al. 2014). Less is known about exposure to such drugs in early development, although antipsychotics administered to mice prior to weaning lead to markedly simpler dendritic structure and reduced spine density in pyramidal neurons across the cerebral cortex.

Diet

Although there is a considerable literature on the effects of early-life nutrition on normal and abnormal behavioral development, relatively little is known about the role of early nutrition and brain plasticity. Most studies have focused on the effects of deficiencies in specific nutrients, such as iron and choline. An intriguing question is whether early brain development might be enhanced by mineral or vitamin supplements. There is evidence that feeding dams diets with enhanced choline or combinations of nutrients increases cognitive function, nerve growth factor in the hippocampus and neocortex, as well as increased dendritic branching and length in cortical neurons in adulthood.

Gut Bacteria

Over the past decade, the idea has emerged that gut bacteria, the microbiome, interacts with the brain and can alter brain plasticity. Manipulation of the microbiome in newborn mice can influence motor and anxiety-related behaviors, and although the authors did not specifically examine brain plasticity, they did describe changes in the turnover of noradrenaline, dopamine, and serotonin in the striatum as well as changes in the production of synaptic-related proteins in cortex and striatum (Diaz Heijtz et al. 2011).

Factors Influencing Plasticity in the Adolescent Brain

There is a much smaller literature examining the factors that alter the adolescent brain, but interest has been growing over the past decade (see Spear 2016). Here I review a range of factors that have been shown to have special effects in adolescents and consider the possibility of others.

Play

Play may be an example of an experience-expectant behavior for the adolescent sensitive period. Although rats engage in play throughout their lifespan, play begins around weaning. Around P25–30, the full complement of play behaviors has emerged, with play reaching a peak about P30–40, followed by a decline beginning about P60. The PFC plays a central role in play behavior, and its development is strongly influenced by play. Bell et al. (2010) manipulated the amount of play that rats could engage in and found a negative correlation between dendritic complexity/spine density in mPFC and the amount of play. In contrast, complexity of neurons in OFC was positively correlated with the number of conspecifics (playmates or adults) present. Burleson et al. (2016) replicated this effect in hamsters and, in addition, found that play deprivation in adolescent hamsters increased the vulnerability to social stress in adulthood. In another study, Himmler et al. (2013) found that play experience changes the effect of nicotine on the same prefrontal neurons affected by play and stress. Taken together, these results suggest that a special adolescent behavior (play) can have a significant impact on the function and organization of the adult PFC.

Sensory and Motor Enrichment

Enriched housing is one of the most powerful experiences that can affect cerebral networks. Although it is generally expected that enrichment would have similar effects in young and older brains, and likely bigger effects in younger brains, this is not the case. We have compared the effects of two months of complex housing in young (P25), young adult (P90), and senescent (P300) rats (Kolb et al. 2003). As expected, the older groups showed increased dendritic complexity and spine density in parietal and visual pyramidal neurons; the young group, however, had a significant decrease in spine density, but an increase in dendritic length. One effect of this arrangement is that it would be easier to add synapses in response to other experiences in the young group.

Given that the young animals were complex housed for the entire adolescent period, this could be another adolescent experience-expectant example in the adolescent brain.

As noted for the infant brain, tactile stimulation is a powerful experience. Given that the adult brain is also affected by tactile stimulation, it would be interesting to determine how the adolescent brain responds to this type of experience.

Drugs

Adolescence is a common time for the initiation of psychoactive drug use (for an extensive review, see Spear 2016). A recent review of MRI studies on adolescent users of alcohol and other drugs (marijuana, nicotine, other stimulants, and a variety of illicit drugs) concludes that the frontal lobe is the most common region showing alterations (Silveri et al. 2016). Overall, the brain alterations appear larger than those observed in adults and, as noted earlier, it has been proposed that the interference with the development of prefrontal circuits increases the risk for cognitive and psychiatric problems, including substance abuse, in adulthood (e.g., Jordan and Andersen 2017). This is likely due, in part, to effects on brain plasticity in adulthood.

There is an extensive literature on the effects of adolescent drug exposure in laboratory animals showing a wide range of neural effects, as summarized in Table 9.1 (Spear 2016). Overall, drug effects are larger in adolescents than in adults. Although these neural effects are correlated with extensive cognitive/ behavioral and affective/social behavior changes, it is not well understood how these effects vary with exposure age, what the underlying mechanisms might be, and how they might influence brain plasticity later in life.

Stress

Although much is known about the effects of stress during adulthood and the perinatal period, surprisingly little is known about the effects of stress on brain plasticity in adolescence. Perinatal and adult stress both alter the structure of neurons in the PFC, hippocampus, and amygdala, although in differing ways at the two time points. Given that the adolescent sensitive period is one of increased plasticity, the adolescent brain may be more vulnerable to stress-related behavioral changes such as anxiety, depression, psychotic episodes, and so on. This is especially so given that the brain regions most sensitive to stress, including the PFC, hippocampus, and amygdala, all continue to mature during adolescence.

Eiland et al. (2012) exposed rats to daily restraint stress from P20 to P41 before looking at behavioral and neuronal changes. In short, they found behavioral changes that differed from similar stress in adulthood but found dendritic retraction in both hippocampus and mPFC and increased dendritic material in

Table 9.1   Effects of repeated adolescent exposure to ethanol (EtOH), nicotine (NIC), cannabinoids (CBs), cocaine (COC), and methamphetamine stimulants (STIM) on neural behavior: impaired/attenuated (↓); enhanced (↑); alterations, often complex, were reported (Y); effect is greater in adolescence than in adults (Adol>Adult); effect is less in adolescence than in adults (Adol<Adult) (after Spear 2016).

	EtOH Adol>Adult	NIC Adol>Adult	CB Adol>Adult	COC ?	STIM Adol<Adult
Neurogenesis	↓				
Cell death	↑	↑			↑
Spines/dendritic branching	↑ (immature spines)	↑			
Electrophysiological alterations	Y			Y	
Neuroimmune activation	Y				
Histone acetylation, epigenetic regulation	Y			Y	
Alterations in:					
Acetylcholine	Y	Y			Y
Glutamate/GABA	Y	Y	Y	Y	Y
Dopamine	Y	Y	Y	Y	Y
Serotonin		Y	Y		Y
CB			Y		
Affected regions:					
Prefrontal cortex	Y	Y	Y	Y	Y
Hippocampus	Y	Y	Y	Y	Y
Nucleus accumbens	Y	Y	Y	Y	
Amygdala	Y	Y	Y	Y	Y

the basolateral amygdala. The dendritic changes in females differ from what is observed in adults, as adult females show dendritic expansion in mPFC neurons that are projecting the amygdala and do not exhibit hippocampal dendritic retraction. Because the animals' brains were examined in adolescence (~P45) it is not known if the dendritic changes persist or change by adulthood.

Finally, although not specifically directed toward the effects of adolescence stress on neural plasticity, there is a growing literature on the association between stress exposure and the altered development of the amygdala, PFC, and ventral striatal dopaminergic systems in human adolescence (for a review, see Tottenham and Galván 2016). The general consensus is that these systems are vulnerable to stress in adolescence, leading to the emergence of a range of abnormalities in affective processing.

Diet

Just as in infant brain development, diet is likely to have an important contribution to changes in the brain during adolescence but few studies have directly studied how brain or behavior might be altered, especially by additives. It seems unlikely that specific nutrients will have large effects but preliminary studies using a combination of vitamin and mineral supplements are suggestive. For example, EmpowerPlus™ is a blend of 36 vitamins, minerals, and antioxidants and includes a proprietary blend of herbal supplements such as gingko biloba and the amino acid precursors for neurotransmitters (choline, phenylalanine, glutamine, and methionine). EmpowerPlus™ has been studied extensively for its effects on a wide range of behavioral problems (e.g., Simpson et al. 2011). In one open label trial, Kaplan et al. (2004) gave children (aged 8–15 yr) with mood and behavioral problems EmpowerPlus™ for several months and found improvement on a variety of outcome measures. In addition, several studies have shown benefits with EmpowerPlus™ and similar concoctions in adults with mood disorders (Rucklidge and Kaplan 2013). Rats fed this diet during development, including adolescence, show significant increases in dendritic length in cortical pyramidal neurons, although no behavioral measures were made.

Cerebral Injury

Although there is an extensive literature on the effects of brain injury in the perinatal brain in rats, cats, and monkeys (for a review, see Kolb et al. 2013), injury in adolescence has largely been neglected in laboratory animal studies. Nemati and Kolb (2011) showed that if the mPFC is damaged at P35 versus P55, there is a dramatic difference in outcomes. Animals with bilateral P35 lesions show remarkably better outcomes than similar lesions at P55 or adulthood, and this is associated with increased dendritic complexity and spine density in pyramidal neurons in adjacent cortical regions. Curiously, in a similar experiment in animals with unilateral motor cortex lesions, the effect was reversed: animals with P35 lesions were as impaired on motor tasks as adults, whereas those with P55 lesions showed good recovery, which was again correlated with dendritic changes in adjacent parietal cortex (Nemati and Kolb 2010). It is not clear if this difference relates to the location of the lesion or to a unilateral versus bilateral difference.

One additional aspect to studies of cerebral injury involves the timing of compensatory changes in adolescence after perinatal injuries. Using rats, we have shown that there are severe behavioral deficits in adults with cortical injuries in the first week of life but very good recovery after similar lesions in the second week. The recovery is correlated with dendritic hypertrophy and/or increased spine density in cortical pyramidal neurons, whereas the absence of recovery is correlated with atrophy of these neurons (for a review, see Kolb et al. 2013). The question is: When do neuronal compensations occur? To this end, we made mPFC lesions on either P1 or P10 (Kolb and Gibb 1991). One set of animals were given behavioral tests on P22–25 and their brains harvested on P28. A second group were given the same tests on P55–58 and their brains harvested on P60. Both the P1- and P10-lesioned rats were severely impaired on the early behavioral tests, but the P10-lesioned rats showed substantial recovery on later tests, which was correlated with dendritic hypertrophy present only in the older brains. It appears that during adolescence the enhanced plasticity in the P10-lesioned animals led to the enhanced behavioral and neuronal compensations. The question remaining, however, is why this occurred. What is different in the P1- and P10-lesioned brains in adolescence? This may be due to some type of epigenetic difference, but this remains to be shown.

One inference that we can draw from the above study is that adolescence may be an especially important time for implementing treatments to remediate early-life negative perturbations on the brain and behavior, such as stress or gestational drug exposure.

Immune System

The immune system intricately interacts with the nervous system throughout life. Various aspects of brain plasticity, including neurite outgrowth and synaptic pruning, are regulated in part by the immune system. Although much is known about the maturation of the immune system in embryonic and perinatal development, very little is known about its maturation during adolescence. Recent studies have shown changes in cytokine expression in rats across adolescence and that earlier exposure to stress can influence this expression (for a review, see Brenhouse and Schwarz 2016). Microglia, the primary immune cells of the brain, become active in response to a variety of perturbations. Their role in adolescence has not yet been studied thoroughly, but they are likely to play a role in the sensitive period and may act in some way to mediate onset/offset as well as neural plasticity.

Conclusions and Unanswered Questions

Adolescence can be seen as a sensitive period in which there is a sharp increase in neural plasticity. Although most plastic changes are likely to be "experience-dependent," there may be "experience-expectant" plastic changes in adolescence, perhaps related to gonadal hormones or the increase in socioaffective behaviors. The onset of the sensitive period is around the onset of pubertal gonadal hormone production, but may or may not be triggered by the hormone release. The offset of the sensitive period may be related to the completion of myelination, which can reduce plasticity. The offset of the sensitive period is likely different in different cortical regions, with the OFC being among the last regions to mature, likely reflecting the continuing impact of adolescent socioaffective experiences. Among the many questions that remain, I highlight the following:

- What is the role of glia in controlling the onset and offset of the sensitive period?
- Are the brakes on the early sensitive periods turned off during adolescence and, if so, how?
- How do experiences in adolescence vary with exposure age, what are the underlying mechanisms of their effects, and how might they influence brain plasticity later in life?
- What are the sex differences in the timing of the sensitive period and the role of experiences in altering brain plasticity during this period?
- Are there experience-expectant plastic changes in adolescence? If so, what are they?
- How do changes in gene expression in adolescence influence the duration of the sensitive period?
- What is the role of the immune system in controlling the onset/offset of the sensitive period and synaptic plasticity during this period?

Acknowledgments

Preparation of this chapter was supported by a grant from the Canadian Institute for Advanced Research.

10

Anesthesia-Induced Brain Oscillations

A Natural Experiment in Human Neurodevelopment

Patrick L. Purdon

Abstract

GABAergic inhibition mediates many crucial aspects of brain development, including the development of structural connections, critical period plasticity, and functional synchronization across large-scale networks via gamma-band oscillations. Disturbances in the development of GABAergic circuits are thought to underlie neurodevelopmental disorders such as schizophrenia and autism. Characterizing the developmental trajectory of these circuits in humans is a vitally important problem, but a challenging one. Current approaches in humans include postmortem studies and noninvasive imaging. These methods can provide highly specific information about GABA circuits in older children, and in-depth functional information in younger children, albeit with only indirect links to GABA circuit function. An ideal characterization in humans would map the continuous trajectory of GABA circuit function from infancy through adulthood with a common set of tools, alongside detailed measurements of sensory, cognitive, and language function. In this chapter it is proposed that studies of anesthesia-induced oscillations could be used to characterize and track the development of GABAergic oscillatory circuits from infancy through adulthood. The most commonly used anesthetic drugs in both pediatric and adult practice are powerful positive allosteric modulators of GABA receptors. These drugs induce large, stereotyped oscillations in the unconscious state that are likely generated by the same GABAergic circuits responsible for gamma oscillations in the conscious state. In the United States alone, these anesthetic drugs are administered to tens of millions of patients each year, under conditions of both neurotypical and atypical development. By harnessing this anesthetic experiment of nature, it may be possible to develop detailed developmental trajectories of GABAergic circuit function in humans.

Introduction

Inhibitory signaling mediated by gamma-aminobutyric acid (GABA) plays a pivotal role in the maturation of the cerebral cortex (Le Magueresse and Monyer 2013). GABAergic interneurons contribute to the development of structural connections within cerebral cortex (Le Magueresse and Monyer 2013), mediate the regulation of critical period plasticity (Fagiolini and Hensch 2000; Takesian and Hensch 2013), and are thought to support functional synchronization of activity across large-scale networks via gamma-band oscillations (~30–70 Hz) (Uhlhaas and Singer 2013). In particular, fast-spiking parvalbumin-positive (PV+) GABA interneurons are required to generate gamma oscillations, as demonstrated in both computational models (Whittington et al. 2000; Börgers et al. 2005) and optogenetic studies (Cardin et al. 2009; Sohal et al. 2009; Cho et al. 2015). Given their essential role in brain maturation and in large-scale network function, disturbances in the development of GABAergic circuits are thought to underlie neurodevelopmental disorders such as schizophrenia and autism. Schizophrenic patients have reduced numbers of PV+ interneurons in the dorsolateral prefrontal cortex (Glausier et al. 2014; Enwright et al. 2016), and also show disrupted synchrony in gamma-band activity (Uhlhaas and Singer 2010). Autistic patients show reduced numbers of PV+ interneurons in medial prefrontal cortex (Hashemi et al. 2017), reduced GABAergic signaling during sensory processing tasks (Robertson et al. 2016), and altered resting-state functional connectivity across multiple oscillatory bands, including the gamma band (Kitzbichler et al. 2015; Vakorin et al. 2017).

Given the pivotal role that GABAergic circuits play in development and developmental disorders, characterization of these circuits in humans is vitally important. However, such characterizations are challenging. Postmortem studies can provide detailed anatomic information on inhibitory neuron circuits (Ariza et al. 2016; Hashemi et al. 2017), but this type of data is difficult to collect, and still more difficult to relate to covarying functional information. Magnetic resonance imaging (MRI) spectroscopy has been used to characterize GABA neurotransmitter levels noninvasively (Robertson et al. 2016), but such methods are challenging to apply in young children or infants. Electroencephalography (EEG) and magnetoencephalography (MEG) have been used to analyze oscillatory activity within large-scale functional networks (Uhlhaas and Singer 2013; Kitzbichler et al. 2015; Vakorin et al. 2017). At present, however, it is difficult to disambiguate gamma-band oscillations from gamma-band power that comes from the "1/f" spectrum generated inherently by neuronal and post-synaptic activity (Buzsáki et al. 2012). As a result, it is difficult to make clear inferences about GABAergic networks relating to gamma oscillations distinct from gamma-band power generated by overall neuronal activity. Event-related potentials have also been used to interrogate specific sensory (LeBlanc et al. 2015) and language (Ortiz-Mantilla et al. 2016) functions in children and infants, in some cases alongside information about gamma-band oscillations

(Ortiz-Mantilla et al. 2016). Although highly plausible inferences can be made about inhibitory networks from these data, such inferences remain indirect in the absence of more specific information related more directly to GABAergic circuits.

Despite the rich information available from these studies, more direct assessments of GABAergic circuit function during human development are clearly needed. An ideal characterization would:

1. Map the continuous trajectory of circuit development from infancy through adulthood using a common set of tools.
2. Explicitly measure specific brain oscillations.
3. Conduct alongside measurements of covarying sensory, cognitive, and language function.

The question is: How could this be accomplished?

In this chapter I propose the idea that studies of anesthesia-induced oscillations could be used to characterize and track the development of GABAergic oscillatory circuits from infancy through adulthood. The most commonly used anesthetic drugs in both pediatric and adult practice are indeed powerful positive allosteric modulators of GABA receptors (Hemmings et al. 2005). These drugs induce large, stereotyped oscillations that coincide with unconsciousness (Purdon et al. 2013, 2015b)—oscillations which are very likely generated by the same GABAergic circuits responsible for gamma oscillations in the conscious state (Börgers et al. 2005; McCarthy et al. 2008; Ching et al. 2010). Moreover, in the United States alone, these anesthetic drugs are administered to tens of millions of patients each year, under conditions of both neurotypical and atypical development. By harnessing this anesthetic experiment of nature, I argue that it would be possible to develop detailed developmental trajectories of GABAergic circuit function in humans.

Anesthetic Mechanisms

General anesthesia is a reversible drug-induced state consisting of unconsciousness, analgesia, amnesia, immobility, and autonomic stability (Brown et al. 2010). Each year, over 20 million patients (Brown et al. 2010), including 6 million children (Sun 2010), receive general anesthesia for surgical and medical procedures. Perhaps one of the most fascinating properties of anesthetic drugs is their ability to produce altered states of arousal and unconsciousness. Anesthetic drugs act by modulating neurotransmitter-gated ion channels in the central nervous system (Hemmings et al. 2005). Propofol, one of the most commonly used anesthetic drugs in both adult and pediatric practice, is a positive allosteric modulator of $GABA_A$ receptors, enhancing GABAergic inhibition (Hemmings et al. 2005; Brown et al. 2011). Sevoflurane, an inhaled ether-derived anesthetic frequently used in pediatric practice, also enhances inhibition via $GABA_A$ receptors (Hemmings et al. 2005). At low doses, both propofol and sevoflurane produce sedation, whereas at high doses they maintain a deep state of unconsciousness from which patients cannot be aroused (Purdon et al. 2015b). Other commonly used anesthetic drugs are known to act through distinct molecular mechanisms to produce other states of altered arousal. For instance, ketamine is an NMDA antagonist that produces analgesia, altered sensory perception, and hallucinations at low doses, and unconsciousness at high doses (Brown et al. 2011; Purdon et al. 2015b). Dexmedetomidine is an α2 adrenergic agonist that produces states of sedation which mimic nonrapid eye movement sleep, from which patients can be aroused with sufficiently strong stimuli (Brown et al. 2011; Purdon et al. 2015b).

Significant progress has been made in recent years to characterize how anesthetic drugs act at a systems level. A key insight has been that anesthetic drugs induce profound brain oscillations, visible in the EEG, whose structure corresponds to the drugs' underlying molecular mechanisms (e.g., GABA, NMDA, α2 adrenergic), as shown in Figure 10.1 (Purdon et al. 2015b). Moreover, for a given anesthetic drug, the structure of the oscillations varies with drug dose, in a way that correlates with a patient's state of altered arousal or unconsciousness (Purdon et al. 2013, 2015b). The neurophysiology and mechanisms of these drugs is becoming increasingly well understood,

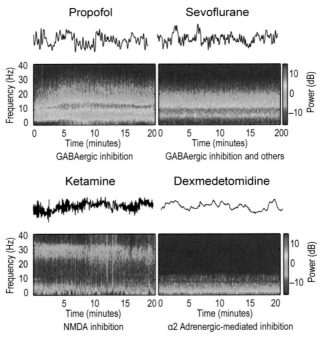

Figure 10.1 The structure of anesthesia-induced EEG oscillations varies by drug class and molecular mechanism (from Purdon et al. 2015b).

making it possible, particularly in the case of the GABAergic drugs, to use the knowledge of anesthesia-induced brain oscillations to help characterize human brain development.

Slow and Alpha Oscillations: Markers of Propofol- and Sevoflurane-Induced Unconsciousness

Propofol-induced unconsciousness is characterized by the presence of large, stereotyped frontal alpha (8–12 Hz) and slow (0.1–1 Hz) oscillations (Figure 10.2; Lewis et al. 2012; Purdon et al. 2013, 2015b). When propofol is gradually administered, a combination of gamma and beta oscillations (~12–35 Hz) appears at a point in time that coincides with reduced responsiveness to external stimuli: a state of sedation (Purdon et al. 2013). At loss of consciousness, defined operationally as the point at which subjects stop responding to stimuli, large slow (0.1–1 Hz) and alpha (8–12 Hz) oscillations develop (Purdon et al. 2013). The propofol-induced alpha oscillations have a frontal predominance, whereas the slow oscillations are present across the scalp. The frontal alpha oscillations are coherent (i.e., spatially correlated at alpha frequencies), whereas the slow oscillations are not (Figure 10.3). General anesthesia maintained with sevoflurane is characterized by coherent frontal alpha oscillations (8–12 Hz), high-amplitude delta (1–4 Hz) and slow (0.1–1Hz) oscillations in the EEG (Akeju et al. 2014b). This pattern, observed when patients are sufficiently anesthetized to conduct surgery, is similar to what is observed during propofol-induced unconsciousness.

Recordings of intracranial neurophysiology in humans—spanning populations of single neurons, local field potentials, and intracranial EEG—show that propofol-induced slow oscillations reflect sustained periods of neuronal silence lasting 1 to 2 seconds, referred to as OFF states, interrupted by brief periods of firing, or ON states (Figure 10.4; Lewis et al. 2012). The slow oscillations become asynchronous or incoherent with increasing distance along the cortex, suggesting that the ON and OFF states in different areas of the cortex are likely to be misaligned in time. Thus, during this propofol-induced slow oscillation, both local cortical neuronal activity, as well as intracortical neuronal interactions are likely blocked, contributing to unconsciousness (Lewis et al. 2012).

Anesthesia-Induced Frontal Alpha: An Indicator of GABAergic Inhibitory Circuit Function

Computational models have been developed to study how propofol-induced oscillations might arise as a result of enhanced GABA inhibition within cortical and thalamocortical circuits. McCarthy et al. (2008) proposed a cortical model featuring inhibitory interneurons and excitatory pyramidal neurons with

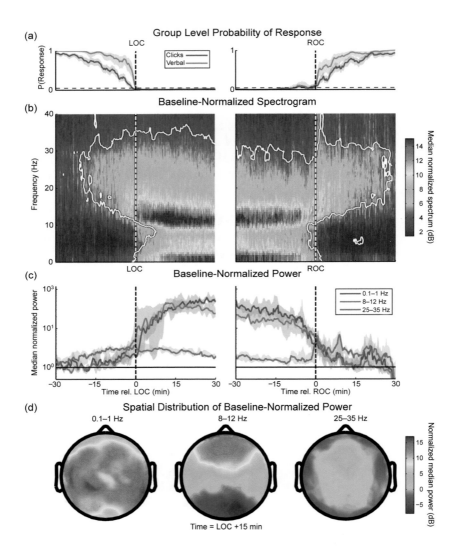

Figure 10.2 (a) Group-level probabilities of response to click (blue) and verbal stimuli (red) relative to loss of consciousness (LOC) and recovery of consciousness (ROC).
(b) Group-level baseline-normalized spectrograms aligned with respect to LOC and
ROC. (c) Group-level baseline-normalized power in the slow (0.1–1 Hz), alpha (8–12
Hz), and gamma (25–35 Hz) bands aligned with respect to LOC and ROC. (d) During profound unconsciousness with propofol, alpha power is concentrated in frontal
channels, whereas the slow oscillations are distributed across the scalp. Reprinted with
permission from Purdon et al. (2013).

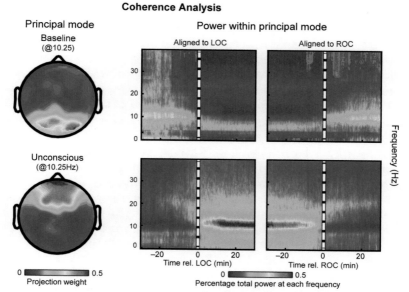

Figure 10.3 Top: In the conscious eyes-closed state, coherent alpha-band oscillations with a posterior distribution are present. Bottom: In the propofol-induced unconscious state, spatially coherent alpha waves are concentrated in the frontal channels. Adapted after Purdon et al. (2013).

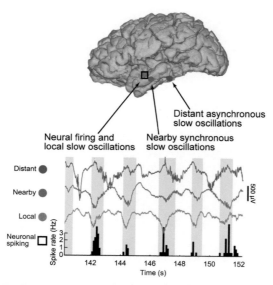

Figure 10.4 During propofol-induced unconsciousness in humans, neurons fire only in a specific phase interval defined by the local slow oscillations. Adapted after Lewis et al. (2012).

Hodgkin–Huxley-style dynamics. The model included both putative PV+ fast-spiking interneurons and low-threshold spiking (LTS) interneurons (McCarthy et al. 2008). They found that enhanced GABA inhibition mimicking propofol, represented by increasing the $GABA_A$ conductance, could produce beta oscillations as well as alpha oscillations at higher levels of $GABA_A$ conductance, corresponding to higher propofol doses. A follow-up study by Ching et al. (2010) characterized thalamocortical loop dynamics in a similar fashion. This model included fast-spiking and LTS inhibitory interneurons and pyramidal excitatory neurons in the cortex as well as thalamic reticular interneurons and thalamocortical relay neurons in the thalamus. Here, propofol's actions were modeled by enhancing $GABA_A$ conductance by up to ~300% in both the cortex and thalamus. At baseline, this network showed gamma oscillations in the ~40 Hz range, driven by fast-spiking interneurons via a pyramidal interneuron mechanism, as described by Börgers et al. (2005). As the $GABA_A$ conductance was increased moderately by ~50% above baseline, corresponding to low propofol doses, oscillations formed in the low gamma (~30 Hz) and beta (13–25 Hz) frequency range (Figure 10.5). However, with increasing $GABA_A$ conductance, the oscillations decreased in frequency into the alpha range. Crucially, at some point the alpha oscillations became coherent (Figure 10.5). In the absence of the thalamic components of the model, the induced alpha oscillations

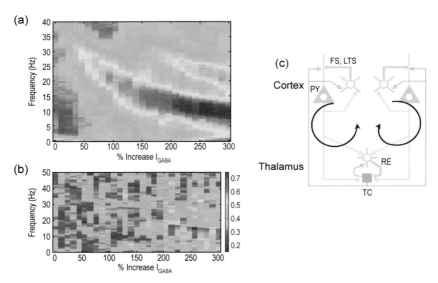

Figure 10.5   (a) Spectrogram and (b) coherence of cortical pyramidal postsynaptic currents from a (c) thalamocortical model as a function of increasing GABAergic inhibition. The dominant frequency decays monotonically, eventually settling into the 10 Hz alpha range at high GABA conductance levels (300% increase). At these levels, the cortical populations display high coherence, mediated by thalamocortical synchronization. Adapted after Ching et al. (2010).

are not coherent (McCarthy et al. 2008). However, when the thalamus is part of the model, thalamic neuronal activity appears to entrain the cortical alpha oscillations, allowing them to become coherent (Ching et al. 2010).

A primary inference from these studies has been that the thalamus plays a key role in generating propofol-induced coherent frontal alpha oscillations. However, these computational studies also suggest that inhibitory cortical circuit components that normally mediate gamma oscillations, namely fast-spiking interneurons within a pyramidal-interneuron network (Börgers et al. 2005), produce beta and alpha oscillations under the enhanced inhibitory influence of propofol (Ching et al. 2010). These computational results suggest that propofol-induced frontal alpha oscillations could be used to quantify cortical inhibitory circuit function, and that coherence in these frontal alpha oscillations could be an indicator of related thalamocortical functional interactions.

Experimental evidence from both human and rodent studies is consistent with the predictions of these computational studies. The model behaviors correspond closely to the EEG dynamics seen in the propofol human volunteer studies described earlier (Purdon et al. 2013). These data have also been corroborated by hundreds of operating room recordings of propofol and other GABAergic anesthetic drugs, such as sevoflurane (Akeju et al. 2014b; Purdon et al. 2015b). Multisite invasive rodent neurophysiological recordings show that propofol-induced alpha oscillations are present in both the thalamus and cortex (Baker et al. 2014; Flores et al. 2017), and that those oscillations are coherent in the alpha band after loss of consciousness (Flores et al. 2017). The specific circuit-level contributions of different cell types (e.g., PV+ fast-spiking interneurons versus LTS interneurons) remain to be studied in detail. Nonetheless, the experimental and computational evidence to date is consistent with the notion that propofol- and sevoflurane-induced frontal alpha oscillations are generated by cortical and thalamocortical GABAergic inhibitory circuits.

Age-Dependent Anesthesia-Induced EEG Oscillations from Infancy to Adulthood

Through studies on human volunteer subjects we have learned a great deal about the clinical neurophysiology of anesthetic drugs (Purdon et al. 2013; Akeju et al. 2014a). Our ability to perform studies on humans safely is a direct consequence of improvements in anesthesia patient safety, developed over several decades, punctuated by the introduction of physiological monitoring standards in 1984, which ensured that electrocardiogram, end-tidal CO_2, blood pressure, and pulse oxygenation were monitored in every patient (Cooper et al. 1984). This allowed anesthesiologists to monitor and precisely manage the cardiovascular and respiratory side effects of anesthetic drugs. These safety improvements address the major concerns that might arise in volunteer studies in adults, but the situation is more complicated in children. First, there are concerns that anesthetic exposure may have lasting neurodevelopmental effects in children (Jevtovic-Todorovic et al. 2013). Moreover, clinical management in children is far more challenging. As a result, anesthesia studies in children are only conducted when surgical or medical procedures require general anesthesia.

Early studies of anesthesia-induced EEG changes in children focused on characterizing (a) quantitative EEG features as a function of age (Schultz et al. 2004), and (b) the behavior of commercial proprietary processed EEG "depth of anesthesia" parameters (Davidson et al. 2005). These studies found that EEG spectral parameters varied with age (Schultz et al. 2004) and that "depth of anesthesia" parameters developed for adults worked differently when applied to children (Davidson et al. 2005). These findings were of value clinically, because they made clear that anesthesia-induced EEG patterns in children differed from those in adults (Schultz et al. 2004), and that existing processed EEG monitors developed for adults could not be reasonably applied to children (Davidson et al. 2005). Their scientific utility, however, was limited by the absence of neural mechanisms that could link drug mechanisms to EEG features. Fortunately, significant progress has been made in recent years to work out the relationships between drug actions, neural circuits, and anesthesia-induced EEG oscillations (Brown et al. 2011; Purdon et al. 2015b). Moreover, as described earlier, there is a plausible and specific link between inhibitory circuits crucial in development and anesthesia-induced oscillations: the putative PV+ fast-spiking interneuron circuits that mediate gamma oscillations slow to beta and alpha frequencies under the influence of GABAergic anesthetic drugs. This new perspective, coupled with improved data analysis methods (Prerau et al. 2017), allows us to gain new insights by analyzing EEG oscillations recorded in children receiving general anesthesia.

A recent study by Lee et al. (2017b) analyzed age-dependent changes in frontal EEG power and coherence in 97 patients, between 0 and 21 years of age, who received propofol to maintain general anesthesia. They found that total power (0–40 Hz) increased from infancy through approximately 7 years of age, subsequently declining to a plateau at approximately 21 years (Figure 10.6). The top panels in Figure 10.6 show EEG spectrograms from representative subjects across the age range. The three older patients, aged approximately 4, 10, and 20 years, show the same combination of slow and alpha oscillations described above for adults. The EEG for the youngest patient, less than 1 year of age, however, has a different structure that lacks a distinct alpha peak. Lee et al. (2017b) also performed a more detailed analysis of alpha (8–13 Hz) and slow oscillation (0.1–1 Hz) power as a function of age (Figure 10.7). They found that alpha power peaked at approximately 7.3 years of age, whereas slow oscillation power peaked at approximately 11.6 years of age. Earlier studies by Akeju et al. (2015) showed similar results for the anesthetic drug sevoflurane. Given the putative GABAergic mechanism for propofol- and

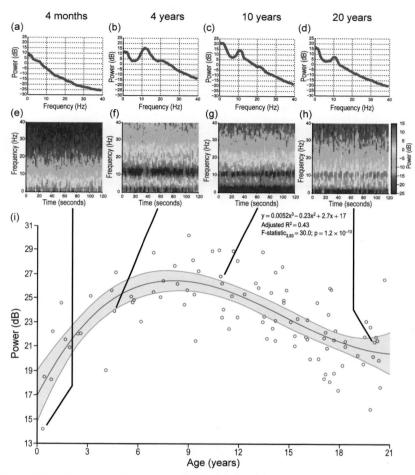

Figure 10.6 Representative frontal EEG spectrograms (a–h) illustrating that slow (0.1–1 Hz) and delta (1–4 Hz) oscillations are present during general anesthesia maintained with propofol. Alpha (8–13 Hz) oscillations appear to emerge after 1 year of age. Total EEG power (1–40 Hz) for each patient is plotted in (i) as a function of age. From Lee et al. (2017b).

sevoflurane-induced frontal alpha oscillations, and the evidence for a medial prefrontal cortical generator (Flores et al. 2017), these results imply that GABAergic inhibition in the medial prefrontal cortex increases in extent from infancy through approximately 7 years of age, declining gradually thereafter into early adulthood. This interpretation is consistent with postmortem studies of age-dependent human (Petanjek et al. 2011) and nonhuman primate synaptic density (Cruz et al. 2003). Human prefrontal cortical pyramidal neuron dendritic spine density peaks at approximately the same age, ~ 7 years (Petanjek et

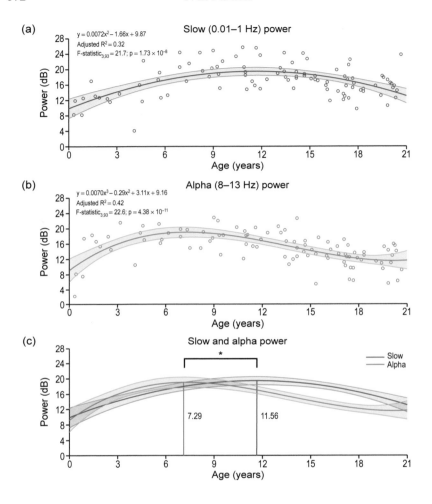

Figure 10.7 (a) Age-dependence of propofol-induced slow (0.1–1 Hz) power. (b) Age-dependence of propofol-induced frontal alpha power (8–13 Hz). (c) The putative GABA-mediated alpha power peaks at 7.3 years of age, distinct from the slow oscillation, whose power peaks at 11.6 years of age. From Lee et al. (2017b).

al. 2011), in accordance with the peak in propofol-induced alpha wave power (Lee et al. 2017b). The dendritic spines correspond to excitatory synapses, but based on studies of nonhuman primate prefrontal cortex, GABAergic inhibitory synapses are thought to develop in parallel with excitatory synapses in early life (Anderson et al. 1995), consistent with the interpretation that the anesthesia-induced frontal alpha power reflects inhibitory signaling.

The EEG spectrogram shown for the youngest representative patient in Figure 10.6 (upper left corner) lacks a distinct alpha wave, suggesting that different circuit dynamics may be active in children less than 1 year of age. Cornelissen et al. (2015) conducted a series of multichannel EEG recordings in infants less than 6 months of age during general anesthesia for surgery maintained with sevoflurane. Their findings reveal that sevoflurane-induced frontal alpha power was absent in children between 0 to 3 months of age (Figure 10.8). At 4 to 6 months of age, sevoflurane induced broad-band power spanning alpha (8–12 Hz) and beta (12–30 Hz) frequencies, albeit without the pronounced peak at alpha frequencies observed in adults (Figure 10.8). The studies by Lee et al. (2017b) showed a similar effect under propofol: patients between 4 months and 1 year of age showed increased power at beta frequencies, but not at alpha frequencies (Figure 10.9). This would suggest that inhibitory networks are still developing in prefrontal cortex in the first year of life. In particular, modeling studies by McCarthy et al. (2008) and Ching et al. (2010) suggest that the capacity for inhibitory signaling is lower in infants compared to adults and older children, such that propofol- and sevoflurane-induced actions are only able to slow cortical pyramidal-interneuron circuits into the beta-frequency range, short of the alpha range as in older children and adults.

Cornelissen et al. (2015) performed a coherence analysis, similar to that depicted in Figure 10.3 for adults (Purdon et al. 2013), which characterized the spatial distribution of coherent activity. Cornelissen et al. (2015) found that, in contrast to adults (Purdon et al. 2013; Akeju et al. 2015; Purdon et al. 2015a), children 6 months of age or younger do not have coherent frontal alpha waves (Figure 10.10). Coherence analysis in children less than 1.5 years old, who received sevoflurane (Akeju et al. 2015), showed that frontal

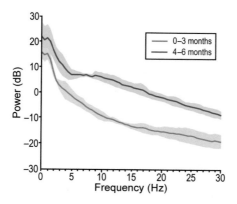

Figure 10.8 Spectra during maintenance of sevoflurane-induced general anesthesia in patients 0–3 months (red) and 4–6 months of age (blue). Unlike adults, infants in both age groups lack a distinct alpha peak. The older infants show increased broad band power spanning alpha (8–12 Hz), beta (12–25 Hz), and gamma bands (>25 Hz). From Cornelissen et al. (2015).

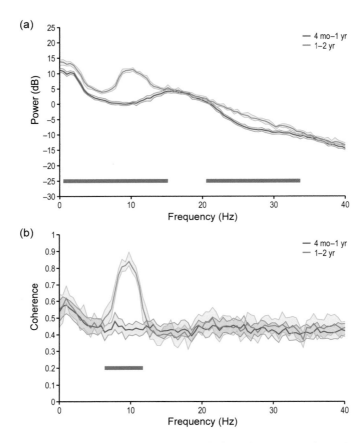

Figure 10.9 (a) The frontal EEG spectrum in infants (blue, 4 months to 1 year of age) and toddlers (red, 1–2 years of age). (b) The frontal EEG coherence in these same groups. The green bars indicate a statistically significant difference (95% bootstrap confidence intervals for the difference do not intersect zero). These data illustrate how propofol-induced coherent frontal alpha oscillations do not fully develop until 1 year of age. From Lee et al. (2017b).

coherence, though absent at 6 months of age, appears at approximately 1 year of age (Figure 10.10). Lee et al. (2017b) showed a similar result for propofol: coherent alpha waves were absent in children less than 1 year of age but appear thereafter (Figure 10.9). These results suggest that in infancy, the functional and/or structural connections required to generate and maintain coherent frontal alpha waves are absent, and only develop later, at approximately 1 year of age. This interpretation is consistent with studies of thalamocortical functional connectivity in infants using functional MRI, which show an absence of frontal thalamocortical connectivity in newborn children that subsequently develops by 1 year of age (Alcauter et al. 2014).

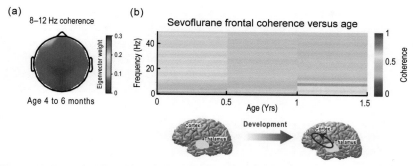

Figure 10.10 (a) Studies of sevoflurane-induced EEG in infants less than 6 months old show that they do not have coherent frontal alpha waves (adapted after Cornelissen et al. 2015). (b) Analysis of sevoflurane-induced EEG in children from 0–1.5 years shows that frontal alpha waves become coherent at approximately 1 year of age (adapted after Akeju et al. (2015).

Discussion

Age-related changes in the EEG of children have been described during both wakefulness and sleep, and show similar trends throughout childhood (Gaudreau et al. 2001; Feinberg and Campbell 2010; Segalowitz et al. 2010; Miskovic et al. 2015). However, age-related anesthesia-induced changes are perhaps unique in that they appear to reflect GABA interneuron-dependent frontal cortical and thalamic function (McCarthy et al. 2008; Ching et al. 2010; Flores et al. 2017). The changes in anesthesia-induced frontal alpha power from infancy through childhood, adolescence, and adulthood might therefore be interpreted as a marker of GABA interneuron circuit development. Our current understanding of the mechanisms for the anesthesia-induced frontal alpha wave has specific implications for the development of fast-spiking PV+ interneuron circuits within the prefrontal cortex. The age-dependent time course for propofol-induced slow oscillations, the other major feature of the EEG under propofol, differs significantly from that of the frontal alpha wave (Figure 10.7). In the future, as we learn more about the mechanisms for the propofol-induced slow oscillations, it may be possible to make inferences about other developing brain circuits. The age-related changes in EEG power shown in Figure 10.6 might also reflect a more general process of development that includes synaptogenesis in early childhood, followed by neural pruning during maturation. Given the putative frontal thalamocortical mechanism for the propofol- and sevoflurane-induced frontal alpha oscillations, the absence of coherent frontal alpha waves during infancy and subsequent appearance at approximately 1 year of age (Figures 10.9, 10.10) likely reflect an underlying development within the thalamus and cortex.

The age dependence of anesthesia-induced EEG oscillations in children has a number of clinical implications. The differences in EEG power across all frequency bands as a function of age suggest that "depth of anesthesia" monitors developed for adult patients are not accurate for use in children. However, the form of the anesthesia-induced EEG oscillations—namely, large, slow, and frontal alpha oscillations—is similar in patients above 1 year of age. Monitoring strategies based on these features might therefore apply equally well in children as they do in adults. In the simplest of forms, this could consist of direct interpretation of the unprocessed EEG and spectrogram during general anesthesia (Bennett et al. 2009; Purdon et al. 2015b). The options for anesthetic drugs are limited, with a limited pipeline of drugs in development. If concerns persist that anesthetic exposure may have neurodevelopmental effects, a strategy to minimize exposure could help mitigate potential harm. EEG monitoring in children might help establish in individual patients an appropriate level of sedation or unconsciousness using the smallest possible anesthetic dose.

The age-dependent features of the propofol- and sevoflurane-induced EEG patterns raise interesting possibilities for future studies: clinical anesthetics could be used to assess cortical inhibitory circuit function in relation to developing sensory or cognitive function, or in relation to developmental neurocognitive disorders. The insights gained from analyses of age-dependent effects in GABAergic anesthetic drugs suggest that similar insights might be gained by analyzing oscillations induced by anesthetic drugs that act at other sites, such as NMDA, $\alpha2$ adrenergic, or μ-opioid receptors. Ketamine is being studied extensively as a model for schizophrenia (Rivolta et al. 2015); studies of the age-dependent effects of ketamine might lead to additional insights from this model. Further investigation of the neural circuit mechanisms for anesthetic drugs could make it possible to gain deeper insights into human brain development viewed through the lens of this clinical anesthetic experiment of nature.

11

Understanding Effects of Experience on Neurocognitive Development through the Lens of Early Adolescence

Marina Bedny, Tomáš Paus, Sam M. Doesburg,
Jay Giedd, Rowshanak Hashemiyoon, Bryan Kolb,
Patrick L. Purdon, Pasko Rakic, and Cheryl L. Sisk

Abstract

How does lifetime experience shape cognitive and neural development? This chapter considers this question from the perspective of early adolescence. Plasticity in the adolescent brain may occur on three possible time courses: adolescent brains may be no more plastic than adult brains; adolescence could mark the end of critical periods which began in infancy; or adolescence might constitute its own, distinct critical period. In humans, all these time courses likely coexist. Many functional properties of the brain continue to change well into adolescence (e.g., regional cortical information selectivity, neural network correlations, oscillatory activity). Determining how these developmental changes are influenced by genes and/or experience in humans is challenging. Some insight comes from studies with individuals who have different developmental histories (e.g., individuals who grow up blind). These investigations suggest that experience plays a fundamental role in determining the functional specialization of cortical networks and patterns of functional connectivity. Studies with animal models provide crucial insight into the causal factors that drive brain development because they allow direct manipulation of experience and genetics as well as more direct measurements of neural function. However, work is needed to bridge the gap between measurements of brain function in animals and humans. At present, work with animal models has focused on plasticity in sensory systems, whereas much of the developmental change that occurs in early adolescence is in higher cognitive systems. Bridging these gaps is an important goal for future research.

Introduction

In this chapter, we consider the time course and mechanisms of experience-based plasticity through the lens of early adolescence. Adolescence is of interest in its own right as a developmental phenomenon. It is a time when humans and other animals undergo profound changes in social behavior, transitioning from a focus on social relationships within the family to relationships with peers. Many important cognitive functions continue to develop into adolescence, including social cognition, aspects of episodic memory, and abstract thinking (Steinberg 2005; Shing et al. 2008; Dumontheil 2014). Adolescence is also a time when many psychiatric conditions, such as depression and schizophrenia, emerge or become exacerbated (Paus et al. 2008). As such, the clinical significance of this time period cannot be overstated.

The study of adolescence provides an opportunity for us to consider some basic questions regarding the effects of experience on neurocognitive development more generally:

- Why might the potential for learning change across the human life span?
- What are critical periods, and does the term "critical period" refer to the same phenomena in the context of sensory and higher cognitive systems?
- Are critical periods that occur in early childhood qualitatively similar to hypothesized critical periods later in development?

Our goal in this chapter is to highlight insights into experience-based plasticity from the perspective of early adolescence.

Adolescence and the Time Course of Experience-Based Plasticity

Before we consider the time course of experience-based plasticity in adolescence, we must first operationalize the term "plasticity." All learning throughout the life span is associated with some neural change, whether this involves modifying long-range tracts during early postnatal development or long-term potentiation of a single synapse. According to an inclusive definition, all neural changes constitute plasticity. On the other hand, one could take a more restrictive view, where the term plasticity is reserved for more substantial and long-term changes. The line between learning and plasticity could be drawn in anatomical terms. For example, changes to individual synapses might not count as plasticity if the overall number of synapses in a particular brain region remains consistent, whereas systemic pruning of large numbers of synapses in a given cortical area would constitute sufficient change to warrant the term.

Here we operationalize the term experience-based plasticity in functional terms, as a substantial change in the representational and processing capacities of a neural system resulting from informational input from the environment. Visual deprivation early in life is a classic example of experience-based plasticity (Hubel and Wiesel 1970). Such deprivation permanently changes the capacity of the visual cortex to support basic visual functions, such as line orientation discrimination, three-dimensional perception, and face recognition (de Heering and Maurer 2012). By contrast, learning a new face would constitute a case of learning without plasticity, according to this definition, since it does not substantially alter the visual system's ability to process or learn other information.

A useful framework for thinking about the time course of plasticity was articulated by Greenough et al. (1987), who distinguished between two broad classes of plasticity with different characteristic time courses: experience-expectant and experience-dependent plasticity. Experience-expectant plasticity occurs in cases where the brain evolved to expect certain types of input from the environment that are nearly ubiquitous for the species (e.g., input from the two eyes, presence of motion, language and social interactions with other agents). Greenough et al. argued that mammalian brains evolved to expect such species-typical experiences during particular temporal windows in development; that is, during critical periods (Figure 11.1a, b). During these windows, such experiences have potent organizing effects on the brain. By

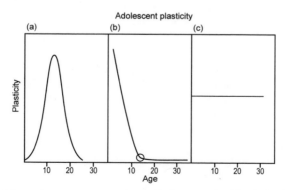

Figure 11.1   Theoretical depiction of how plasticity (i.e., responsivity to experience) changes over the life span: (a) adolescence may be its own independent critical period that is not contiguous with childhood; (b) adolescence could mark the end of a critical period that began in infancy or early childhood; (c) the brain is able to change in response to experiences and is stable over the entire life span.

contrast, before and after the critical period, the same type of experience has little to no effect on the same neural systems.

In contrast to experience-expectant plasticity, experience-dependent plasticity results from experiences that, in general, vary widely across species members. Only some of us will learn to ride a bicycle or play the piano, and literacy levels vary widely across humans. Evolution could not have prepared the brain to expect such experiences. Greenough et al. (1987) proposed that unlike experience-expectant plasticity, which is restricted to critical periods, the capacity of the brain to change in response to experiences is stable throughout the life span (Figure 11.1c). Thus the organism is capable of this type of learning to a similar degree early and late in life. An example of this kind of learning is the capacity of humans to add words to their vocabulary throughout the life span.

Within this framework, we can ask whether the human brain is intrinsically more plastic during adolescence: Has the brain been prepared by evolution to expect particular experiences during this period of life? If so, we would expect the adolescent brain to be differentially sensitive to the influences of environmental input relative to the adult brain. Consistent with the idea of enhanced sensitivity, humans undergo large-scale behavioral changes during adolescence. However, such changes could be driven by external environmental changes that occur during this time of life (e.g., intensification of interactions with peers), rather than due to enhanced potential for change within the brain (Figure 11.1c) (Fuhrmann et al. 2015). They could also purely be due to maturational factors that do not reflect an enhanced sensitivity to the environment.

Alternatively, adolescence could be a special time of sensitivity to environmental input (i.e., a critical period). Adolescence could mark the end of critical period(s) that began in infancy or early childhood (Figure 11.1b). An example of such an effect has been postulated in the context of acquiring the grammar of language. A seminal study examined the acquisition of a second language by immigrants from China and Korea to the United States. Here, Johnson and Newport (1989) reported that individuals who entered the country before approximately age 7 attained native-like abilities in the use of English sentence-level grammar, with proficiency falling off linearly until at approximately age 15 before it plateaus. The presence and precise timing of such a plateau in second language learning remains controversial. The critical period may be earlier and sharper for the acquisition of the first language (Friedmann and Rusou 2015; Mayberry 1998). Nevertheless, in principle, such a pattern corresponds to a prolonged critical period that terminates in adolescence. There is also evidence that the brain undergoes anatomical changes during the adolescent period. Gray matter volume thins in much of the cortex from early childhood throughout the late teens, and the total amount of white matter in the human brain increases, abating only in young adulthood (Gogtay et al. 2004; Giedd and Rapoport 2010; Walhovd et al. 2017). This long-lasting neural immaturity could be associated with sensitivity to environmental influences on the brain. Such temporal trajectories are examples of adolescence marking the end of a very long critical period lasting the first two decades of life (Figure 11.1b).

It is worth noting that the protracted duration of putative critical periods that terminate in adolescence distinguishes them from the classic critical periods that have been described in sensory systems. For example, the period of ocular dominance plasticity in mice is estimated to last 1–2 months (Morishita and Hensch 2008). The critical period for monocular deprivation in humans lasts for the first few years of life, peaking sometime before the second year and ending by approximately year six (Berardi et al. 2000; Maurer and Hensch 2012). In general, animals with longer life spans have longer critical periods. Thus the critical period for monocular deprivation in mice is shorter than that of monkeys, which is in turn shorter than that of humans (Berardi et al. 2000). Nevertheless, even relative to the human life span, critical periods which end in adolescence are protracted. Why might putative critical periods be so long?

One factor that could contribute to long-lasting critical periods in humans is the relatively late maturation of higher cognitive systems as compared with sensory systems. Even within the visual system, visual functions that mature slower (vision for high spatial frequencies) have more prolonged critical periods and mature later (vision for low spatial frequencies) (Maurer and Hensch 2012). Analogously, critical periods in higher cognitive systems may be longer because higher cognitive systems mature more slowly. From an evolutionary perspective, such a protracted period of sensitivity to external information may provide an advantage by enabling humans to achieve maximal adaptation to a variety of changing environments. There may have been a specific evolutionary pressure for the human brain to remain flexible later into development to enhance learning and adaptability (e.g., Thompson-Schill et al. 2009). Whether such long critical periods are mediated by similar neurophysiological mechanisms, as the shorter-lasting critical periods in sensory systems, remains to be determined. Irrespective of the mechanism, however, in humans, adolescence may mark the end of a number of protracted critical periods that began in early childhood.

An alternative nonmutually exclusive possibility is that adolescence is its own independent critical period that is not contiguous with childhood (Figure 11.1a). If so, we might expect the potential for plasticity to begin rising in early adolescence and fall off by adulthood. Consistent with this possibility, there is some evidence from studies with humans and animal models that adolescents are more sensitive than juveniles or adults to experiences of social stress (Fuhrmann et al. 2015). Such enhanced sensitivity to experience could be mediated by adolescence-related hormonal changes. Alternatively, the same or analogous neural mechanisms that render the brain especially sensitive to the influence of hormonal changes in adolescence could independently enhance sensitivity to experience.

Studies with animal models provide extensive evidence for the idea that hormone exposure during adolescence affects the brain, social behavior, and learning in ways that are distinct from hormonal exposure outside of the adolescent period. In rodents, like in many other species, puberty is associated with a surge in testosterone. If male rodents are gonadectomized prior to puberty, then adolescent brain development proceeds in the absence of the normal influences of testicular hormones. Under these conditions, hamsters fail to acquire a wide range of male social behaviors typically learned during adolescence, and these behaviors are compromised in adulthood. Prepubertally gonadectomized male Syrian hamsters display lower levels of sexual behavior compared with male hamsters that are gonadectomized in adulthood. The deficits resulting from prepubertal gonadectomy are not reversed either by prolonged testosterone replacement therapy or sexual experience in adulthood (Schulz et al. 2004). Other male typical adult behaviors that are organized by pubertal testosterone include aggression, scent marking, play fighting, and nonaggressive social interactions (reviewed in Schulz et al. 2009a; Schulz and Sisk 2016). Thus, the absence of testicular hormones during adolescence results in long-lasting impairments of sociosexual behaviors. Conversely, the presence of testicular hormones during adolescence masculinizes neural circuits underlying sociosexual behaviors and programs enhanced activational responses to testosterone in adulthood.

Research suggests that pubertal testosterone specifically affects social learning—the ability to make behavioral adaptations as a function of social experience (De Lorme et al. 2013; De Lorme and Sisk 2013, 2016). By contrast, pubertal testosterone does not affect performance or motor execution of sociosexual behaviors per se, because males deprived of testosterone during adolescence do display the consummatory components of sexual behavior, aggression, and scent marking, albeit at lower levels compared with males that did experience testosterone during adolescence. For example, normal male hamsters gain social proficiency over the course of repeated encounters with another male in a neutral arena. During the first social encounter between two unfamiliar males, an aggressive interaction occurs initially, and a dominant-subordinate relationship is established within a few minutes. In subsequent encounters, there is little aggression but the dominant–subordinate relationship is maintained through flank marking by both males. This experience-dependent behavioral pattern is disrupted in males deprived of testosterone during adolescence: these males display low overall levels of flank marking, even if they are dominant, and the dominant–subordinate relationship is not maintained by flank marking, but instead is reestablished via aggression in subsequent encounters (De Lorme and Sisk 2013).

Thus, during adolescence, the brain appears to "expect" pubertal testosterone, and this testosterone exposure organizes neural circuits that govern social cognition, the mental processes by which an individual encodes, interprets, and responds to sensory information from a conspecific. Crucially, the adolescent brain appears to be specifically sensitive to the influence of testosterone, thus suggesting a critical period of plasticity where hormones have a unique potential to influence the brain. Whether similar sensitive periods exist for the influence of experience on the brain during adolescence, apart from the experience of hormones, remains to be tested. A further interesting possibility is that hormones interact with experience, such that deprivation from specific social experiences during adolescence interferes with the typical hormonal effects.

In sum, the potential for plasticity in the adolescent brain could theoretically follow one of three types of time courses. First, adolescent brains could be no more plastic than adult brains. Second, adolescence could mark the end of critical periods that began in infancy. Third, adolescence could be its own critical period, with a beginning and end that is distinct from critical periods occurring in early childhood. Given the available evidence, it seems probable that all of these time courses coexist in the human brain. As was pointed out by Greenough et al. (1987), the brain does not follow a single sensitive period: each neurocognitive system has its own time course of development. Some neurocognitive systems may be stable in their plasticity throughout the life span, others may begin their sensitive periods early in life and taper off during adolescence, and still others may have a specific critical period of sensitivity spanning adolescence itself (e.g., sensitivity to social stress). Future research is needed to uncover the time course of plasticity across different neurocognitive systems.

Animal Models for the Study of Experience-Based Plasticity

It is not possible to understand the causal variables that drive human brain development based on studies with humans alone. Currently, noninvasive imaging approaches in humans are severely limited in their ability to measure specific neural mechanisms, since the measurements of neural activity in humans are indirect and limited both in their temporal and spatial resolution. Functional magnetic resonance imaging (fMRI) and light-based hemodynamic measures (near infrared spectroscopy) infer neural activity through metabolic markers that are substantially divorced from the biological mechanisms of primary interest. Furthermore, such studies lack temporal precision. While neural events occur on the timescale of milliseconds, hemodynamic responses stretch over seconds. Electroencephalography (EEG) and magnetoencephalography (MEG) provide millisecond timing accuracy but have poor spatial resolution. They are unable to disentangle unambiguously the specific neurophysiological generators of activity even at the level of cortical region and are insensitive to certain neural sources, depending on their depth and orientation in the brain. Even the comparatively "high" spatial resolution of fMRI (millimeters) is far lower than what is required to disentangle the neurophysiological mechanisms that drive developmental change at the level of circuits (e.g., specific neural subpopulations, neurotransmitters, receptors, synapses). Thus, at present, studies in humans are limited to measuring coarse network properties. While these measurements are informative in their own right, they represent only a fraction of what we need to know about human brain development.

A further challenge to conducting research with humans is that we are unable to manipulate experimentally genes or experience. This limits our ability to disentangle the contribution of these factors to developmental change. By contrast, the experience of animals can be precisely controlled during specific windows of time. This includes both studies of deprivation (e.g., dark-rearing or monocular deprivation), studies of enrichment, and more fine-grained changes such as exposure to motion in a specific direction and deprivation from occluded objects (e.g., Hubel and Wiesel 1970; Sale et al. 2007; Vallortigara et al. 2009; Arcaro et al. 2017). Such controlled rearing studies continue to be a bedrock of developmental science. Over the past decade, it has become possible to manipulate the genes of animals. This enables not only testing the effects of specific genes, but also dissecting the neurophysiological mechanisms that govern critical period plasticity (Hensch 2005). Studies of animals, therefore, provide crucial insights into the causal factors that drive brain development.

There are some inherent challenges, however, in leveraging insights gained from animal studies to answer questions about development in humans and in adolescents, in particular. At present, most of what we know about the biological mechanisms of experience-based change in animals is circumscribed to early development in primary sensory systems (e.g., effects of monocular deprivation on V1; Hensch 2005). The degree to which similar mechanisms mediate development during early adolescence is not known. A key feature of learning and plasticity in early adolescence is that higher cognitive systems are involved. The neurocognitive systems that continue to develop into early adolescence include those which support language, social cognition, executive function, memory, emotion, and decision making. Analogously, psychiatric disorders that emerge during adolescence implicate these higher cognitive systems (e.g., depression). Since higher cognitive systems are more developed in humans compared to other species, the creation of adequate animal models to study plasticity in higher cognitive domains poses a key challenge.

Most work in animals has concentrated on sensory systems, and comparatively few studies with animals have even attempted to look at experience-based change in higher cognitive systems. One study by Yang et al. (2012) examined the neural basis of music preference learning in juvenile mice. The acquisition of such preference is associated with changes in medial prefrontal cortex, rather than the auditory cortices, consistent with the idea that such learning is mediated by higher cognitive systems. This study found that mice have a critical period for developing music preferences, and this critical period is influenced by neurochemical modulators similar to those which affect visual system development (valproic acid). This result is consistent with the possibility that experience-based plasticity in higher cognitive systems is mediated by similar neurophysiological mechanisms as those that govern critical period plasticity in primary sensory systems (e.g., excitatory–inhibitory balance) (Yang et al. 2012).

Notably, although the above study examined neural changes in prefrontal cortices, the nature of the learning experience itself was more similar to the passive sensory-based learning in the visual and auditory systems than to the higher cognitive experiences of human adolescents. In other words, the mice either did or did not experience a type of music. By contrast, higher cognitive learning in humans is more complex, both in terms of the knowledge acquired (e.g., the grammatical structure of sentences or the causal relationship between people's mental states and their behaviors) and the nature of the learning experience itself (e.g., constrained by previous knowledge, self-directed, socially situated).

In this regard, play is a key example of a learning experience that has a complex character representative of the type of learning that occurs in higher cognitive systems. During the first years of life, children spontaneously engage in object-directed and social play. There is evidence that such play provides children with crucial information about how the world works. For example, children actively test hypotheses about the causal mechanisms that govern how toys work (Cook et al. 2011). They systematically seek out evidence both by physically manipulating the objects and seeking out information from other social agents that they perceive to be reliable (Gweon et al. 2014). Although the play behavior of humans is likely to have a distinctive character, other animals, including rodents, also engage in play in the wild. In their studies of the neurobiological mechanisms of social play learning in rats, Kolb and colleagues found that prefrontal cortex plays a central role in play behavior and, conversely, that the development of prefrontal cortex is strongly influenced by play. For example, the complexity of neurons in the medial prefrontal cortex of rats and hamsters is related to the amount of play behavior (Bell et al. 2010; Burleson et al. 2016). Furthermore, the number of conspecifics that an animal plays with during development influences the pruning of the orbitofrontal cortex (Bell et al. 2010). These studies provide an example of how animal models could be used to understand the effects of higher cognitive experience on the developing brain.

For animal model work to inform maximally our understanding of plasticity in humans, further work is needed that examines the neurophysiology of experience-based learning in higher cognitive systems, perhaps using naturalistic behaviors of the species in question, such as play. Nevertheless, since there is a qualitative gulf between the cognitive repertoire of humans and other species, it will ultimately be necessary to measure higher cognitive plasticity directly in humans and to link these measures to markers of experience-based learning in animals. One attempt at such linking is discussed at the end of this chapter.

Neural Markers of Development and Experience-
Based Change in the Human Brain

The human brain continues to change substantially throughout early adolescence and even into young adulthood (e.g., Uhlhaas et al. 2010). Below we highlight some examples of the type of neural markers of human brain development that have been identified, focusing on measurements of functional development. In most cases, directly linking these markers to the influence of experience per se, as opposed to maturation, has remained out of reach. However, some work on cross-modal plasticity provides an in-principle demonstration of how the effects of experience on the brain can be studied in humans (discussed in the following section).

One course of action in the study of development of human cortical functions is to compare the amount of response in a given cortical area during a cognitive task across groups. For example, if we wish to know whether working memory circuits are developing in human children, we could compare the amount of response in frontoparietal networks during a working memory task in adults and children. Such comparisons are complicated, however, by the fact that the overall amount of neural activity in a given system reflects not only the stable properties of that system but the degree of their engagement during a particular cognitive task. Thus greater activity in children during a working memory task might reflect the immaturity of the working memory system. Alternatively, it could reflect the relative difficulty of the working memory task for children and therefore the increased processing required within this cognitive system.

One alternative to measuring the overall amount of neural activity during a task is to measure information selectivity; that is, the degree to which a given brain area responds selectively to one type of information over another (Saxe et al. 2009). Although information selectivity measures are not immune to difficulty, they are arguably less prone to reflecting merely the degree of difficulty of the current task for the age group in question. One example of such work comes from the field of social cognition. There is evidence that cortical areas that support social cognition increase their information selectivity into late childhood/early adolescence. In adults, a subset of socially relevant cortical regions are specifically involved in reasoning about the internal mental states (e.g., their beliefs, desires, and goals) of social agents (Saxe and Kanwisher 2003). For example, the right temporoparietal junction (RTPJ) is highly active when participants comprehend stories or view images that require representing an agent's mental state (e.g., Rachel thought it was going to rain) but not stories about agents that do not require "mentalizing" (e.g., Rachel went to the store to buy milk) or stories about physical events (e.g., the bridge is going to fall down). Studies with children show that this preference for mentalizing emerges slowly over the course of development and is not adult-like until late childhood/early adolescence. In children, unlike in adults, the RTPJ responds similarly to all stories about social agents. Furthermore, across children, the degree of specialization in the RTPJ correlates with performance on mentalizing tasks outside the scanner, even when age is factored out (Gweon et al. 2012). Whether this change emerges as a result of social/linguistic experience, intrinsic maturation, or both is currently under investigation. It is worth noting that delays in theory of mind performance have been observed in deaf children who grow up without access to sign language. By contrast, deaf children who grow up with early access to sign language perform similarly to hearing controls (Schick et al. 2007). This suggests that theory of mind development is influenced by linguistic experience during childhood and early adolescence.

A further aspect of human brain function that continues to develop well into adolescence is spontaneous as well as task-driven synchronized rhythmic neural activity, so-called neural oscillations (see Vakorin and Doesburg 2016). In humans such oscillations can be measured using electrophysiological recordings such as EEG and MEG. Changes in spontaneous neural oscillations begin in infancy and continue throughout childhood and adolescence. They include not only a restructuring of neurophysiological synchronization among brain areas but also (a) reduction in overall power in oscillation, (b) acceleration of the peak frequency of alpha oscillation, and (c) reduction in lower-frequency oscillations (< ~10 Hz) and increase in higher-frequency oscillations (> ~10 Hz) (Miskovic et al. 2015; Vakorin et al. 2017). Development of such coordinated neurophysiological activity is atypical in neurodevelopmental disorders such as autism, and is associated with symptomatology (Vakorin et al. 2017).

Task-related oscillations also develop throughout childhood and adolescence. For example, increased activity and interelectrode synchronization have been reported in beta- and gamma- (>30 Hz) frequency ranges during Gestalt visual perception (Rodriguez et al. 1999). This percept-dependent network synchronization has also been shown to change throughout child and adolescent development, with qualitative maturational shifts suggesting late restructuring of neurophysiological networks during adolescence (Uhlhaas et al. 2009b). Such developmental changes in task-dependent neurophysiological synchronization may contribute to cognitive and behavioral maturation, as age-dependent increases in interregional synchronization during performance of a language task have been shown to correlate with individual differences in language abilities (Doesburg et al. 2016). Neural oscillations could play an important role in experience-dependent plasticity, since the synchrony of neural firing profoundly affects the nature of the resulting plastic changes (Uhlhaas et al. 2009b).

Although substantial evidence suggests that cortical oscillations change during development in humans and in animals, there are important challenges in determining the neurobiological and cognitive significance of measured oscillations. For example, not all oscillatory activity measured by MEG and

EEG reflects oscillatory neural activity; alternatively, it is not clear whether non-oscillatory neural activity is measured as oscillation by MEG and EEG. This limitation is particularly severe when measuring high-frequency oscillations (e.g., in the gamma band). Non-oscillatory, nonsynchronized activity (spikes, EPSPs, IPSPs) gives rise to signals in the high-frequency band and mimics, when band pass filtered, gamma oscillations. Because it is high-frequency activity, it is often characterized as gamma activity but does not necessarily reflect true oscillatory process. Special analysis procedures are required to distinguish true oscillatory signals from other causes of EEG and MEG oscillations. A key goal in outstanding research is to link measurements of brain development in humans (e.g., changes in oscillations observed in EEG) to neurophysiological processes measured in studies with animal models.

Adolescent development is also associated with changes in the synchronization of spontaneous activity across regions and networks, as measured by correlating the spontaneous fluctuations in hemodynamic activity in fMRI (Biswal et al. 1995). This is sometimes referred to as "resting-state connectivity" or "functional connectivity." In adults, functionally related areas express coordinated fluctuations in activity, even in the absence of a task. For example, activity is correlated among areas within a network, such as the dorsal attention network, the salience network, and default mode network, and is less correlated between networks. These are examples of intrinsic connectivity networks (ICNs) that have been identified using functional connectivity measures with fMRI.

The correlation structure of ICNs changes over the course of development, including during adolescence. Specifically, there is a shift toward progressively stronger interactions among functionally related areas and progressively weaker interactions with anatomically proximal but functionally unrelated areas (Fair et al. 2008, 2009). This can be conceptualized as a strengthening of "within network connectivity" and a weakening of "between network connectivity" (Dosenbach et al. 2010). This refinement of the correlation structure of hemodynamic activity during adolescence is a continuation of a process that begins in the perinatal period (Smyser et al. 2010).

Developmental changes in spontaneous neurophysiological network, as measured by fMRI (ICNs), may be related to some of the spontaneous oscillatory activity measured by MEG and EEG. The anatomical structure of correlations in cross-frequency coupling and the envelope of alpha (8–12 Hz) and beta (15–30 Hz) activity recapitulate ICN topographies in MEG (Brookes et al. 2011; Florin and Baillet 2015). Age-related increases of neurophysiological amplitude correlations in ICNs during childhood and adolescence are also strongest in the alpha- and beta-frequency ranges (Schäfer et al. 2014), suggesting that spontaneous MEG and fMRI correlations may capture overlapping aspects of organization in coordination of large-scale activity.

Linking Markers of Human Brain Development
to Changes in Experience

The developmental changes in neural activity described above have not been directly linked to experience as opposed to maturation. Doing so based on studies with typically developing children is challenging. However, comparisons across populations with different developmental experiences makes it possible to tease apart these variables in human development. One example of such research comes from studies of blindness. These studies suggest that both the informational selectivity within a cortical area and functional coordination of activity across regions are influenced by developmental experience.

Studies of visual cortex function in individuals who are congenitally blind demonstrate that the basic functional properties of cortical networks can change dramatically as a result of developmental experience. In individuals who are blind from birth, so-called "visual" cortices respond to auditory and tactile stimuli (Sadato et al. 1996; Gougoux et al. 2005). Remarkably, visual areas appear to take on higher cognitive functions, including language and mathematical reasoning (Röder et al. 2002; Amedi et al. 2003; Bedny et al. 2011; Kanjlia et al. 2016). For example, a subset of visual areas responds more to spoken sentences than noise, more to sentences than lists of words, and more to grammatically complex than grammatically simple sentences (Röder et al. 2002; Lane et al. 2015). By contrast, dorsal regions within visual cortex of blind individuals participate in numerical tasks: they are active when participants solve spoken math equations and activity scales with equation difficulty (Kanjlia et al. 2016). The information selectivity of visual cortices, therefore, changes dramatically as a result of blindness (Bedny 2017).

The functional coordination of visual cortex activity with other regions also changes in blindness. In particular, activity in "visual" areas becomes more correlated with frontoparietal networks (Deen et al. 2015). Furthermore, different subregions of visual cortices become functionally coupled at rest with distinct higher cognitive frontoparietal networks, and resting-state specialization in functional connectivity is related to specialization in task-based responses. Number-responsive visual areas (e.g., middle occipital gyrus and foveal V1) become coupled with a frontoparietal network that is involved in numerical processing. By contrast, language-responsive visual areas (e.g., lateral occipital and posterior fusiform regions, peripheral V1) become coupled with inferior-frontal regions involved in linguistic processing (Figure 11.2) (Bedny et al. 2011; Lane et al. 2015; Kanjlia et al. 2016). Together, this evidence demonstrates that both information selectivity within a cortical area and the functional coordination of cortical networks are profoundly influenced by developmental experience.

Studies with individuals who became blind as adults further suggest that the capacity of cortex to respond to changes in experience is qualitatively different in childhood and adulthood. Although visual cortices show responses to

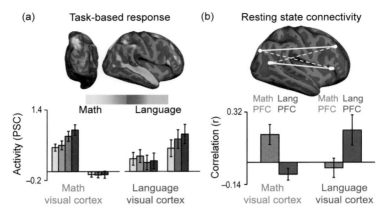

Figure 11.2   (a) BOLD activity during sentence comprehension task (blue) and math equation task (red) in congenitally blind participants. Contrast depicts math task > sentence task (P < .05, corrected). White circles highlight visual cortex responses that are absent in those who are sighted. Bottom graphs show percent signal change (PSC) in two visual cortex areas with different functional profiles in blind individuals: a math-responsive visual cortex region and language-responsive visual cortex region. Red bars show responses to equations of varying difficulty; darker reds indicate the hardest equations. Blue colors show sentences of differing grammatical complexity; darker blues indicate more complex sentences. Gray bars show responses to nonword lists. (b) Resting-state connectivity patterns in blind individuals show that while the math-responsive visual cortex region is more correlated with math-responsive prefrontal cortex (PFC), the language-responsive visual cortex region is more correlated with language-responsive inferior PFC areas.

nonvisual stimuli, even in those who become blind as adults, there is evidence that such changes are less functionally specific and less behaviorally relevant (Cohen et al. 1997; Bedny et al. 2012; Collignon et al. 2013). Studies of sensory loss thus suggest that developmental experience plays a unique role in determining the functional architecture of the human brain.

Conclusions

Childhood and adolescence are periods of particular sensitivity to the influences of experience on the human brain. Studies with animal models have uncovered local circuit neurophysiological mechanisms that govern these periods of enhanced sensitivity. In humans, many functional properties of the brain continue to change throughout childhood, adolescence and beyond, including regional cortical information selectivity, neural network correlations, and oscillatory activity. Evidence from studies of plasticity in sensory loss suggests that such functional properties of cortical networks can be fundamentally altered by experience and that sensory experience during development has especially potent effects on cortical function. Two key challenges for future research are (a) to link neurophysiological mechanisms of critical period plasticity identified in animals to developmental changes in human behavioral and brain function and (b) to understand how complex experience (such as play) influences the function of higher cognitive systems in both animals and humans.

Emerging Adulthood

12

Modifications in Brain Coordination and the Emergence of Brain Disorders during Late Adolescence

Peter J. Uhlhaas

Abstract

Late adolescence is associated with the emergence of major brain disorders, such as schizophrenia and affective disorders, thus raising the question of the underlying biological vulnerability and mechanisms that confer risk for psychopathology. This chapter presents evidence which shows that during late adolescence, dynamic brain coordination undergoes major modifications in specific circuits involved in cognition and affective regulation. These data are consistent with emerging findings from physiology and anatomy that physiological and anatomical underpinnings of brain coordination are characterized by profound changes during the transition from adolescence to adulthood. This chapter posits that the expression of psychopathology may be intimately linked to ongoing modifications in brain coordination, which occur during adolescence, and that these could confer a biological vulnerability for disturbances in affect and cognition.

Introduction

Until recently, an essential dogma of developmental neuroscience was the assumption that fundamental properties of cortical networks were formed mainly *in utero* as well as in the early postnatal years. From this perspective, later developmental stages, such as adolescence, were viewed as having little or virtually no effect on the functional characteristics and anatomical layout of cortical and subcortical networks. Only scant systematic evidence existed on the fundamental properties of neural circuits and their associated neuronal dynamics during later developmental periods. This position has recently been overturned by evidence, from a range of disciplines, showing that brain coordination undergoes profound reorganization; many important features of large-scale networks, such as rhythmic activity and functional integration of distributed neural responses, mature only fully during late brain development.

In this chapter, I outline evidence which suggests that the reorganization of circuit properties during late adolescence constitutes a critical period for development and confers crucial vulnerability for the emergence of major brain disorders. I summarize studies that implicate the transition of adolescence to adulthood in the expression of schizophrenia and disorders involving affective dysregulation. I also review data from developmental cognitive neuroscience that reveal important changes in oscillatory dynamics and functional connectivity during brain maturation. In conclusion, I discuss candidate mechanisms for changes in brain coordination, such as changes in neurotransmitter systems, that have been involved in the generation of fast neuronal dynamics as well as modifications in the layout of anatomical connections.

Adolescence and Brain Disorders

Adolescence represents a paradoxical stage of human development whereby increased mental and physical fitness is accompanied by vulnerability for severe and enduring mental disorders (Kessler et al. 2005; see also Figure 12.1). Although the troubling nature of adolescence has been described since antiquity, cognitive neuroscience and clinical psychology have only recently started to investigate systematically the changes in brain function during adolescence that are likely to constitute the vulnerable developmental window from which adult mental disorders emerge (Paus et al. 2008; Lee et al. 2014). The fundamental question of which biological and psychological mechanisms give rise to this phenomenon still remains to be answered. The hypothesis that I would like to advance here is that the expression of psychopathology is intimately linked to ongoing modifications in brain coordination, especially during adolescence, that confer a biological vulnerability for disturbances in affect and cognition.

The expression of psychopathology is of paramount importance for current health systems, as the onset of 75% of all mental disorders can be traced to this age period (Kessler et al. 2005) and of these, one in four to five adolescents develops a mental disorder with severe impairment across their lifetime (Lee et al. 2014). If we are ultimately to improve mental health, research and treatment approaches require a fundamental reframing. We need to move beyond the current focus on adulthood—the stage at which mental disorders are already established, symptoms fully expressed, and associated disabilities clearly visible. Indeed, a paradigm shift is emerging that highlights the importance of early intervention in adolescents at risk for psychosis and affective disorders (McGorry 2010). However, a crucial prerequisite for the implementation of the early intervention paradigm is a coherent understanding of the trajectory

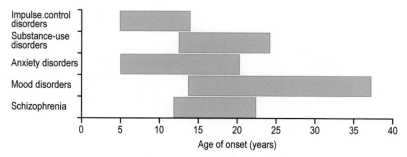

Figure 12.1   Emergence and duration of mental disorders during adolescence. Mood disorders, anxiety, substance abuse, and schizophrenia typically begin in adolescence (after Paus et al. 2008 with permission).

and mechanisms of adolescent brain development that could support an improved understanding of developmental vulnerabilities and novel treatment approaches. As Lee et al. (2014:547) state: "...understanding neurodevelopmental changes and their roles in both emergence of mental disorders and how they affect treatment efficacy is imperative."

Schizophrenia

Brain disorders closely associated with adolescence include disturbances in reality testing and affective regulation. Psychotic disorders, such as schizophrenia, typically emerge during the transition from adolescence to adulthood. In schizophrenia research, analysis of the biological mechanisms that underlie clinical symptoms and cognitive deficits has focused for a long time on the contribution of circumscribed brain regions, such as the prefrontal cortex (PFC). Contrasting this view, which was largely inspired by findings from clinical neuropsychology, current research indicates that anatomical alterations involve a large number of cortical and subcortical regions; this suggests that schizophrenia, and perhaps other mental disorders, are likely to constitute systemic disturbances involving essentially a disruption of brain coordination (Uhlhaas and Singer 2012). This position is supported by extensive evidence from electro- and magnetoencephalography (EEG/MEG) studies which show impairment of the amplitude and synchronization of oscillatory activity, in particular at gamma-band frequencies (Uhlhaas and Singer 2010). Significantly, these dysfunctions are present prior to illness onset and independent of medication status, suggesting that readouts of neural oscillation could be used as biomarkers for early intervention and diagnosis.

Importantly, this fundamental disruption of large-scale neuronal dynamics is supported by data that has consistently implicated disturbances in excitation–inhibition (E–I) balance parameters, especially in disturbances of gamma-aminobutyric acid (GABA)ergic interneurons and N-methyl-D-aspartate (NMDA) receptors (Uhlhaas and Singer 2012). This has important implications for the development of novel treatments. From this perspective, disturbances in E–I balance parameters during brain development give rise to abnormal dynamics and disturbed coordination which, in turn, lead to cognitive deficits that provide the underlying vulnerability from which the more complex manifestations of the disorder emerge. Currently, however, it is not known at which point disturbances in E–I balance emerge during development, and whether they are causally involved in the expression of aberrant neural dynamics. Testing such relationships requires careful manipulations of circuit properties during development in animal models and could eventually provide more mechanistic insights into these relationships.

Cognitive impairments across several domains are detectable several years before the onset of psychosis, which typically emerges during late adolescence or early adulthood (Insel 2010). The onset of clinically manifested symptoms during this developmental period may thus be due to the ongoing changes in oscillatory dynamics and the underlying neurobiological mechanisms responsible for coherent mental states. This perspective is consistent with current thinking that perceptual and cognitive impairments are the earliest indicators for an at-risk mental state, although subthreshold and full psychotic symptoms emerge only later during development (Fusar-Poli et al. 2013).

The development of certain positive symptoms, such as delusions and hallucinations, could represent an adaptive response aimed at attributing significance to the abnormal states that result from disturbed dynamic coordination of distributed networks. One of the functions of the dopaminergic system is to reward consistent brain states, such as those likely to be associated with the eureka effect that accompanies solutions of perceptual tasks and the confirmation of predictions. One might expect an upregulation of dopaminergic signaling if such states are rare. This, in turn, could cause normally neutral external and internal experiences to be judged as meaningful or salient, which is a hallmark of positive symptoms. The transition from adolescence to adulthood could constitute a critical period for such a scenario, since it is only during this late developmental phase that dopamine signaling is fully matured (Tseng and O'Donnell 2007).

Affective Dysregulation

In addition to the emergence of impaired reality testing and cognitive deficits, adolescence is also associated with the emergence of affective dysregulation. Intense and frequent emotions are common during the early adolescent years and are likely to constitute the earliest indications for the emergence of a pathological state of mood in adulthood. Indeed, evidence has emerged that mood disorders and anxiety peak in prevalence during adolescence (see Figure 12.1). More recently, borderline personality disorder, which involves profound affective dysregulation and instability of emotional experience, has been conceptualized as a life-span developmental disorder which has its roots in adolescence. As a result, diagnostic criteria have been developed that are similar to risk symptoms of psychosis and could potentially identify adolescents at risk for the development of disorders involving altered states of mood and affective regulation (Chanen et al. 2007; Bechdolf et al. 2014).

Vulnerability for psychosis and affective dysregulation during adolescence, however, should not be considered separately, since evidence suggests that a large percentage of at-risk participants for psychosis develop a range of affective disorders (Lin et al. 2015). This highlights the importance of developing novel approaches for risk prediction and stratification for diverse health outcomes.

One potential mechanism that could underlie the changes in affective experiences and the emergence of its disorders during adolescence is emotion regulation. In this context, emotion regulation can be broadly defined as conscious or unconscious strategies to start, stop, or otherwise modulate the trajectory of an emotion. These two forms of emotion regulation depend on interactions between prefrontal and cingulate control systems and cortical and subcortical emotion-generative systems. In recent years, deficits in emotion regulation have emerged as a putative maintaining factor and promising treatment target for a broad range of mental disorders in adulthood, such as depression, bipolar disorder, and borderline personality disorder (Berking and Wupperman 2012).

Recent fMRI/MRI and behavioral studies indicate profound modifications in the ability to regulate emotional states during adolescence (Heller and Casey 2016). This could explain the high prevalence of mood disorders and instability of emotional experience during this period. Accordingly, understanding the mechanisms of emotion regulation, its developmental modifications as well as its impairments in adolescence may have wide-ranging consequences for mental health and brain development.

Do Modifications in Brain Coordination Increase Vulnerability for Emerging Brain Disorders?

The evidence outlined above suggests that adolescence is associated with (a) the expression of dysfunctions in both cognitive processes and emotion regulation and (b) important windows of vulnerability for the development of brain disorders. This hypothesis is supported by recent evidence (discussed below) that brain coordination undergoes important modifications during the transition from adolescence to adulthood.

Neural Oscillations and E–I Balance Parameters

Developmental findings on neural oscillations indicate that cortical circuits during late brain maturation are accompanied by profound modifications in

202   P. J. Uhlhaas

the amplitude and synchrony at theta, beta, and gamma frequencies (Uhlhaas et al. 2010). Uhlhaas et al. (2009b) examined the development of induced oscillations in the 4–80 Hz frequency range in children, adolescent participants, and young adults during the perception of Mooney faces (Figure 12.2). During

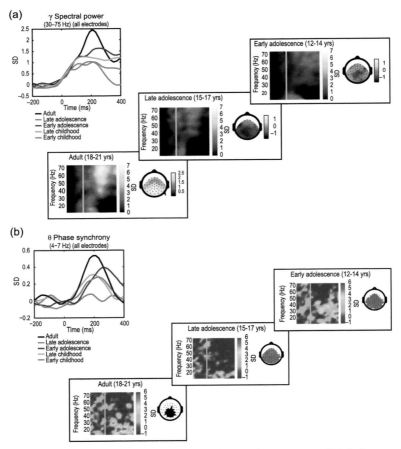

Figure 12.2 Development of induced oscillations and emergence of high-frequency oscillations and synchrony during the transition from adolescence to adulthood. (a) Spectral power in the beta- and gamma-frequency bands during the perception of Mooney faces across all electrodes in the 13–75 Hz frequency range for adult (18–21 years), late adolescence (15–17 years), early adolescence (12–14 years), late childhood (9–11 years), and early childhood (6–8 years). Topography for the 30–75 Hz frequency band lies between 100–300 ms. All values are expressed in standard deviations in reference to the baseline. Group comparison across all electrodes of spectral power in the 30–75 Hz range between 100–300 ms. (b) Phase synchrony in the beta and gamma bands in the face condition averaged across all electrode pairs in the 13–75 Hz frequency range. Group comparison for all electrodes of phase synchrony in the 13–30 Hz frequency range between 100–300 ms. Modified after Uhlhaas et al. (2009b).

adolescence, important changes in the amplitude and synchrony of theta- and/or gamma-band oscillations occurred that correlated with improved detection rates and reaction times. In particular, phase synchrony in the beta and gamma bands increased until age 14, followed by a reduction during late adolescence (15–17 years) before synchrony increased again sharply in 18- to 21-year-olds. This nonlinear development of phase synchrony was accompanied by a reorganization in the topography of phase-synchrony patterns in the beta band. Finally, phase synchrony between frontal and parietal circuits was significantly increased during adolescence.

Developmental modifications during adolescence in rhythmic activity are supported by several studies that have investigated the maturation of basic sensory-evoked oscillations, revealing improved phase consistency and amplitude of gamma-band activity during auditory and visual stimulation (Uhlhaas et al. 2010; see also Figure 12.3). These preliminary data suggest that changes in stimulus-locked activity frequently follow a linear trajectory: earlier maturation is associated with power and phase values of oscillatory activity, not with

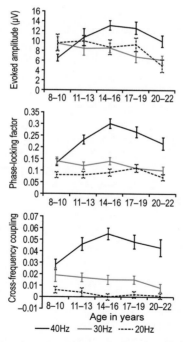

Figure 12.3 Maturation of sensory-evoked oscillations show improved phase consistency and amplitude of gamma-band activity during auditory stimulation. Age-related changes are shown in evoked amplitude, phase-locking factor, and cross-frequency coupling during auditory steady-state stimulation: 20, 30, and 40 Hz. All measures for the 40 Hz condition follow an inverted U trajectory. Responses to 30 and 20 Hz stimuli show a flat or decreasing trend with age. Used with permission from Cho et al. (2013).

nonphase-locked, induced oscillations. The reason for this is that the latter correspond more closely to the internal, self-generated dynamics characteristic of large-scale networks that govern higher cognitive functions. Accordingly, one hypothesis is that brain coordination processes that give rise to these phenomena are more likely to reveal developmentally sensitive signatures of developmental processes during the transition from adolescence to adulthood which, in turn, can be disrupted by aberrant brain development and existing pathophysiological events.

One possibility is that maturational changes in oscillatory activity are closely linked to the development of circuit properties that are linked to the emergence of neural oscillation and their synchronization. While the number of GABAergic cells undergoes only small modifications during adolescence, axons of parvalbumin-containing basket and chandelier neurons seem to undergo more extensive modifications (Hoftman and Lewis 2011). Changes in GABAergic neurotransmission also comprise modifications in the subunit composition of GABA receptors. Hashimoto et al. (2009) described a decrease of $GABA_A$ receptor α2 subunits and an increase of α1 subunits with age in the monkey dorsolateral PFC. This change is accompanied by marked alterations in the kinetics of induced pluripotent stem cells (iPSCs), including a significant reduction in the duration of miniature iPSCs in pyramidal neurons. The shift in GABAergic subunit expression could lead to an increase in the precision of temporal patterning, as the time course of iPSCs is an important determinant for the frequency at which a network can oscillate.

In addition, there are changes in excitatory and modulatory systems that lead to a modification of inhibitory processes, such as alterations of the dopaminergic modulation of prefrontal interneurons (Tseng and O'Donnell 2007) and the reconfiguration of NMDA and AMPA receptors in fast-spiking interneurons (Wang and Gao 2009). Together, these data suggest that the transition from adolescence to adulthood is accompanied by circuit modification that supports the emergence of high-frequency activity and precise synchronization and will likely have an important impact on the functional properties of large-scale networks. It should be noted, however, that such relationships are exceedingly complex: the different types of oscillation parameters (evoked vs. induced) and distinct frequencies most likely involve different generating mechanisms whose detailed developmental modifications remain to be elucidated.

These modifications also allow for the impact of environmental factors, such as cannabis, to exert detrimental effects on brain development. Previous work has indicated that tetrahydrocannabinol (THC)—the principal psychoactive constituent of cannabis—dysregulates the E–I balance of cortical circuits through its impact on cannabinoid 1 receptors (Robbe et al. 2006). Because E–I balance parameters are important for rhythmic activity, THC leads to a disruption of neural oscillations (Morrison et al. 2011), constituting a potential candidate mechanism for the generation of cannabis-induced cognitive deficits and possibly psychosis. This is supported by emerging evidence in mice, where chronic THC administration during adolescence, but not in adulthood, was found to reduce the strength of *in vitro* oscillations in PFC circuits (Raver et al. 2013).

Maturation of Emotion Regulation Circuits

Studies that have looked at changes in circuits involved in emotion regulation during adolescence have found similar modifications in brain coordination that point to crucial modifications in the properties of large-scale networks. In the adult brain, functional and anatomical studies indicate that interactions between the PFC, in particular the ventromedial PFC, and the amygdala are fundamentally involved in emotion regulation (Hare et al. 2008). In adolescent participants, converging evidence indicates heightened amygdala activity in response to threat-related stimuli; this correlates with levels of trait anxiety when modulatory feedback from the ventromedial PFC is decreased. Resting-state fMRI data also provide support for a developmental switch in this pathway (Gee et al. 2013), thus suggesting the existence of positive connectivity between the amygdala and PFC in early childhood and a switch to negative functional connectivity during the transition to adolescence.

Moreover, studies in rodents have shown that fear extinction learning is attenuated during adolescence relative to preadolescent and adult animals (Pattwell et al. 2011), indicating a sensitive window for the development of affective disorders during adolescence. Together with evidence for increased risk-taking and neural signatures of elevated activity in reward structures (Galván et al. 2006), these findings suggest a unique nonlinear developmental trajectory of adolescent brain networks, characterized by a transient destabilization involving elevated activity in subcortical versus cortical regions (Heller and Casey 2016).

These modifications in brain coordination can be linked to ongoing development in anatomical pathways that are critical for the corticolimbic interactions. Although *in vivo* anatomical studies in humans are relatively scarce, tracing studies suggest that amygdala-to-PFC projections emerge earlier than PFC-to-amygdala projections, and that these connections continue to develop through adolescence in rodents (Cunningham et al. 2002).

Summary and Perspective

Current data suggest that important modifications in brain coordination occur during the transition from adolescence to adulthood: improved generation of rhythmic activity and its synchronization at low and high frequencies as well as changes in the functional interactions between brain regions that underlie emotion regulation. Because of the coincidence of these changes with the emergence of brain disorders that are characterized by profound disruption of reality testing and emotional experience, it is likely that these maturational changes provide windows of vulnerability during which favorable conditions give way to the expression of dysfunctions that then lead to behavioral anomalies. Here I have applied this framework of understanding to schizophrenia and affective disorders but suggest that it can also be applied to other mental disorders that emerge during the transitional period from adolescence to adulthood, such as substance abuse and anxiety disorders (Lee et al. 2014). Moreover, developmental modifications of circuits involved in major mental disorders are also vulnerable to the effects of stress and its epigenetic regulation (Niwa et al. 2013), thus providing an important link to environmental factors.

Important characteristics of these modifications in brain coordination are (a) the nonlinear trajectory of developmental changes, supported by EEG and fMRI studies, and (b) the close relations shown with the underlying neurobiological parameters (Heller and Casey 2016). This indicates that brain coordination during the late adolescent period is characterized by unique properties that transiently disrupt large-scale brain dynamics and are possibly required for the emergence of fully mature brain coordination. From a dynamical systems perspective, the nonlinear developmental trajectory of functional networks during adolescence is consistent with the idea that phase transitions between different states of a system are characterized by critical fluctuations. In the adolescent brain, the transient reduction in large-scale synchronization of cortical networks and the concomitant increase of subcortical input could be a condition that favors critical fluctuations. If these become supracritical when the developing system undergoes the phase transition toward the adult state, the brain could remain in a faulty bifurcation and fail to accomplish the final development steps:

1. Increase in the precision of synchronized, high-frequency oscillations
2. Integration of frontal and subcortical activity patterns
3. Shift in the balance between local and global coordinated brain states

These insights have implications for the development of interventions designed to target large networks and brain coordination mechanisms. In view of the converging evidence for a disturbed E–I balance and the resulting changes in brain dynamics caused by alterations in GABAergic and glutamatergic neurotransmission, we need to intensify the search for drug targets to restore E–I balance. Consideration should be given to interventions that modulate brain dynamics. There is increasing evidence that transcranial magnetic stimulation and transcranial direct-current stimulation could be used to modulate neuronal oscillations and large-scale synchrony in a frequency-specific fashion (Thut et al. 2011). Finally, to correct and improve central psychological processes, such as reality testing and affect regulation, we should consider using psychological interventions, such as cognitive behavioral therapy and cognitive remediation. Such applications could play an important role in early intervention and prevention of mental disorders that emerge during adolescence.

13

Plasticity beyond Early Development

Hypotheses and Questions

Ulman Lindenberger

Abstract

Applying insights from research on critical periods in early development, this chapter outlines a life-span research agenda on human plasticity and uses it as the conceptual foundation for a set of research hypotheses and open questions. *Plasticity* is defined as the capacity for lasting changes in brain structure associated with expansions in behavioral repertoire. As a complement to plasticity, *flexibility* refers to the instantiation and reconfiguration of the existing behavioral repertoire during periods of stability that are characterized by the absence of structural change. Mammalian and avian brains evolve through cycles of plasticity and stability, with a general trend toward stability. Animal work on critical periods in motor and sensory development substantiates three hypotheses that can serve as guideposts for research on plasticity in later age periods: First, likelihood, rate, and magnitude of plastic changes decrease after maturity. Second, when triggered, plastic changes often entail an overproduction of new synaptic connections, followed by pruning. Macroscopically, this sequence is associated with a pattern of gray matter volume expansion, followed by renormalization. Third, earlier plastic changes provide a structural scaffold for later learning. These hypotheses await empirical testing in humans, engender research design recommendations, and are related to fundamental open issues in research on human plasticity.

A Life-Span View of Structural Brain Plasticity

Life-Span Gradients in Plastic Potential

The term plasticity is often used interchangeably with learning, maturation, or adaptation. To avoid ambiguity that arises from such usage, I follow Lövdén et al. (2010) and define *plasticity* as the brain's capacity to respond to environmental demands by triggering and implementing long-lasting structural changes that alter its functional and behavioral repertoire. I refer to the exploitation of that repertoire, or range of functioning, as *flexibility* (Lövdén et al. 2010). At the behavioral level of analysis, the distinction between plasticity and flexibility can be traced back to the Swiss psychologist and epistemologist Jean Piaget. Piaget argued that cognitive development alternates between phases of structural change, in which new structures and relations are created, and phases of elaboration, in which the implications of these structures and relations are explored and instantiated (Piaget 1980).

I posit that plasticity is triggered by a *mismatch between the current range of functioning and experienced demands*. To trigger and direct plastic changes, this mismatch between demand and supply needs to exceed the scope of the current repertoire while still being representable by the organism's nervous system (Figure 13.1). In addition, I postulate that plasticity is characterized by inertia. A central nervous system in a permanent state of plasticity-induced renovation would not be able to develop a coordinated set of habits and skills, and would constantly drain a large amount of precious metabolic resources (Kuzawa et al. 2014). Hence, demand–supply mismatches have to surpass some threshold of intensity and duration to trade the goal of stability for that of plasticity. This dynamic equilibrium shifts with age.

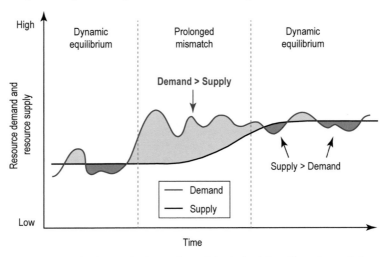

Figure 13.1 Supply–demand mismatch model of plasticity. The mismatch between functional supply and experienced environmental demands can be caused by primary changes in demand (shown here) or by primary changes in functional supply (not shown). Functional supply denotes structural constraints imposed by the brain on function and performance, and permits a given range of performance and functioning. Mismatches between supply and demand need to be present for some period of time to overcome the system's tendency toward stability (sluggishness) and to push the system away from its current dynamic equilibrium. Adapted after Lövdén et al. (2010).

Plasticity forms a necessary part of ontogeny in birds and mammals and establishes individuality (Freund et al. 2013). The evolving brain strikes a balance between plasticity and stability that supports the construction, modification, and maintenance of behavioral repertoires from early ontogeny into late adulthood. Over the course of their lives, humans acquire a rich model of the world that enables flexible deployment of established behavioral repertoires. For this reason alone, the number of situations requiring a plastic response is likely to decrease with advancing adult age. In addition, putting a premium on stability also favors continuity of social structures, which in turn may facilitate the deployment of plastic potential in the next generation (Lindenberger 2014). Finally, the metabolic costs of plasticity are likely to be amplified in neural systems that have accumulated damage, reflecting evolved limitations in somatic maintenance, as is the case for brains in later adulthood, when senescent changes become dominant. Primarily for these reasons, it can be assumed that the brains of older adults are both less capable and less in need of reacting to a supply–demand mismatch with a plastic response, as compared to the brains of typically developing children and adolescents. Hence the set point of the plasticity–stability equilibrium follows an overall life-span trend, moving from a greater relative emphasis on plasticity to a greater relative emphasis on stability (Lindenberger 2014).

There is evidence to support these claims. For a long time, plasticity was assumed to peak during critical periods early in life and to be absent thereafter. In contrast, early work in motor and auditory domains (Recanzone et al. 1993) as well as more recent studies have confirmed that plasticity is present throughout ontogeny, but to varying degrees and in different ways (Hensch 2005; Uhlhaas et al. 2010; Kempermann 2011; Hübener and Bonhoeffer 2014). In particular, there is accumulating evidence for experience-dependent plastic changes in the structure of the adult brain (Hübener and Bonhoeffer 2014), and these changes are large enough to be captured by magnetic resonance imaging (MRI) in adult humans (Draganski et al. 2004; for a review, see Lövdén et al. 2013). Using T_1-weighted MRI, gray matter alterations have been observed following extensive behavioral interventions, such as several months of juggling training, intensive studying for medical exams, foreign language acquisition studies, spatial navigation training (Lövdén et al. 2012; Wenger et al. 2012), playing video games (Kühn et al. 2014), and tracing with the nondominant hand (Wenger et al. 2017b). Other studies have reported gray matter changes after two weeks of mirror reading, seven days of juggling training, a few days of signature writing with the nondominant hand, and even after only two sessions of practice in a complex whole-body balancing task, or mere hours of training on color subcategories (for references, see Lövdén et al. 2013). Taken together, these results suggest that plastic changes in gray matter volume can emerge quite rapidly in adults. Note, however, that the method most commonly used to delineate these changes, voxel-based morphometry, does not permit firm conclusions about their physiological basis.

It is assumed that the dynamic interplay between mechanisms promoting plasticity and mechanisms promoting stability organizes behavioral development into alternating, sequentially structured periods that support the hierarchical organization of cerebral function and higher-order cognition (Figure 13.2). The canonical example is the sequence of critical periods that drive sensory and cognitive development from infancy to adolescence (Shrager and Johnson 1996; Hensch 2005; Hübener and Bonhoeffer 2010). Adopting knowledge about critical periods in early ontogeny may prove useful in understanding and, if deemed desirable, overcoming the greater inertia of the adult brain (Bavelier et al. 2010). Moreover, it is assumed that plasticity decreases further from early to late adulthood, reflecting senescent alterations of the brain involving reductions in energy metabolism, gray matter volume, white matter integrity, receptor densities, and neurotransmitter availability (Lindenberger 2014). Behavioral evidence is consistent with the prediction of life-span age gradients in plasticity (Brehmer et al. 2007, 2008; Schmiedek et al. 2010, 2014; see Figure 13.3). Eliciting plasticity in the adult brain may require shifting the excitatory–inhibitory circuit balance closer to levels present during critical periods in early ontogeny (Bavelier et al. 2010). In line with this notion, recent studies that modulated the excitability of motor cortex in adult humans with anodal transcranial direct-current stimulation have revealed improved learning (Hashemirad et al. 2016) along with reductions of the inhibitory neurotransmitter GABA (Stagg et al. 2009).

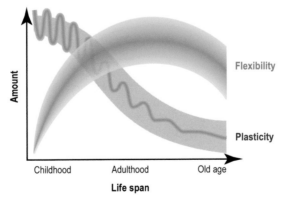

Figure 13.2 Plasticity and flexibility across the life span. *Plasticity* refers to long-lasting alterations in the brain's chemistry, gray matter, and structural connectivity in support of behavior. *Flexibility* denotes the capacity to optimize performance within the limits of the current functional supply. The dynamic interplay of mechanisms promoting plasticity versus stability, illustrated by the oscillating pattern of the plasticity trajectory, organizes behavioral development into alternating, sequentially structured periods that permit the hierarchical organization of cerebral function and higher-order cognition. The range of the functions at any give age denotes between-person differences and within-person modifiability. Reprinted with permission from Kühn and Lindenberger (2016).

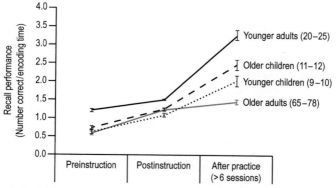

Figure 13.3 Life-span age differences in skilled memory performance. Individuals of different ages were instructed and trained in the Method of Loci, an imagery-based mnemonic technique. Recall performance is based on a ratio score of correctly recalled items over encoding time. Postinstruction scores for younger adults cannot be interpreted because of ceiling effects; all other data points can be interpreted. Error bars represent standard errors. Figure adapted after Brehmer et al. (2007).

In summary, converging evidence suggests that cortical plasticity is less easily activated through novel sensory and motor interactions with advancing age. Surprisingly, an experimental test of this hypothesis at the neural level in typically developing humans has thus far not been undertaken. Work by Brigitte Röder and others, however, provides supporting evidence from individuals with congenital sensory impairments (Nava and Röder 2011).

Dynamics of Plasticity: Overproduction–Pruning Model

Animal models have helped to uncover the dynamics of the molecular mechanisms that promote or suppress plasticity (Hensch 2005). In 1894, Santiago Ramón y Cajal proposed that mental activity might induce "novel intercellular connections through the new formation of collaterals and protoplasmic expansions." He then raised an intriguing question (Ramón y Cajal 1894:466; Azmitia 2007):

> One objection immediately presents itself: How can the volume of brain remain constant if there is a multiplication and even new formation of terminal branches of protoplasmic appendices and nerve collaterals?

On the ontogenetic timescale, the *pruning model of brain maturation* was proposed to offer an answer to Cajal's question (Changeux and Dehaene 1989). According to this model, an increase in the number of synapses is followed by experience-dependent selective stabilization of behaviorally relevant connections and the elimination of those connections that prove to be functionally irrelevant. Recent animal work provides a mechanistic basis for integrating microgenetic and ontogenetic timescales (Yang et al. 2009a), suggesting that overproduction followed by pruning may point to a set of mechanisms that is common to all forms of plasticity.

Macroscopically, the pruning model leads to the expectation that plasticity is accompanied by an initial phase of volume expansion, followed by a phase of volume renormalization (Lindenberger et al. 2017; Wenger et al. 2017a). Animal work indicates that structural MRI methods should be capable of capturing such changes (Scholz et al. 2015). To observe the pattern of expansion and renormalization with greater precision in humans, voxel-based morphometry needs to be augmented by structural MRI methods that assess the thickness of the more heavily myelinated cortical laminae *in vivo*.

This expansion–renormalization pattern of plastic change is predicted by Darwinian accounts of cortical plasticity (Edelman 1987; Kilgard 2012) and neural development (Changeux and Dehaene 1989). The hypothesized pattern is also consistent with microscopic evidence which shows that plastic changes in sensory and motor cortex are marked by the rapid formation of new dendritic spines, followed by a slower process of spine elimination, almost returning the overall number of spines to pretraining levels (Hübener and Bonhoeffer 2014). For example, such rapid formation of new dendritic spines was observed in mice being trained to perform a reaching task (Xu et al. 2009). The rapid increase was followed by a slower process of elimination of spines that had existed before training, bringing the overall number of spines almost back to pretraining levels, while performance on the trained task remained high. Similarly, monkeys and rats learning to retrieve food showed training-related gray matter volume expansion that partially renormalized while behavioral performance remained stable (Molina-Luna et al. 2008; Quallo et al. 2009). Effects of exercise on progenitor cell proliferation have also been shown to follow an inverted U-shape (Kronenberg et al. 2006).

In relation to human data, we acquired up to 18 structural MRIs over a 7-week period while 15 right-handed participants practiced nondominant, left-hand writing and drawing (Wenger et al. 2017b). After four weeks of practice, increases in gray matter in both left and right primary motor cortices relative to a control group were observed; another three weeks later, these differences were no longer reliable. Time-series analyses revealed that gray matter in both primary motor cortices expanded during the first four weeks *and then partially renormalized, in particular in the right hemisphere, in the presence of continued practice and increasing task proficiency* (Figure 13.4).

It is worth noting that the present considerations call for a radical change in research designs to address plasticity in humans (Lindenberger et al. 2017). The pretest–posttest design, which implicitly equates structural plasticity with monotonic growth, has to be replaced by designs that capture nonmonotonic structural changes accompanying functional reorganization. Specifically, only research designs with multiple observations in the course of plastic change are able to detect and test the expansion–renormalization pattern, which posits a pattern of initial growth (e.g., overproduction of synaptic connections)

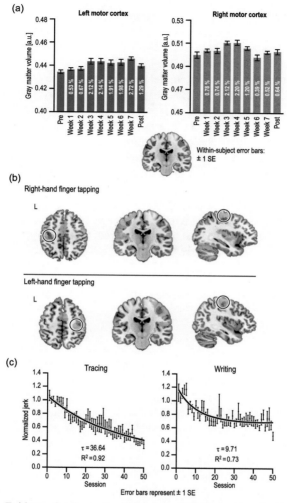

Figure 13.4 Evidence for the expansion–renormalization pattern during human skill acquisition. (a) In the course of left-hand training, gray matter volumes (measured in arbitrary units, a.u.) in the left and right primary motor cortices show initial expansion, followed by partial renormalization. (b) There is high spatial congruence between anatomical hand knobs, indicated by circles, and structural change as a function of left-hand training, displayed in blue and red for the left and right motor cortices, respectively. The functional activation maps during right-hand finger tapping and left-hand finger tapping are depicted in green. (c) Gains in left-hand tracing and writing during training. Normalized jerk is an index of movement smoothness. Individual training trajectories were fitted to exponential curves. Effects were quantified by the time constant τ of the exponential fit, indicating how fast participants approached the estimated asymptote, as well as relative improvement, expressed as R^2. Data shown here are averaged across all participants and displayed with error bars representing one standard error (SE) of the mean. Adapted after Wenger et al. (2017b).

followed by renormalization (e.g., pruning of these connections). As stated above, this model is based on physiological evidence, inspired by Darwinian concepts of cortical plasticity and neural development, and has never been tested in typically developing humans of different ages.

The overproduction–pruning model also speaks to concomitant changes in neural activation patterns and plasticity-induced network reorganization that are currently not well understood (see also section, Plasticity beyond Early Development: Open Issues). In line with Darwinian accounts of plasticity, task-related *functional* activations in cortical areas undergoing plastic reorganization should *increase* during the initial period of cortical expansion and *decrease* in the course of renormalization, when the pruning of new connections is likely to have led to sparser coding of task-relevant perception–action links, or schemata (Gdalyahu et al. 2012). Thus, the metabolic cost of plastic change is balanced by the benefit that a more efficient, metabolically less costly task representation is eventually achieved; more energy needs to be invested during the plastic episode to reach a metabolically more efficient state.

Primary Cortex Organization and Plasticity: Two Examples

Motor Cortex

Early research using low-intensity electrical stimulation led to the discovery of a somatotopically ordered representational map, or "homunculus," that resembled a distorted cartoon of the body. Later evidence confirmed the existence of functional subfields for legs, arms, and the head, but questioned the topographical representation of all body parts. As a prominent and undisputed part of the map, the cortical representation of motor hand functions is located in the superior part of the precentral gyrus, in a region labeled M1. Functional MRI work has delineated the human motor hand area as a knob-like field on the precentral gyrus. Typically, this field has an inverted omega shape and an extension of about 1.4 cm in the sagittal plane. Importantly, this topographical view of map organization has been complemented by the discovery of another organizing principle: a map of complex, meaningful movements or "ethological action maps" (Graziano 2016). The action map organization has been found in primates, prosimians, and rodents using a variety of stimulation, lesion, and neuronal recording methods.

Thus, the organization of M1 reflects the structures of both the body and its movement repertoire. Accordingly, motor skill acquisition consistently engages M1 (Dayan and Cohen 2011). It has been found that synaptogenesis induced by motor learning occurs in the same region in which learning-dependent alterations of the cortical map take place, indicating that motor skill acquisition is marked by the co-occurrence of functional reorganization and structural plasticity (Kleim et al. 2002). In support of this view, we observed high spatial congruence between structural change, anatomical hand knobs, and functional activation patterns (Wenger et al. 2017b; see Figure 13.4b). Evidence from string players, correlated with the age at which they began to play their instruments, suggests that the plasticity of human motor cortex decreases from childhood to adolescence (Elbert et al. 1995). Further, experimental behavioral evidence indicates that motor skill acquisition decreases with advancing age (Ghisletta et al. 2010).

Primary Auditory Cortex

The tonotopic place–frequency code of auditory cortex originates in the inner ear's organ of Corti and is comparable to retinotopic and somatotopic representations. The core of the human auditory region comprises two fields that jointly fold across the transverse superior temporal gyrus, also known as Heschl's gyrus. The shape and size of these two fields varies between individuals (Gaser and Schlaug 2003). Both fields are organized by V-shaped tonotopic best-frequency gradients that can be mapped with fMRI using jittered tone sequences (Langers et al. 2014). Auditory plasticity decreases with age, but auditory maps retain some degree of plasticity throughout life. In adult animals, persistent exposure to random, band-limited, moderately loud sounds leads to changes in auditory cortex that are similar to those observed after restricted hearing loss (Pienkowski and Eggermont 2011). I hypothesize that learning to discriminate increasingly small pitch intervals may lead to a more fine-grained representation of frequencies in primary auditory cortex, and hence to changes in best-frequency gradients that are discernible with fMRI-based tonotopic mapping.

A Note on Plastic Changes in Primary Cortical Areas

Learning-induced improvements in perceptual thresholds and motor skills have often been attributed to an increase in the extension of sensory or motor cortical maps. However, it is worth noting that plastic changes *within* cortical maps are not necessarily associated with changes in the *extension* of this map. From the perspective of the expansion–renormalization model (Lövdén et al. 2013; Wenger et al. 2017a), an overproduction of neural connections can lead to a *transient local thickening of deeper cortical layers which may or may not trigger changes in the size of the map*. Recently proposed MRI protocols gauging the thickness of myelinated layers of the cortex are likely to capture such localized plastic changes in humans.

Plasticity Effects on Later Learning

An intriguing finding in plasticity research concerns the physiological basis of the effects of earlier episodes of plasticity on later learning. It has been observed that the malleability and later preservation of postsynaptic spines on apical dendrites of pyramidal neurons in layer V serve as mechanisms to encode and store new experiences in cortical circuits (Yang et al. 2009a; Hofer and Bonhoeffer 2010; Hübener and Bonhoeffer 2010; Meyer et al. 2014). A remarkable example is the formation and elimination of dendritic spines during motor skill acquisition in rodents (Yang et al. 2009a). Importantly, Yang et al. (2009a) found that a small fraction of new spines were preserved and appeared to provide the structural substrate for memory retention throughout the animal's lifetime. Thus, plastic changes during skill acquisition form lifelong memories that are stored in stably connected neural networks, or plasticity-induced engrams (Hofer and Bonhoeffer 2010; Tonegawa et al. 2015).

This microscopic evidence on engram formation is of great importance and warrants the hypothesis that the *expansion–renormalization pattern indicative of plasticity will be reduced or absent when a previously acquired skill is reactivated and learning is resumed, after a break, on a task that has previously elicited plastic change.* In terms of our theoretical framework (Lövdén et al. 2010), this would mean that skill acquisition elicits plasticity, whereas skill reactivation, in its extreme form, draws on existing structures, and requires flexibility. I hypothesize that the lack of a need for plasticity during skill reactivation, and hence the minimization or absence of the expansion–renormalization pattern, depends on three factors:

1. The age at which the skill has been originally acquired, in the sense that plasticity at younger ages, or during age periods falling into a critical period, leads to more durable engrams, hence reducing the need for plasticity when the skill is reactivated at an older age (see Figure 13.5).
2. The age at which the skill is reactivated, in the sense that plasticity is reduced at older ages (e.g., in late adulthood).
3. The time period elapsing between initial acquisition and reactivation, in the sense that plasticity-induced engrams may deteriorate with age, in particular when they were acquired during later periods of ontogeny.

In line with these considerations, human behavioral evidence suggests that the positive effects of plasticity on later learning decline with age (Brehmer et al. 2008; Schmiedek et al. 2014).

Individual Differences in Plasticity

Individual Differences in Skill Acquisition

The primary cortices form part of a structured learning architecture (Chein and Schneider 2012). Networks that generate and monitor new behavioral routines and action sequences also belong to this architecture and contribute to individual differences in skill acquisition. It is assumed that a network, sometimes

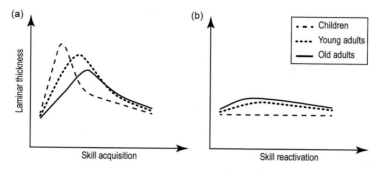

Figure 13.5 Hypothesized age differences in expansion and renormalization during (a) skill acquisition and (b) skill reactivation. Both the degree of plasticity and the durability of plasticity-induced engrams are assumed to decline with age. Hence skill reactivation after a constant time period is in greater need of plasticity if the skill was originally acquired at a later age. Predictions are based on the expansion–renormalization model, and on animal models indicating that stably maintained dendritic spines in layer V of primary cortex serve as a substrate for memory.

designated as the metacognitive system (Chein and Schneider 2012), notes and keeps track of the mismatch in supply and demand that triggers plasticity. It follows that reductions in that mismatch due to improvements in task performance should be accompanied by decreasing activations of prefrontal and temporal brain areas, which are critically involved in the metacognitive system. Repeated MRI scans during skill acquisition, as carried out by Wenger et al. (2017b), would permit researchers to quantify age group differences, as well as individual differences within age groups, in the expression of the expansion–renormalization pattern, as a quantitative index of plasticity in primary cortical areas, along with associated changes in functional activation patterns during task performance.

Individual Differences in Plasticity

Plasticity differs greatly among people of the same age (Brehmer et al. 2007; Mårtensson et al. 2012; see also Figure 13.6). However, the physiological predictors of between-person differences in plasticity remain poorly understood. Below, I outline some antecedents and correlates of individual differences in plasticity that await further study.

First, given that the formation of new neural connections is metabolically costly (Kuzawa et al. 2014), one may expect that general individual differences in brain metabolism predict differences in plasticity; a metabolically more resourceful brain should be more likely to shift from a stability to a plasticity regime than a metabolically less resourceful brain. Second, individual differences in plasticity within specific domains, such as auditory pitch discrimination, are likely linked to preexisting differences in brain anatomy,

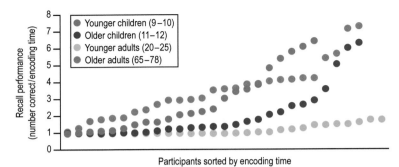

Figure 13.6   Individual differences in plasticity within age groups. The figure refers to performance after practice, illustrated in Figure 13.3, and shows the adaptively determined encoding times needed by individual participants to recall an average of 57 out of 96 words distributed across 6 lists of 16 words. Individuals are sorted by encoding time. In children, encoding times ranged from 1–7.2 seconds per word. Figure adapted after Brehmer et al. (2007).

such as the morphology of primary auditory cortex (Gaser and Schlaug 2003). Furthermore, a reliable portion of individual differences in plasticity is associated with genetic differences; for instance, behavioral and genetic evidence indicates that genetic variation in individuals' musical abilities affects both the ability and the inclination to practice, such that music practice in itself does not add unique variance to individual differences in musical ability (Mosing et al. 2014). At the molecular levels, a range of single nucleotide polymorphisms (SNPs) has been tentatively related to individual differences in plasticity (Lövdén et al. 2011; Bellander et al. 2015). Candidate gene or genome-wide association approaches could be used to examine the unique and interactive effects of SNPs that result in individual differences in transmitter systems known to affect excitatory–inhibitory balance in primary cortical areas, such as glutamate, GABA, and dopamine. Finally, in terms of epigenetics, it seems worthwhile to examine differences in DNA methylation affecting the expression of plasticity-related genes during skill acquisition versus skill reactivation, with the caveat that DNA methylation markers obtained outside the brain (e.g., via buccal swabs) may not be indicative of DNA methylation in the brain.

Below I will outline a set of research hypotheses that derive directly from the framework just outlined. This will be followed by a discussion of open issues, speculative in nature and only loosely connected to the framework.

Plasticity beyond Early Development: A Set of Research Hypotheses

Life-Span Differences in the Plasticity of Primary Cortices

Hypothesis 1: Life-span gradients in plasticity. The plasticity of motor and auditory cortices in response to experience is greatest in children and smallest in older adults, with younger adults falling in between (Figure 13.5a). This hypothesis corresponds to a long-standing claim of life-span psychology (Baltes et al. 2006; Lindenberger 2014) that has thus far not been tested experimentally in typically developing humans.

Delineating the Expansion–Renormalization Pattern of Plastic Change

Hypothesis 2a: Plasticity is expressed by nonmonotonic volume changes in primary cortices. The task-relevant deeper laminae of primary cortices will show a pattern of initial expansion, reflecting the formation of new neural connections, followed by renormalization, reflecting subsequent pruning. This hypothesis is consistent with microscopic findings in animals and informed by Darwinian models of plasticity and development (Changeux and Dehaene 1989; Kilgard 2012).

Hypothesis 2b: Plasticity leads to sparsification of the neural code. During later periods of a plastic episode, brain activation will decrease while performance will continue to improve, reflecting sparsification of the neural code. This hypothesis is based on the assumption that plastic change, if successful, results in a metabolically efficient neural representation of the trained task.

Testing the "Memory of Plasticity" or the Effects of Plasticity on Later Learning

Hypothesis 3a: Plasticity-induced structural alterations are partially preserved and serve as a scaffold for later learning. The expansion–renormalization pattern will be attenuated or absent when a previously acquired skill is reactivated at a later point in time, reflecting the stability of previously formed neural connections.

Hypothesis 3b: The preservation of plasticity-induced structural alterations decreases with age. We assume that the mechanisms that stabilize newly formed synaptic connections in cortical circuits (Meyer et al. 2014; Tonegawa et al. 2015) decrease in efficiency with advancing age. Hence the reduction of the expansion–renormalization pattern when a sensory or motor skill is reactivated after a constant period of time will be greatest in children and smallest in older adults, with younger adults falling in between (Figure 13.5b). This finding would corroborate the widely held but untested claim that the positive effects of plasticity on later learning decline in the course of ontogeny.

Individual Differences in Plasticity and Skill Acquisition

Hypothesis 4a: Individual differences in primary cortex plasticity, learning rates, and final task proficiency form a positive manifold and contribute to gene–environment correlations (Beam and Turkheimer 2013); see also Hypothesis 4d.

Hypothesis 4b: The contribution of the metacognitive network to task performance will decrease with practice. Individual differences in this decrease will be positively related to skill acquisition, reflecting less effortful and more efficient performance.

Hypothesis 4c: Individual differences in the expression of the expansion–renormalization pattern will show positive associations with skill acquisition. It is assumed that initial overproduction and pruning both contribute to plastic change, akin to mutation and selection on an evolutionary timescale.

Hypothesis 4d: Individual differences in metabolic, anatomic, and genetic markers predict differences in plastic responses to experience. The following exemplary predictions emanate from this generic hypothesis:

1. Individual differences in brain metabolism can serve as trait markers of plastic potential.
2. Preexisting differences in the anatomy of primary auditory cortex selectively predict plasticity in the auditory domain.
3. SNPs related to individual differences in GABA and glutamate expression are associated with experience-induced shifts in excitatory–inhibitory balance, which in turn predict laminar expansion and renormalization.
4. Genes identified previously as being relevant for plasticity, such as the BDNF gene and dopamine-related genes, predict individual differences in plasticity within age groups, especially in old age, when brain resources are scarce (Papenberg et al. 2015).
5. Periods of skill acquisition marked by expansion–renormalization of primary cortical areas will be associated with changes in DNA methylation status (Guo et al. 2011).

Plasticity beyond Early Development: Open Issues

The purpose of this final section is to raise issues of general importance for the age-comparative study of human plasticity. Many of these are not yet sufficiently well understood to permit the operational definition of testable hypotheses. Rather, they are meant to serve as guideposts for future conceptual and empirical efforts in the study of human plasticity.

Issue 1: We need to be aware of the almost ubiquitous and often unavoidable confound between age and experience whenever we wish to make claims about age differences in plastic potential. Individual development reflects age-graded changes in the interactions among maturation, residues of earlier learning (e.g., memories of all kinds), senescence, and new learning. Hence, manifest age differences in plasticity may not solely reflect age-graded differences in plastic potential but rather a variety of additional influences that interact in unknown ways with plastic potential. In particular, memories, or the sediments of past learning, accumulate with age. Even a system whose plastic potential does not decline with age would experience fewer and fewer episodes of plastic change with increasing age for the simple reason that the accumulation of experience makes it increasingly difficult to experience an environmental demand characteristic for the first time.

Issue 2: We need a better understanding of how age-based changes and between-person differences in large-scale network topography affect the context for local plastic change. Cognitive development from childhood to adulthood is accompanied by profound changes in structural and functional connectivity (Uhlhaas et al. 2010), presumably associated with declines in general synchronizability and increases in controllability (Tang et al. 2017). Hence the cerebral context for acquiring any given skill changes with age, and these changes may affect the plasticity of cortical areas in unknown ways. In particular, the maturation of the parietal and prefrontal cortices during childhood and adolescence and the growing number of available skills and bodies of declarative knowledge (see Issue 1) lead to an increase in top-down strategic control over perceptual processes with advancing age. This increase in control may help specify the supply–demand mismatch and hence direct attention to specific aspects of behavior that are in need of plastic change. This increase, however, may also hinder local plastic change through an excessive strategic guidance of local overproduction–pruning dynamics. For instance, although directing attention toward a to-be-acquired skill may generally be helpful, overly precise knowledge about what should be done to acquire it may lead to an "over-instruction" of local circuits that hinders local plastic change.

Issue 3: We need a mechanistic account of the plasticity of higher-order cognition. In this chapter, research on critical periods of perceptual and motor skills during early development was used as a template to delineate a research program on human plasticity across the life span. This strategy reflects the premise that human research on plasticity needs a strong connection and a firm grounding in animal models. In particular, optogenetic tools and two-photon microscopy have provided insights into the molecular dynamics of plastic changes that need to be brought to bear upon research in humans. At the same time, we need to be cautious. We do not know the extent to which plasticity observed in primary sensory and motor cortices in the context of perceptual and motor skill acquisition offers a viable analogy to the role of the association cortices in the context of higher-order cognitive abilities such as episodic memory, working memory, task-set switching, and fluid intelligence. Similar to the plasticity of perceptual and motor skills, the plasticity of these abilities, if present, is likely to require a mixture of local plastic change (e.g., akin to cortical map extension in the primary cortices) and more global changes such as myelination of relevant white-matter tracts in the service of network reorganization (e.g., to improve synchronization of posterior and anterior regions). However, with the exception of the role of the hippocampus in memory and related functions, the relative importance of local plastic change for higher-order cognitive abilities is not well understood. Can we target a higher-order cognitive function, such as efficient switching between task sets, and identify a cortical area, or set areas, that shows the expansion–renormalization pattern when this function is trained? Or is improvement of higher-order cognitive abilities, especially in adulthood, generally more a matter of flexibility than of plasticity, in the sense that the behavioral repertoire available to the system is exploited more fully and reconfigured more efficiently without any structural change?

Issue 4: We lack neural theories of generalization and transfer to predict consequences of plastic change. In recent years, the issue of transfer of training attracted great scientific and public interest (Simons et al. 2016), with much of the debate focused on whether the effect sizes of transfer of training do, in some cases, differ from zero. Unfounded claims about real-life benefits of "brain jogging" abound[1] while attempts to use mechanistic accounts of plastic change for deriving hypotheses about generalization and transfer (Dahlin et al. 2008) are scarce. Clearly, a better understanding of plasticity in humans is a prerequisite for arriving at hypotheses about transfer gradients and generalization (Lindenberger et al. 2017). This is especially true for higher-order cognitive functions (see Issue 3), where we lack evidence on the processes associated with plastic change.

Issue 5: We need a better understanding of the relationship between brain size and neural efficiency. General cognitive ability shows a weak to moderate positive association with brain size (Luders et al. 2009). For instance, a recent meta-analysis found that larger prefrontal cortex volume and greater prefrontal cortex thickness are associated with better executive performance (Yuan and Raz 2014). Apparently, then, both neural code efficiency and brain size determine an individual's effective "functional cerebral space" (Kinsbourne and Hicks 1978). However, the recursive relations between size and efficiency are not well understood. For instance, size may enable efficiency, or efficiency may alleviate the effects of smaller size.

Recommendations for Future Research

What all these open issues have in common is that they can only be tackled successfully if research on plasticity in humans goes beyond the pretest–posttest design (Lindenberger et al. 2017). To understand plastic change in humans, we first need to observe it using imaging protocols that facilitate mechanistic interpretation. As a critically important step toward this goal, we need research

---

[1] For a critique, see "A Consensus on the Brain Training Industry from the Scientific Community," Max Planck Institute for Human Development and Stanford Center on Longevity, http://longevity.stanford.edu/a-consensus-on-the-brain-training-industry-from-the-scientific-community-2/ (accessed Oct. 11, 2017).

with rodents and nonhuman primates using optogenetic and MRI methods in combination to inform the interpretation of MRI results obtained in research with humans (Lerch et al. 2017). Investigating the dynamics and temporal progression of plastic change in humans requires experiments with multiple imaging sessions during # that include a selection from a wide range of imaging modalities (e.g., structural and functional MR as well as MR spectroscopy) to assess metabolites such as GABA, glutamate, and creatine, as well as electrophysiological recordings (e.g., electroencephalography, magnetoencephalography, and intracranial recordings) to assess related changes in oscillatory patterns. If motivated by a search for mechanisms of plastic change, the repeated application of these methods *in the context of age-comparative high-intensity training studies* bears great promise for the future of plasticity research.

These empirical efforts will benefit from close interactions with work on neuromimetic computational architectures and machine-learning algorithms (LeCun et al. 2015; Mnih et al. 2015). Building on pioneering work by Terry Sejnowski, Jay McClelland, Mark Johnson, and others (Sejnowski et al. 1990; Elman et al. 1996; McClelland 1996; Shrager and Johnson 1996), such artificial systems enable researchers to observe the dynamics of plastic change within and between the layers of artificial neural networks, and hence can guide them in formulating and testing hypotheses about developing biological systems.

Acknowledgments

The author thanks Julia Delius for her editorial contributions as well as Simone Kühn, Martin Lövdén, Naftali Raz, and Elisabeth Wenger for valuable discussions that contributed to the conceptual framework and helped to improve this chapter. Preparation of this chapter was facilitated by a Fernand Braudel Senior Fellowship of the European University Institute to Ulman Lindenberger.

14

What Factors Determine Changes in the Adolescent Brain?

Adriana Galván

Abstract

Over the past two decades there has been an explosion of research on the human ado-lescent brain. This research has demonstrated that the brain continues to mature during the second decade of life, due to ontogeny and experience. The majority of this work has focused on changes that occur in regulation and affective circuitry; in particular, on how these neurobiological changes relate to characteristic adolescent behavior. This chapter summarizes existing understanding and speculates about agents of change that impact neurobiological development in the adolescent brain. It begins with a discus-sion of what adolescence refers to and reviews the prevailing neurobiological models of adolescent brain development. Factors are considered that contribute to adolescent brain development (e.g., puberty, sleep, social relationships, adolescent risk-taking). Open questions are posed to aid further consideration and research.

What Is Adolescence?

Across the world and across different species, adolescence refers to an impor-tant function in development: it is the time when individuals move from a state of dependence on caregivers to one of relative independence. This transitional period lends itself to many changes in physical growth and biological devel-opment, as well as cognitive sophistication and psychosocial skills. These changes, in turn, drive ongoing development of the brain.

The Beginning and End of Adolescence

The determination of when adolescence "begins" and "ends" is currently a question of intense debate. Most scientists pinpoint adolescence as "the gradual period of transition from childhood to adulthood" (Spear 2000) that begins at the onset of puberty and ends as individuals attain adult roles, responsibilities, and rights. The range of age at which this occurs, however, varies according to cultural and historical circumstances. In the United States, for example, adolescence begins at approximately 10–12 years of age and ends in the late teenage years (approximately 18–19 years of age). In this chapter, the work that I review and the speculative comments that result refer to these age boundaries.

Adolescence across the Globe and across Species

Adolescent-related behaviors are observable worldwide, across different cultures (Schlegel 2001) and species (Spear 2000). A recent study of sensation-seeking and self-regulation in more than 5,000 individuals from 11 countries in Africa, Asia, Europe, and the Americas found that sensation-seeking peaked in late adolescence and that self-regulation increased linearly until the midtwenties (Steinberg et al. 2018). One explanation for this global phenomenon is that regardless of cultural experience, there are particular neurobiological and hormonal changes that arise during adolescence that are common to typically developing young people. Yet despite these observed general trends, there are vast individual differences in the extent and manner in which adolescent risk-taking manifests itself in different areas of the world. Nonetheless, these findings lend support for current working models (reviewed below) of neurobiological development in brain regions that underlie these behaviors. Juvenile rodents, immediately prior to and following sexual maturation, exhibit behavioral changes that are similar to those commonly observed in human adolescents: increased peer-directed social interactions; occasional increases in fighting with parents; increases in novelty-seeking, sensation-seeking, and risk-taking; increased consummatory behavior; and greater per occasion alcohol use. The increased proclivity toward drug use observed in human adolescents is also observed in adolescent rats (Brenhouse and Andersen 2008; Torres et al. 2008) and nonhuman primates. These data suggest that some of the characteristic adolescent behaviors observed in humans may be embedded in our evolutionary past, and that they emerge to facilitate behaviors important to the developing organism. Indeed, rapid progress is being achieved across laboratories (Crone and Dahl 2012; Varlinskaya and Spear 2008), from studies involving both animals and human adolescents, which show that neural changes in systems that underlie motivational, affective, and behavioral regulation influence the processing of and response to events in the environment in ways that bias behavior.

Theoretical Models of Adolescent Brain Development

Currently, research is guided by four neurobiological models of adolescence: the dual systems model, the triadic model, the imbalance model, and the fuzzy trace theory. These models reflect the differences in maturation rates of brain systems implicated in emotion, social, and reward processing from those that are important for regulation of behavior.

In the domains of sensation-seeking and risky decisions, Steinberg (2010) described adolescent behavior in terms of a *dual systems model*. According to the model, risky decision making in adolescence is the product of an interaction between two neurobiological systems: (a) the socioemotional system, comprised of limbic regions including the amygdala, ventral striatum, orbitofrontal cortex, and medial prefrontal cortex (PFC), and (b) the cognitive control system, comprised of the lateral prefrontal and parietal cortices. Around the time of puberty, the surge in dopaminergic activity within the socioemotional system leads to increases in sensation-seeking and risky decision making, outpacing the development (and engagement) of the cognitive control system. This temporal gap leads to heightened vulnerability to these behaviors during adolescence.

To explain motivated behavior in adolescent decision making, Ernst et al. (2006) proposed the *triadic model*. This model attributes the determinants of motivated behavior to three functional neural systems (the PFC, the striatum, and the amygdala) and focuses on how the maturational timing of each region contributes to age-related differences in motivated behavior as people mature. The PFC is implicated in the regulation aspect of motivated behavior, the striatum in motivational aspects of the model, and the amygdala in the emotional components of behavior. Together, these three nodes and their associated constructs serve (a) to coordinate the calculation of whether to approach (engage in) or avoid a particular behavior and (b) to regulate the resulting calculation. This model has been used to describe typical adolescent behaviors, including cognitive impulsivity, risk-seeking, emotional intensity, and social orientation.

The *imbalance model*, developed by Casey et al. (2008), emerged from empirical studies that examined the developmental transition in humans—from childhood through adolescence and into adulthood—and translated the results across species (nonhuman primate and rodent). According to the model, developmental changes in the neurochemical, structural, and functional composition of the brain proceed on distinct time lines: some brain regions exhibit changes earlier in development than other brain regions. This leads to an imbalance in how these regions bias behavior due to differential engagement across different stages of development (see also Uhlhaas, this volume). For instance, the model has been used to explain nonlinear changes in behavior during adolescence because regions implicated in reward (e.g., striatum) exhibit greater engagement—in terms of striatal activation and behavioral bias toward reward—relative to regions critical for behavioral regulation (e.g., PFC). Importantly, unlike models that focus on specific brain regions, the imbalance model aims to attribute adolescent behavior to the coordinated integration of multiple brain circuits.

Reyna and Farley (2006) have applied *fuzzy trace theory* as an explanatory framework for adolescent risk behavior. This model posits that sophisticated judgment and decision making are based on simple mental representations of choice ("fuzzy" memory traces) as opposed to more detailed, quantitative representations (verbatim memory traces). Accordingly, decision making becomes less computational and more intuitive as development proceeds. Specifically, risky decision making involves a focus on precise calculations (e.g., determining whether the exact amount of fun or money gained will outweigh the exact amount of risk involved in achieving the fun or money) earlier in development as compared to a "fuzzier" calculation that simply ranks the options (e.g., ranking the potential rewards against the risk involved to get the reward) as individuals get older.

Factors that Determine Change in the Adolescent Brain

Human brain development is a prolonged process compared to nonhuman animals. The developmental periods of early childhood and adolescence, in particular, exhibit protracted development in humans as compared with most other species. Although the majority of changes and growth in the brain occur postnatally during the first few years of life, the brain undergoes another period of major development during adolescence.

The past two decades has witnessed an explosion of research on the adolescent brain aimed at examining adolescent brain development (Galván 2014). What triggers changes in the brain during adolescence? Similar to many other developmental milestones, these changes are a product of ontogeny as well as environmental or experiential input. In this regard, the adolescent brain is not unique. What distinguishes development in the brain during adolescence, however, is the sensitivity of particular circuitry, namely regions in frontostriatal circuitry, to the changing social and cognitive landscape.

Physiological Changes

Puberty

Much has been written about the role of pubertal hormones in inciting neurobiological change in the adolescent brain (see, e.g., Sisk, this volume), so in the interest of brevity, discussion here will be brief.

Puberty is the result of a series of hormonal events during which young adolescents undergo the physical and neuroendocrine changes required to reach sexual maturity. Three characteristics describe puberty:

1. It is controlled and sustained by hormones.
2. It involves changes in body height, weight and shape.
3. It is associated with changes in behavior and mood.

What is perhaps most fascinating about puberty is that although the physical manifestations occur at a discrete point in development, puberty is actually a long process that is influenced by many factors, some of which occur much earlier in life. This has implications for the various roles and influences pubertal hormones have on the developing brain.

The beginning of puberty is marked by the activation of the hypothalamic-pituitary-gondal (HPG) axis, when the brain starts to communicate with the gonads (sex glands). One brain region, the hypothalamus, plays a central role in this process. Generally the hypothalamus is responsible for monitoring basic human needs (e.g., eating, drinking, sex), but at the onset of puberty, it plays a special role in governing the pituitary gland through gonadotropin-releasing hormone (GnRH) neurons. The pituitary gland produces the hormones, called gonadotropins, necessary to stimulate the release of sex hormones from gonads. The level of sex hormones that need to be released from the gonads is regulated by two hormones secreted from the pituitary gland: the follicle-stimulating hormone stimulates sperm production in males and follicle development in females, whereas the luteinizing hormone regulates testosterone production in males and estrogen secretion and ovum development in females.

Adrenarche, an early stage of sexual maturation, typically begins in humans around 6 to 8 years of age. During adrenarche, the adrenal glands secrete adrenal androgens, such as dehydroepiandrosterone and dehydroepiandrosterone sulfate. Their secretion leads to androgen effects, including the emergence of pubic hair and body odor due to changes in sweat composition, and appears to play a role in changes in the oiliness of the skin that lead to acne.

Gonadarche begins typically around 8 to 10 years of age, but there is considerable variability among individuals as to its onset. Gonadarche is the period most commonly recognized as puberty because it involves the maturation of observable sexual characteristics. In females, menarche (the first menstrual period) occurs in the middle to late stages of gonadarche whereas in males, spermarche (the first ejaculation of semen) occurs in the early to middle stages.

What triggers puberty? Decades of research in animals and humans have not identified any one hormone, event, age, or environmental experience that induces puberty. Instead, all of these factors converge to signal that the organism is healthy and physically mature enough to permit sexual reproduction. These factors have been called "permissive signals" because they permit (or stop inhibiting) pubertal onset (Sisk and Foster 2004). These signals include changing levels of melatonin, body fat, and leptin, all of which are related to weight and energy balance. It is generally held that individuals do not go through puberty until they are energetically and metabolically capable of doing so.

Sleep

Sleep is essential for survival and plays an important role in supporting healthy development. Growing public and scientific concern have raised awareness of adolescent sleep patterns and led to what is called a "sleep deprivation epidemic" among human adolescents (National Sleep Foundation 2014). Adolescent sleep deficiencies are rooted in biological and psychosocial changes that occur during this developmental period (Carskadon 2011).

Biological alterations during puberty, including brain coordination in hormonal circuitry, contribute to delay the sleep phase—the body's internal clock shifts—making it more difficult for adolescents to go to sleep earlier (Hagenauer and Lee 2012). This delay (a biological factor) pushes adolescent bed times later while school starting times (an environmental factor) force early waking times (Hagenauer and Lee 2012). Figure 14.1 shows that as children transition from grade school to high school, they go to bed at increasingly later times yet rise at roughly the same time in the morning to attend school. As a result, adolescents regularly experience insufficient sleep. Some studies report that *only 15% of adolescents sleep the recommended 8–10 hours on weekdays* (National Sleep Foundation 2014). Additional environmental factors, such as socializing and studying, contribute to sleep loss among adolescents.

Sleep is integral for various functions, ranging from restorative purposes and memory consolidation to removal of neurotoxic waste. Persistent sleep deficiency and subsequent sleepiness negatively impact adolescent health and safety, including increased risk of suicide and substance use (Owens 2014). Many studies indicate that insufficient sleep is associated with poor emotional functioning in adolescents. Less sleep in adolescents is associated with more depressive symptoms, feelings of hopelessness, and greater anxiety (Fredriksen et al. 2004). Emerging research suggests that insufficient sleep is also detrimental to brain function: poor sleep is associated with less dorsolateral PFC activation during cognitive control (Telzer et al. 2013a) and lower white matter integrity longitudinally (Telzer et al. 2015).

Despite the dwindling time spent asleep, studies suggest that the sleep "need" per se does not undergo dramatic changes during adolescence. An early longitudinal study, which followed adolescents yearly from age 10–12 until age 15–18, found that when given the opportunity to sleep ten hours, adolescents slept an average of approximately 9.25 hours, irrespective of age or maturational stage (Carskadon 2011). Wahlstrohm et al. (2014) noted that early morning school schedules contribute significantly to lower the sleep times of adolescents. When school start times are delayed, sleep is increased, enrollment rates and attendance improve, students fall asleep in class less, symptoms of depressed mood are reduced, and automobile crash rates in teen drivers are lower (Wahlstrohm et al. 2014).

Observing brain activity during sleep may provide a unique window into adolescent cortical maturation and complement waking measures (Tarokh et al. 2016). Recent studies suggest that sleep not only offers an opportunity to measure otherwise unperturbed brain activity, it may also play an active role in sculpting the adolescent brain. Using two-photon microscopy in adolescent mice, for example, Maret et al. (2012) found that synaptic spine elimination

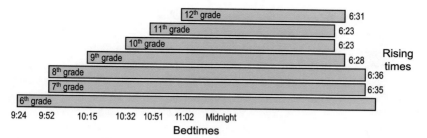

Figure 14.1 Bedtimes (p.m.) and rising times (a.m.) for youth between sixth and twelfth grade in the United States. This figure, adapted from Carskadon (2011), illustrates the increasingly late bedtime in youth as they transition from grade school to high school.

was higher during sleep than during waking in adolescent but not adult mice, suggesting a distinctive role for sleep in the adolescent brain. Correlational studies in humans have also found associations between sleep behavior and brain development. One study examined structural magnetic resonance imaging (MRI) scans in 290 children and adolescents between the ages of 5 and 18 years and found that self-reported sleep duration was positively correlated with bilateral hippocampal gray matter volume (Taki et al. 2012). Another study found an association in adolescents between variability in sleep duration across fourteen days and white matter integrity, as measured with diffusion tensor MRI a year later (Telzer et al. 2015). Although this line of research is in its nascent stage, evidence for a role of sleep in brain development is emerging.

Uy and Galván (2017) recently published research showing that the relationship between insula response and risky behavior was exacerbated in individuals who reported that they regularly slept less than the 7 hours per night (currently recommended by the National Sleep Foundation). In a separate study from the same group, Tashijian et al. (2017) found that variability in sleep quality, not sleep duration, was predictive of immature development of neural connectivity in the default mode network. Furthermore, their data suggest that stronger neural connectivity buffers the relation between sleep variability and impulsive behavior.

Social Relationships

Adolescence is a period of social reorientation during which young people begin to develop the identities that will define their adult relationships, interests, and social roles. As a part of the process of social identity formation, adolescents must integrate the perspectives of others with their own to create a unique, coherent sense of self, independent from others. This task can be challenging because while adolescents continue to value input from those they admire, they are also generating their own ideas, values, and behaviors, particularly as they become increasingly aware of the identity bestowed upon them by society (e.g., in terms of gender or ethnicity). Parents remain a crucial source of feedback and authority, but sensitivity to peer attitudes becomes increasingly essential to the adolescent. The plasticity and flexibility of the adolescent brain may render it particularly sensitive to social input because all of the brain regions that populate the "social brain network" undergo significant maturation during the adolescent period (Blakemore and Mills 2014). The relative importance of parental and peer perspectives seems to shift over the course of adolescence, at least in some domains. Parental influence is not likely to be replaced by peer influence during adolescence (Brown et al. 1993), as is commonly believed.

In one recent study, Welborn et al. (2016) used fMRI to investigate the neural basis of peer and parental influence on adolescents' subjective evaluations of artwork. We reasoned that works of art would provide a neutral domain, with potentially flexible attitudes that are not already saturated with influence from either group. While undergoing scanning, participants received information regarding their own peers' or parents' actual attitudes (i.e., there was no deception) and immediately provided their own evaluation of the artwork stimulus. Shifts in participants' attitudes toward those of their peers (i.e., peer influence) or those of their parents (i.e., parental influence) were assessed based on participant ratings of each stimulus acquired prior to the scanning session. Adolescent participants shifted their attitudes to indicate significant influence by both peers and parents. As shown in Figure 14.2, there was a significant relation between the level of social influence from both peers (left bottom)

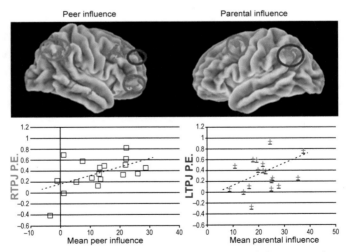

Figure 14.2 Relation between brain activation and social influence. Brain activation in the temporoparietal junction correlates with peer (left) and parental (right) influence: right temporo-parietal junction parameter estimates (RTPJ P.E.) and left temporo-parietal junction parameter estimates (LTPJ P.E.). From Welborn et al. (2016) with permission of Oxford University Press.

and parents (right bottom) and the extent of activation in regions typically associated with "mentalizing," including the temporoparietal junction as well as precuneus and ventrolateral PFC during both peer (left top) and parental (right top) influence. This suggests that peers and parents often play distinct roles in adolescence, but that parents continue to exert an important influence.

Adolescence: A Sensitive Period for Romantic and Sexual Development

Neurodevelopmental models have identified the onset of adolescence, marked by the biological transition into puberty, as a period in which profound changes occur in motivation, cognition, behavior, and social relationships. However, despite the emergence of many excellent models which highlight the importance of puberty for neural development and new, adaptive learning (e.g., reviewed above), these models give limited consideration to the importance of adolescence as a sensitive period for romantic and sexual development. In the few developmental models that did consider romance and sexuality, sexual development was characterized as negative risk behavior (i.e., a risk framework of sexual behavior) (Victor and Hariri 2015). It is equally important to consider normative, healthy aspects of sexual and romantic development and the neurodevelopmental underpinnings of learning about romantic and sexual behavior. As young people enter adolescence, one of their primary tasks is to gain the knowledge and experience that will allow them to take on the social roles of adults, including engagement in romantic and sexual relationships. As such, my coauthors and I argue that the psychological, social, and hormonal changes associated with becoming a sexual being help trigger neurobiological changes in the adolescent brain (Suleiman et al. 2017).

Young people's romantic relationships—from primary school crushes (where two people might interact to a limited extent) to relationships that involve significant investment of emotion, time, and energy—are often dismissed as insignificant. In fact, these relationships serve important developmental purposes and form the primary context for young people to explore their sexual identity and gain sexual experience (Furman and Shaffer 2003). In hopes of gaining social status and winning the companionship of desirable partners, adolescents are highly motivated to learn how to navigate the complex social interactions involved with establishing and maintaining romantic relationships. A person's ability to engage in behaviors that will facilitate intimate relationships and create opportunities for sex and reproduction is the normative developmental outcome of puberty.

Although puberty motivates mating and sexual behavior, only limited research has explored the emergence of sexual behavior in adolescent humans. In contrast, pubertal research on other species has included in-depth exploration of the onset of sexual and mating behavior associated with puberty, acknowledging that the emergence of these novel behaviors requires immense coordination of developmental transitions in the brain, endocrine system, and nervous system. As such, animal researchers perceive early sexual experiences not only as behavioral outputs, but also as physiologic inputs that shape neural and hormonal function and development (e.g., Nutsch et al. 2014). The dearth of knowledge about the learning and the reciprocal feedback loops involved in the onset of human mating and sexual experiences highlights important oversights in existing models of human adolescent development. At the same time, while animal models offer important insights into understanding sexual developmental trajectories, they do not expand our understanding of romantic relationships and experiences in humans, nor do they identify developmental changes relevant to these important social milestones. Moreover, the mating framework of animal models offers solely a heterosexual framework for sexual development, thus limiting our understanding of the diversity and fluidity of attraction, behavior, and identity in human sexuality.

The animal literature serves as a critical reminder of the biological purpose of puberty and the reciprocal feedback loops involved in romantic and sexual experiences, which have been largely ignored in models of human adolescent development. Unfortunately, animal models and the limited human research on this topic have done little to explore how puberty shapes the opportunities for learning about the *meaning* of romantic and sexual behaviors (Fortenberry 2014). On one hand, a basic capacity for procreative behavior can be achieved with relatively little skill, knowledge, or experience; on the other, from an evolutionary perspective, social competition in attracting a mate and success in coupling relies heavily on mastery of a complex set of social and emotional skills and behaviors. The learning relevant to acquiring these skills and knowledge necessary to navigate the intertwined social and sexual motivations that emerge with puberty is central to the normative trajectory of social, affective, and cognitive development in humans. Therefore, pubertal maturation (and the natural increase in social motivation, including interest in sexual and romantic behavior) is likely to represent a normative window of learning—not simply about the mechanical aspects of sexual behavior, but also about the complex emotional and social cognitive processes that are part of navigating the charged, high-intensity emotions involved in developing an identity as a sexual being.

In our research (Suleiman et al. 2017), we explored how cognitive and socioaffective development that occurs at puberty creates a unique window of opportunity for adolescents to engage in developmentally appropriate learning opportunities relevant to navigating romantic and sexual experiences. We propose that changes in underlying neural circuitry associated with social and emotional processing may open a second developmental window (after the one in early childhood) for learning about love and attachment relationships. Further, we hypothesize that these learning processes begin with the pubertal physical and neurobiological transitions that influence motivation, yet are highly dependent on context and interpersonal relationships during this time (Suleiman et al. 2017).

The onset of puberty seems to reorient greater attention and salience toward social and emotional information-processing streams. More specifically, puberty leads to the development of novel social behaviors and responses to newly emerging social contexts (Brown et al. 1993). Young people begin to spend increasingly more time with their peers and, at the same, experience new, sexualized feelings of attraction that motivate relationship-facilitating behaviors. Given that the biological purpose of puberty is to achieve reproductive maturity, it makes sense that the balance between plasticity and stability in this unique peripubertal neural system would create a window of opportunity for learning and motivation relevant to romantic and sexual behavior. Consider the skills that an adolescent must learn in this domain: coping with emotions related to finding someone attractive, building the communication skills required to ask someone out on a date, experiencing sexual arousal with a stranger, navigating the social consequences of dating someone more or less popular, coping with rejection or break up, and balancing the biological desire to have sexual experiences with the complex emotions associated with maintaining a romantic relationship.

Although it has been established that many of the neural systems involved in romantic love and sex undergo significant structural, connectivity, and functional transformation during puberty, little is known about how this intersects with a normative romantic and sexual developmental trajectory. Integrating what is known about the neural underpinnings of romantic love and sexual desire/arousal in adults with the literature on pubertal neurodevelopment points to some intriguing questions. While it is beyond the scope of this paper to summarize this body of literature, adolescent neurodevelopmental models have clearly demonstrated significant sex-specific restructuring of the brain during puberty (Dennison et al. 2013). Beginning with puberty, the developmental transitions in brain networks involved in motivation, reward, and social-emotional processing likely create a unique inflection point for romantic love and sexual arousal to be experienced as positive rewards.

Both love and sexual desire are dopaminergically mediated motivation states that can globally affect cognition (Diamond and Dickenson 2012). Given the developmental transitions that occur during adolescence related to emotional processing and cognitive control, it has been proposed that adolescence is an opportune time to explore the cognitions and emotions associated with romantic relationships (Collins 2003). These new motivational states significantly increase in salience at the same time that youth develop an increased capacity for self-regulation of other appetitive behaviors (Fortenberry 2014). Therefore, it makes sense that physical maturation is accompanied by increased neural plasticity and a heightened motivation to seek out a range of highly arousing, slightly scary, highly rewarding, novel experiences, and that increases in sensation-seeking make adolescents more likely to find these high-intensity experiences, such as having a first crush, enjoyable. The co-release of dopamine and oxytocin associated with repeated interactions with a specific partner contributes to additional reward-driven learning about romantic behaviors. Once a young person has a crush and begins to build a relationship with someone, they develop a conditioned partner response in which the dopaminergic reward that is expected and experienced is greatest with that specific bonded partner (Ortigue et al. 2010). Because of the neural development that occurs in puberty, a partner-specific response in early romantic relationships, when both the emotional and physical intimacies are novel, makes them particularly exciting, rewarding, and satisfying.

Risk-Taking

Risk-taking behavior—for instance, elevated rates of experimentation with alcohol, cigarettes, and illicit drugs; higher rates of risky sexual activity, petty and violent crime, and reckless driving—increases in adolescence. Parents, educators, and policy makers have long wondered why this occurs and before the advent of neuroimaging, this phenomenon was primarily attributed to shifts in pubertal hormones. Although such behavior can result in negative consequences, I argue that "healthy" risk-taking, or *exploration*, serves an adaptive developmental process. A prevailing narrative in developmental cognitive neuroscience is that the increase in risk-taking behavior can be attributed to changes in frontostriatal circuitry that occur during adolescence, in particular in the mesolimbic striatum. It is certainly likely that neurobiological changes induce behavioral changes. However, the opposite may also be true: risk-taking behavior may help shape the brain in a similar fashion that experience more generally helps refine the brain across development.

Is Adolescent Risk-Taking Adaptive?

Some adolescents engage in risky behaviors that are a threat to healthy development while others pass through this developmental window relatively unscathed. Scientists have long wondered what the adaptive aspects are of a brain that is hyperexcitable, responsive to the social environment, and primed for learning.

The brain is built this way to facilitate the important task of transitioning from a state of dependence on caregivers to one of relative independence. Imagine if such a period in life, when individuals actively sought out autonomy from their parents, did not exist. There would be minimal exploration of the environment, a lack of thirst for learning new things, and little curiosity to meet new people. Taken to an extreme, without this development stage, it is doubtful that our species would have survived. Growth of the human species is dependent on the motivation of individuals to innovate, create, and procreate utilizing a diverse gene pool. At no other time in life is there greater intrinsic motivation to explore the world than during adolescence. A model by Crone and Dahl (2012) emphasizes the positive and negative trajectories that can result from increased flexibility in the adolescent brain. Their model (Figure 14.3) has become a cornerstone of a nuanced depiction of adolescent brain development. It spans the period from puberty onset until the transition to adulthood, showing that goal-driven behavior is increasingly influenced by social input as well as cognitive flexibility because of the ongoing maturation of the social brain and cognitive control networks, respectively. Together, these changes have the possibility of yielding positive (e.g., adaptive exploration) as well as negative growth trajectories (e.g., mental health issues).

Youth are often at the forefront of new ideas; they are impassioned defenders of ideals, fervid leaders, and the ones having the most "fun" in their quest for autonomy. Despite possessing better cognitive, intellectual, and reasoning abilities than children, adolescents are not simply "mini-adults" nor are they overgrown children, despite immature emotion regulation, inexperience, and dependence on caregivers (Galván 2014). Instead, adolescents are in a distinct developmental stage that facilitates the creativity, rebellion, and progressive thinking that characterizes this period. Puberty jump-starts this process by giving individuals the biological means to procreate. This is followed by a few years of activities and behaviors that facilitate, and in some cases expedite, the move away from caregivers to establish independence—a period often marked by increased conflict with parents, more time spent with peers, frequently engaging in risk-taking behavior, and a greater desire for romantic partners. By late adolescence, independence is achieved. In the United States, for example, marriage often marked this move in years past, according to the U.S. Census

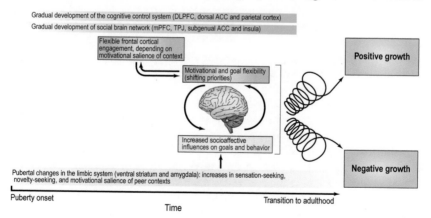

Figure 14.3   Crone and Dahl's model of neurodevelopment illustrating the potential neurobiological mechanisms by which goal-driven behavior is increasingly influenced by social input and cognitive flexibility. Examples of trajectories that would lead to positive growth include adaptive exploration, mature long-term goals, and social competence. Examples of negative growth trajectories include diminished goals (e.g., depression, social withdrawal) and excessive motivation to achieve negative goals (e.g., substance use, excessive risk-taking). After Crone and Dahl (2012) with permission of Macmillan Publishers Ltd.

Bureau. Nowadays, however, marriage plays a lesser role as many individuals simply leave their familial home to live with friends or romantic partners or to attend college.

Humans are not the only species to undergo this characteristic shift in seeking independence. In fact, risk-taking around the time of sexual maturation is also not unique to humans. Like their human counterparts, adolescent rats demonstrate a significant increase in the amount of time spent in social interactions with peers and at play (Varlinskaya and Spear 2008). American psychologist Jerome Bruner proposed that the function of being "immature" is so an organism can engage in experimental play, without serious consequences, and is able to spend considerable time observing the actions of skilled others in conjunction with oversight by and activity with its caregiver (Bruner 1972). Further, he suggested that this type of play helps the species practice and perfect imitative acts, such that "re-interpretive imitation" leads to innovation through extensive exploration of the limits on one's ability to interact with the world. Some have argued that this extended period of immaturity may serve the adaptive purpose of extending the period of neural plasticity (Steinberg 2014).

The Role of Peers in Risk-Taking

Risk-taking is a very social behavior, particularly during adolescence. Teenagers spend an astonishing amount of time with their friends, not to mention the time spent planning or yearning to be with their friends. It is thus not surprising that most risky behavior occurs in the presence of friends. Could it be that being with friends amplifies the excitability of the mesolimbic system to an even greater degree than it already is in puberty? Although this is an interesting question, studying it faces major challenges, given the difficulties of capturing brain activity while teens are with their friends.

One approach, developed by Laurence Steinberg and Jason Chein, involves a clever fMRI experiment in which participants played a risk-taking video game in the presence of their peers while undergoing fMRI (Chein et al. 2011). Three groups of research participants—adolescents (14–18 years of age), college students (aged 19–22), and adults (aged 24–29)—were recruited to the study and each one played the "Stoplight Game." This game is a first-person task wherein participants must advance a car through a series of street intersections to reach a finish line as quickly as possible to receive a monetary reward. The risky component mimics real-life driving circumstances in which each intersection contains a stoplight that turns yellow as the car approaches: participants must decide whether to make a risky choice by running the yellow light or take a nonrisky choice of stepping on the brakes (and thus incurring extra time to get to the finish line). Each participant in the study played the game alone or in the presence of two same-aged, same-sex friends.

College students and adults exhibited the same behavior (i.e., they made the same number of risky and nonrisky choices) regardless of whether there was a peer watching them. Adolescents, however, made significantly more risky choices when a peer was present than when they were alone (Chein et al. 2011). This finding is especially interesting because risky behavior in adolescents did not differ from the other age groups in the alone condition. Interestingly, risky decisions in the presence of peers elicited greater activation in the mesolimbic circuitry (specifically in the ventral striatum) only in the adolescent group. This study provides compelling evidence that in the company of friends, reward sensitivity in adolescents is amplified when confronted with a risky choice. Similar studies using a "virtual" peer found similar results: the presence of a peer yielded worse cognitive control in the adolescent group, but not in young adults or adults (Breiner et al. 2018). This peer effect has also been observed in mice. A sample of mice raised in same-sex triads and tested for alcohol consumption, either as juveniles or as adults, showed that the presence of "peers" increased alcohol consumption among adolescent mice, but not adults (Logue et al. 2014).

The Role of Family in Risk-Taking

Recent work has extended the study of how social relationships influence risk-taking and brain development by examining the role of the family unit on this behavior. The changing nature of family relationships during adolescence can have significant implications for risk-taking and associated health consequences, such as substance use and externalizing problems. Family obligation—the importance of spending time with the family, high family unity, family social support, and interdependence for daily activities (Fuligni et al. 1999)—is a key aspect of family relationships that may have significant consequences for adolescents' health.

Family obligation may reduce risk-taking because it is a meaningful activity that increases adolescents' motivation to control their own impulses and desires for the sake of their family, thus providing adolescents opportunities to practice engaging in self-control. For example, adolescents who value family obligation report greater negative consequences for engaging in risky behavior because risk-taking reflects poorly on their family (German et al. 2009). The negative consequences of risk-taking may be more consequential for these youth, and thus risk-taking becomes comparatively less rewarding. Likewise, adolescents who value family obligation may be more motivated to engage in self-control to avoid risky behaviors. To test this hypothesis, Telzer et al. (2013b) examined whether family obligation related to neural markers of risk-taking. Participants performed the Balloon Analogue Risk Task, a computerized assessment of risk-taking, while undergoing fMRI to derive measures of family obligation values and self-reported risk-taking behavior. Results suggest that adolescents with greater family obligation values show decreased activation in the ventral striatum when receiving monetary rewards and increased dorsolateral PFC activation during behavioral inhibition. Reduced activation in the ventral striatum correlated with less real-life risk-taking behavior and enhanced dorsolateral PFC activation correlated with better decision-making skills. Thus, family obligation may decrease reward sensitivity and enhance cognitive control, thereby reducing risk-taking behaviors (Telzer et al. 2013b).

Conclusions and Unresolved Issues

In humans, the adolescent brain changes significantly until at least the mid-twenties. The next frontier for adolescent neuroscience research is to discover the factors that contribute to this significant period of neurobiological maturation. The influx of gonadal hormones at puberty certainly plays a significant role in refining and, in some cases, reforming the developing brain. However, we need to gain traction on other social, psychological, and physiological factors that contribute to adolescent brain development and dynamic brain coordination. To help anchor future directions in this area, I wish to highlight several promising areas of inquiry.

What is adolescence? As discussed, there are many issues involved in defining adolescence. An increasing number of scholars have argued that age-based boundaries of adolescence limit our understanding of the adolescent experience, the factors that mark the beginning and "end" of adolescence, as well as the policy, law, and education-relevant implications of this research. To progress, it may be useful instead to define adolescence based on neurobiological criteria, psychosocial responsibilities, and/or skill-based capabilities. Defining adolescence via these factors may help disambiguate who is considered an adolescent. To enable this, however, neurobiology needs to coalesce around the parameters of each of these operational definitions. For instance, which brain metric should be used to determine a "mature" versus an "immature" brain? Which skills are necessary for reaching maturation? These questions reveal an interesting set of issues that would benefit from discussions among scholars from interdisciplinary fields.

Social influence: The increasing importance of social relationships in adolescence is clear. What remains unknown, however, are the neurobiological and psychological mechanisms by which social relationships may serve to "sensitize" or trigger change in the adolescent brain. Research may profitably focus on the circumstances under which parental and peer influence may diverge, or the various factors that render adolescents more susceptible to influence from parental or peer sources. For example, parents may exert profound influence on adolescents' choices when values or moral concerns are made salient, whereas peers might be more influential in shaping adolescents' social activities and relationships at school.

The positive attributes of adolescent brain maturation: The majority of early studies on the adolescent brain focused on the negative or problematic attributes of neurobiological "immaturity" during adolescence. Fortunately, scientists have increasingly rectified this perception through empirical research which shows that the ontogenetic changes in the adolescent brain are adaptive for the individual and beneficial for society. Some have even argued that this extended period of immaturity may serve the adaptive purpose of extending the period of plasticity (Steinberg 2014). Like development itself, the science on the adolescent brain is a dynamic process. With every study, methodological advance, and collaboration with nonscientists, knowledge grows. By appreciating that the adolescent brain is sponge thirsty and receptive for new knowledge, rather than problematic, awareness of this significant period of life will continue to grow. New studies have begun to focus on the power of the adolescent brain to learn (e.g., Davidow et al. 2016), to engage in prosocial behavior, and to explore the environment in a healthy way. Greater research into the positive attributes of the brain and its dynamic coordination during this key developmental window is warranted.

Acknowledgments

The author thanks B. J. Casey, Lawrence Steinberg, members of the Galván Lab, and the Executive Committee of the Center on the Developing Adolescent for lively discussions that are represented in this document.

15

Late Adolescence

Critical Transitions into Adulthood

Jennifer N. Gelinas, Sylvain Baillet, Olivier Bertrand,
Adriana Galván, Thorsten Kolling, Jean-Philippe Lachaux,
Ulman Lindenberger, Urs Ribary, Akira Sawa,
and Peter J. Uhlhaas

Abstract

Adolescence is a critical stage of brain development prior to the attainment of a more mature state. The neurobiological underpinnings of this transition have been difficult to characterize, contributing to the challenges in diagnosing, treating, and preventing the neuropsychiatric diseases that commonly emerge during this developmental epoch. This chapter proposes a multidisciplinary approach with a focus on the changing patterns of both physiologic and pathologic brain dynamics across adolescence. The intellectual merit and scientific promises of combining multiple research modalities are discussed: longitudinal studies in humans and animal models are encouraged as are potential for contributions from computational models, including artificial neural systems. Adolescence represents a nonlinear, discrete period of perturbation during which specific brain systems for higher cognitive, emotional, and social functions are highly, and often irreversibly, modified. Identifying the neural processes that underlie these developmental modifications will help facilitate their normal expression during adolescence and ultimately prevent their disruption and onset of neuropsychiatric disease.

Introduction

Most of us remember adolescence as a kind of double negative: no longer allowed to be children, we are not yet capable of being adults.—Julian Barnes

Group photos (top left to bottom right) Jennifer Gelinas, Sylvain Baillet, Akira Sawa, Olivier Bertrand, Urs Ribary, Thorsten Kolling, Jean-Philippe Lachaux, Ulman Lindenberger, Adriana Galván, Sylvain Baillet, Peter Uhlhaas, Patrick Purdon and Adriana Galván, Urs Ribary, Jennifer Gelinas, Akira Sawa, Olivier Bertrand, Thorsten Kolling, Ulman Lindenberger, Jean-Philippe Lachaux, Peter Uhlhaas, Adriana Galván

The transition from immaturity and dependence on caregivers to maturity and autonomy has been experienced throughout human history, and elements of this key developmental phase can be identified in organisms across the evolutionary spectrum. Understanding late adolescence and the subsequent emergence of adulthood, including the neurobiological basis of this transition, is crucial for better diagnosing and treating neuropsychiatric disorders that may arise from disturbances in these brain processes. Here, we propose a research framework for late adolescence based on development and plasticity of brain dynamics. Describing concepts and methods currently and prospectively available to investigate this framework, we address potential ways of translating them into improved diagnostics and therapeutics for the neuropsychiatric disorders prevalent in this developmental period. Specifically, the following questions guided our discussions:

- How might the neurobiological transition from adolescence to adulthood be defined?
- Does brain plasticity end with adolescence?
- What tools and models are needed to help investigate the dynamical processes of the adolescent brain?
- Why and how do specific abnormalities of brain coordination predominantly arise and remit in adolescence?
- Can information about adolescents' neural networks improve diagnosis, treatment, and prevention of adolescent brain disorders?

Toward a Neurobiological and Multidimensional Description of Late Adolescence

The end of adolescence cannot be defined by a discrete event; it gradually emerges from a complex combination of societal and biological influences. From a societal perspective, the age range considered to encompass adolescence varies with cultural and historical circumstances. Currently, in Western cultures and societies, adolescence begins at approximately 11 to 13 years of age and ends in the late teenage years (approximately 18–19 years of age). Early adolescence typically encompasses the period associated with middle school education and includes most pubertal development. Late adolescence refers approximately to the period following the pubertal transition. Significant psychosocial and cognitive changes occur during this time, including increases in orientation toward peers, romantic interests, and identity exploration, as well as more sophisticated cognitive abilities, such as abstract thought, future planning, goal setting, and career exploration.

Adolescence is therefore essentially recognized as a distinct developmental period in which children begin to transition into adults. Typically, this occurs through the adoption of increasingly "adult-like" behaviors (e.g., getting married, moving away from the family, bearing children). Anthropologists note, however, that the extent to which adolescence is acknowledged and the way each society characterizes the transition from childhood to adulthood varies greatly by culture. In some traditional societies, public ceremonies are used to commemorate the transition from child to adult social status. Modern industrialized societies, by contrast, rarely acknowledge adolescence publicly, in part because there are several developmental milestones (at different ages) that are considered critical to the transition from child to adult, including completion of secondary schooling, age of legal status, getting a job, getting married, or becoming a parent. We suggest that neurobiology can be leveraged to articulate a definition of the end of adolescence, as key features of this transition are reflected in measurable brain processes.

Neurobiological evidence supports the hypothesis that adolescence does not exist solely as a linear chronologic connector between childhood and adulthood. Instead, we assert that adolescence is an identifiable period of perturbation with unique hormonal, neurophysiological, and experiential features that combine to provide adaptive advantages as well as vulnerabilities. Studies have examined the developmental modifications in neural circuits central to emotion regulation and reward prediction, as well as phase synchronization of neural oscillations during the transition from adolescence to adulthood. Emotion regulation circuits involve interactions between several subregions of the prefrontal cortex (PFC) and limbic structures that dynamically change through late adolescence. In adolescent participants, there is converging evidence for heightened amygdala activity toward threat-related stimuli that correlates with levels of trait anxiety, while modulatory feedback from the ventromedial PFC is decreased (Hare et al. 2008). Resting-state fMRI data provide support for a developmental switch in this pathway (Gee et al. 2013), suggesting a positive amygdala–PFC connectivity in early childhood that changes to negative functional connectivity during the transition to adolescence. These findings can be directly linked to anatomical changes observed in long-range connections that occur in late adolescence between amygdala and medial PFC (Cunningham et al. 2002). Similar nonlinear changes have been observed in reward sensitivity and prediction-error signaling during adolescence that could account for age-specific elevated risk-taking. Adolescents exhibit greater striatal activation relative to other age groups in response to the following reward scenarios: monetary (Ernst et al. 2005; Galván et al. 2006; Geier et al. 2010), decision making (Jarcho et al. 2012), social (Chein et al. 2011), prediction error (Cohen et al. 2010), and primary reward tasks (Galván and McGlennen 2013). Longitudinal assessments, in which over 200 participants between the ages of 10–25 years were scanned twice, confirmed that the striatum shows peak activation during the adolescent period in response to reward and risk-taking (Braams et al. 2015).

A nonlinear trajectory of brain coordination was also observed for the development of phase synchronization of high-frequency oscillations. Data presented by Uhlhaas (this volume) show that phase synchrony in the beta and gamma band increases until the age of 14 years, followed by a reduction during late adolescence (15–17 years), before synchrony increases sharply again in 18- to 21-year-olds. This nonlinear development of phase synchrony was accompanied by reorganization in the anatomical topography of phase synchrony in the beta band.

The increasing use of functional connectivity techniques to examine the development of networks in the human brain has been useful in identifying important maturational changes that characterize adolescence. For instance, a study comparing network connectivity between children, adolescents, and adults found that connectivity of networks associated with social and emotional functions exhibited the greatest developmental effects, whereas connectivity of networks associated with motor control did not differ between the three groups (Kelly et al. 2009). These findings confirm a long-hypothesized organizational principle of development: the maturation of sensory and motor systems precedes those underlying higher cognition (Chugani et al. 1987). This idea reflects the self-organizing principle of dynamic systems theory in that complex systems, such as the maturing brain, develop through hierarchical, nonlinear processes (Johnson and Shrager 1996).

Several resting-state studies have demonstrated that during the development of large-scale brain systems, functional connectivity shifts from a local to a distributed architecture. For example, intrahemispheric connectivity within local circuits precedes the development of large-scale interhemispheric connectivity (Fransson et al. 2007). Others have found that nodes within the default mode network are sparsely connected in children and strongly functionally connected in adults (Fair et al. 2008). One group collected short (5 min) resting-state scans from typically developing subjects across a range of ages to predict each individual's brain maturity across development (Dosenbach et al. 2010). The best predictive feature of individual brain maturity in this study was the strengthening of and segregation between the adult brain's major functional networks.

Together, these maturational patterns provide support for the notion that brain coordination within large-scale networks during late adolescence shows profound modifications that frequently involve nonlinear trajectories and can be considered as developmental perturbation, and thus may facilitate the emergence of novel principles of large-scale interactions. It should be noted that these observations do not apply to all system-level observations during adolescence. More research is thus required to delineate the functional significance of these changes for understanding of brain coordination during development.

Although the concept of developmental perturbation may help define adolescence, the patterns of these changes across different brain processes observed at multiple scales follow significantly variable trajectories. Several examples serve to illustrate this point:

- *Molecular changes*: dopaminergic projections and neural concentrations of dopamine increase during adolescence, and subsequently decline throughout adulthood.
- *Synaptic changes*: there is a reciprocal relationship between the number of excitatory and inhibitory synapses in the PFC, with inflection points occurring throughout adolescence.
- *Structural measures of neural networks*: gray matter volume decreases monotonically from middle childhood to old age (Douaud et al. 2014), but there is a set of brain regions comprised of lateral PFC, frontal eye field, intraparietal sulcus, superior temporal sulcus, posterior cingulate cortex, and medial temporal lobe that peaks in volume late during adolescence and then shows accelerated degeneration in old age compared with the rest of the brain (e.g., in accordance to the last-in, first-out notion, also termed Ribot's law); white matter tracts increase throughout childhood and adolescence, reaching a plateau between the fourth and sixth decades.
- *Functional measures of neural networks*: the power of postsynaptic potentials and the percentage of low-frequency activity, as measured by electroencephalography (EEG), follow a relatively linear trend throughout adolescence; it is speculated that this monotonic decrease in total power and magnitude of evoked brain responses may reflect a life-span transition from rate coding to temporal coding (Müller et al. 2009), allowing brain processes to involve less but more coordinated activity.
- *Cognitive development*: "fluid" abilities that represent individual differences in the speed and coordination of elementary processing operations show their life-span peak in late adolescence, followed by gradual decline accelerating in old age; "crystallized" abilities, which depend on acquired bodies of knowledge, peak later in age and show a long plateau that extends into old age.

It is unlikely that any one of these processes, or numerous others that can be assayed, individually reflects maturation. Therefore, we propose that late adolescence be defined as a transitional period in relation to a combination of inflection points and ranges of linear trends across multiple structural, molecular, neural network, and cognitive measures (Figure 15.1). The confluence of these measures should lead to a more versatile definition of late adolescence, facilitating translation between data points obtained from different individuals and strengthening correlations between chronological age ranges that represent adolescence and maturity across species.

Given the profound biological and experiential changes that occur during adolescence and the relative expansion of this phase during mammalian evolution, as the brains and bodies of organisms increased in complexity, it is likely that this developmental phase serves an evolutionary purpose. At no other time

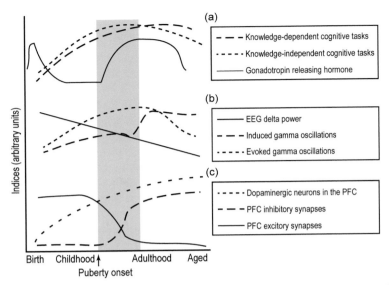

Figure 15.1 Variable trajectories of multidimensional neurobiological measures that can define a window for the transition from adolescence to adulthood: (a) functional measures, (b) network-level measures, and (c) molecular measures. Gray area defines the putative period of late adolescence.

in life is there greater intrinsic motivation to explore the world than during adolescence (Crone and Dahl 2012). Adolescents are in a distinct developmental stage that facilitates all of the creativity, rebellion, and progressive thinking that characterize this period. From the perspective of brain processes, adolescence may represent an experience-expectant window, during which a wide range of novel experiences is actively sought out to broaden an individual's model of the world, and hence improve the accuracy of predictions about future experiences. It is perhaps for this reason that the increased frequency of "surprising" events, or unexpected uncertainty (Yu and Dayan 2005), is more welcomed in adolescence than in any other time in life. As adolescents forage for new experiences, their brains may become more accurate Bayesian predictors (Friston 2010) that are better able, in the long run, to minimize unexpected and potentially harmful responses to actions. It is perhaps this extended period of flexibility and adaptability that has allowed our species to flourish, often at the expense of less adaptable organisms.

Does Plasticity End with Adolescence?

Neural plasticity processes importantly shape development across the life span, and we sought to explore which of these processes are at play during the transition from adolescence to adulthood. Plasticity takes multiple forms.

Structural plasticity includes the formation or elimination of long- and short-range synaptic connections. Synaptic plasticity includes the alteration of receptors, channels, and other synaptic proteins to modify synaptic weights, with the knowledge that long-term synaptic plasticity can subsequently initiate local structural plasticity.

A different form of plasticity, characterized by critical periods, has been identified during early development. These critical periods involve a confluence of neural processes that create a unique epoch during which experience can fundamentally shape neural networks and their functional capabilities, sometimes irreversibly (see also Hensch, this volume). It is unclear, but interesting to consider, whether the concept of a critical period can be extended to adolescence. It is known that specific circuits are fundamentally modified during normal adolescence, including higher association, prefrontal, and limbic regions. There is also evidence to suggest that depriving rodents of social interaction during adolescence can lead to different effects than similar deprivation at earlier or later time points during development, and that certain forms of extinction learning are temporarily attenuated during adolescence (Baker et al. 2016; Lander et al. 2017). We suggest that it could be mechanistically relevant, and potentially clinically significant, to investigate whether adolescence represents the last normative critical period.

In a similar vein, in our discussions it was tentatively proposed that the typical human brain does not change its overall organization after the end of adolescence. Hence, plasticity beyond adolescence is increasingly less likely to involve reorganization of neural circuitry, and more likely to be restricted to structural changes at the local level and modification of synaptic weights. However, it seems likely that more generic mechanisms related to maturation, learning, and senescence cannot be confined to specific age periods, and that some of the mechanisms known to regulate critical period plasticity also are operating during later forms of plasticity (Takesian and Hensch 2013).

At the more local level, available experience clearly shows that structural plasticity continues to be present after adolescence. Pretest–posttest comparisons in adults have revealed increases in gray matter after several months of juggling training, intensive studying for medical exams, foreign language acquisition, spatial navigation training, playing video games, and tracing with the nondominant hand (see Lindenberger, this volume). Similar changes have been observed after two weeks of mirror reading, a few days of signature writing with the nondominant hand, and even after only two sessions of practice in a complex whole-body balancing task. In all of these cases, plasticity is specific to the trained skill and shows a narrow transfer gradient, if any.

Age-graded differences in plastic change deserve to gain center stage and need to be delineated through age-comparative studies (Lövdén et al. 2010). Cognitive development from childhood to adulthood, for example, is accompanied by an increasing dominance of top-down control processes over bottom-up mechanisms. This shifting balance may facilitate some aspects of plastic change and hinder others. On a related note, local plastic change needs to be studied in a global context. For instance, the primary cortices form part of a structured, complex learning architecture. Networks that generate and monitor new behavioral routines and action sequences belong to this architecture and contribute to individual differences in skill acquisition. Higher-order regions like prefrontal and temporal brain areas are likely to signal and keep track of the mismatch between the current range of functioning and experienced demands. We expect reductions in mismatch due to increasing task proficiency to be accompanied by decreasing activations in these areas.

Evidence at ontogenetic and microgenetic timescales supports an *overproduction–pruning model of plasticity* (see also Lindenberger, this volume). The model posits an increase in the number of synapses at the beginning of the plastic episode, followed by experience-dependent selective stabilization of behaviorally relevant connections and the elimination of those connections which prove to be functionally irrelevant. Using two-photon microscopy and optogenetic tools, the overproduction–pruning sequence has been observed in behaving animals with unprecedented precision (Hübener and Bonhoeffer 2014). Plastic changes in the sensory and motor cortices are marked by the rapid formation of new dendritic spines, followed by a slower process of spine elimination, returning the overall number of spines close to pre-intervention levels. The dendritic spines that have been newly formed and retained during a plastic episode show a remarkable degree of structural stability over time, and they may function as the physiological substrate for skill retention and reactivation. This process appears to be specific to the practiced skill, with different skills encoded in different dedicated sets of synapses.

Macroscopically, the overproduction–pruning model leads to the hypothesis that plasticity in the human brain, regardless of the development phase, is accompanied by an initial phase of gray matter volume expansion, followed by a period of volume renormalization. To test this hypothesis in adolescents, Wenger et al. (2017) recently acquired 18 structural MR image volumes over a 7-week period in 15 right-handed young adults who practiced nondominant left-hand writing and drawing. After four weeks of practice, increases in gray matter in both left and right primary motor cortices, relative to a control group, were observed; another three weeks later, these differences were no longer reliable. Time series analyses showed that gray matter in both primary motor cortices expanded during the first four weeks and then partially renormalized, particularly in the right hemisphere, in the presence of continued practice and increasing task proficiency.

Task-related functional activations in cortical areas undergoing plastic reorganization are likely to increase during the initial period of cortical expansion, and decrease in the course of renormalization, when the pruning of new connections has led to sparser coding of task-relevant perception–action links. In fact, one may speculate that the transient increase in metabolic load at the beginning of a plastic episode gives way to a more efficient, metabolically less costly task representation at its completion. These mechanisms are likely present throughout development, but may be more easily invoked at earlier developmental stages. This hypothesis merits further investigation with longitudinal studies of plasticity to similar stimuli across development.

Reinforcement learning from prediction errors, mediated primarily by the striatum and hippocampus, is a plastic process of particular importance during adolescence. Reinforcement learning theory is couched in the notion that we learn by interacting with our environment. Specifically, reinforcement learning is learning how to maximize reward through trial-and-error based actions, which can also include a search for cost minimization (either physical or cognitive/emotional). Learning from the environment occurs via the neural computation of a *prediction-error signal*, which is derived directly from the Rescorla-Wagner model of classical conditioning. The discovery that the prediction-error signal is coded by dopamine neurons points to the central role of the dopamine system in reinforcement learning (Schultz et al. 1997).

Prediction errors occur when outcomes do not match expectations. This mismatch provides new information for the organism, which then learns from this information. A positive prediction error refers to when the outcome is better than expected. For example, if an adolescent expects her weekly allowance of $50 and instead receives $60, she experiences a positive prediction error of +$10. If she instead receives $25, then she experiences a negative prediction error of –$25. One study used a learning task to violate such expectations from participants: the outcomes of the task were unpredictably better or worse than expected. When better than expected, the adolescent group (13–19 years) showed an elevated positive prediction-error signal in the striatum compared to children (8–12 years) and adults (25–30 years) (Cohen et al. 2010). With training, all participants became faster and more accurate at responding to predictable stimuli; only the adolescent group (14–19 years) responded more quickly to stimuli associated with a higher reward value compared with small rewards. In addition, compared with children and adults, the adolescent group exhibited higher ventral striatum responses to higher, unpredicted reward. This suggests that responsiveness to dopaminergic prediction error is higher in adolescents, which might contribute to elevated reward-seeking in this age group.

An alternative notion is that a greater sensitivity to prediction errors in adolescents facilitates learning. Indeed, a study that tested the ability of adolescents and adults to learn simple associations between cues and outcomes found that adolescents outperformed adults (Davidow et al. 2016). This is a remarkable finding because on many other cognitive tasks, adults tend to outperform adolescents. Adolescents showed better memory for positive reinforcement events than for negative reinforcement events, whereas adults' memory did not differentiate between positive and negative events. Congruently, on prediction-error tasks, the brains of adolescent subjects exhibited more activation in the hippocampus than adults, as well as significant functional connectivity between the hippocampus and striatum, which correlated with memory for positive reinforcement events (Davidow et al. 2016).

A related study found that following positive prediction errors, there was stronger connectivity between the striatum and medial frontal cortex in adolescents and young adults (age 13–22 years) than in children (age 8–11 years) (van den Bos et al. 2012). Similar studies have also found that adolescents, compared to adults, are more responsive to unpredictable outcomes in terms of modifying behavior in response to new information (Van Duijvenvoorde et al. 2012). These studies suggest that prediction-error signals help adolescents learn about the environment and, importantly, to adjust their behavior flexibly in response to the dynamic nature of life experiences. We suggest that this flexibility is possible because of the malleability of activation in striatal and frontal networks during adolescence.

Furthermore, this malleability is likely to be affected by changes in societal and cultural norms. We note that a significant portion of the lives of adolescents in many societies is being spent increasingly in virtual or online environments that have developed their own unique contingency sets and social norms. For example, online dating, where potential romantic partners can first interact anonymously, carries reduced risks of damaged self-confidence, potentially allowing expression of a more daring, diverse set of behaviors. Frequent use of text messaging with multiple members of a social group and posting of personal information to the online environment are now prevalent, and offer a very different framework for prediction testing among social peers. We posit that the novel opportunity to receive frequent feedback for a more extended range of behaviors with reduced risk and effort might accelerate the learning process that leads to adulthood. Similarly, socially assistive robots are increasingly being evaluated for use in clinical settings for patients with disorders such as autism or dementia (Rabbitt et al. 2015). We can safely speculate that interactions with robotic agents and machines could supplement, or even substitute, human social interaction at all ages. It is hard to anticipate, however, the nature of new issues (and opportunities) for social interactions that may arise from such societal change. How this may affect or facilitate the trajectory of learned social behaviors in adolescents should undoubtedly become the focus of future research.

Earlier, we raised for consideration the possiblity that adolescence may function as the last normative critical period of human ontogeny, characterized by a shifting balance in the expression of different forms of plasticity (long-range structural, local structural, and synaptic). We then explored reinforcement learning from prediction errors, observing that this form of learning exhibits a key nonlinearity across development, with increased responses to positive deviations from expectation during adolescence compared to life periods both preceding and following it. Hence, we hypothesize that detailed investigation of plasticity in frontal-hippocampal-striatal networks, specifically including changes in the dopaminergic modulation of reward, is likely to be a critical starting point for attempts to provide a mechanistic account of a critical period during adolescence.

Tracking the Dynamics of the Late Adolescent Brain

A major obstacle to a better understanding of the neurobiological changes that occur during the transition from adolescence to adulthood is the relative lack of large-scale, longitudinal data from multiple modalities in both human participants and animal models. Although many challenges exist to efficient gathering of such data, we propose several key considerations and potential solutions to these issues.

One benefit of research targeting this developmental phase is that the breadth of brain processes that occur in parallel allows multiple methodological tools to have potential utility. Indeed, the combination of results from various scales of measurement and methodologies is particularly critical to obtaining a complete picture of adolescence and its trajectory into adulthood. Several of these tools have been effectively used, and hold promise for future studies:

- *Molecular profiling*: postmortem brain histology of adolescent victims of sudden death allows molecular profiling of neural tissue; histochemical analyses of tissue and body fluids can determine neurotransmitter levels; a polygenic risk score for psychiatric disease can be generated based on a peripheral blood sample to address how genetic predisposition affects the molecular landscape of adolescence and its trajectory into adulthood at the individual levels.
- *Electrophysiology*: EEG and magnetoencephalography (MEG) permit noninvasive measurements of brain oscillations, with the possibility of using transcranial magnetic/electric stimulation to modulate these oscillations; intracranial EEG/electrocorticography, though restricted to a small number of patients undergoing neurosurgical procedures, can assay electrophysiological responses at higher spatiotemporal resolution.
- *Structural neuroimaging*: changes in both gray and white matter volumes, as well as white fiber tract density can be determined.
- *Functional and molecular neuroimaging*: positron emission tomography, magnetic resonance spectroscopy, and functional MRI can investigate task- or group-specific brain activation, metabolism, and the presence of specific metabolites.
- *Cognitive/behavioral testing*: various higher perceptual, reward-based, and cognitive tasks can delineate patterns of cognitive function.
- *Epigenetic measures*: changes in methylation status of various genes can be used to explore contribution to disease risk.

Importantly, the contribution of electrophysiological, structural, and functional imaging measures is amplified when combined with behavioral data, thus providing a fine-grained, multifaceted picture of developmental changes in neural function and associated behavior. In light of the multidimensional nature of adolescent changes, we propose that studies involving human participants, animal models, and neural network models can all make contributions to our neurobiological understanding of adolescence.

Human Participant Studies

Properly designed human longitudinal studies are needed to increase our understanding of how neurodevelopment in humans relates to behavioral and psychological change over time as well as to characterize trajectories of development that span childhood, adolescence, and adulthood. Decisions about the spacing and frequency of measurement occasions in many longitudinal studies are often based on the practicalities of human subject research rather than on theoretical considerations about appropriate temporal sampling. This may limit the interpretability of the results obtained. Tools are available to optimize the statistical power of longitudinal designs at detecting effects of interest, such as individual differences in change (Brandmaier et al. 2015). Likewise, continuous time modeling methods yield parameters that generalize across studies that differ in the spacing of measurement occasions (Voelkle 2015).

Since adolescence is also characterized by wide population heterogeneity, studies need to have a large number of participants to be appropriately powered to detect relevant effects. Obtaining behavioral, demographic, or societal data from large (>1000) populations of adolescents can be challenging. One option that has been successful is to initiate collaborations with schools or museums, as they provide consistent access to many adolescents. However, agreement from parents and teachers is necessary to ensure a mutually beneficial interaction for the adolescents and researchers. Another emerging option is the development of dedicated apps on smartphones to test participants on behavioral measures repeatedly throughout the day (Killingsworth and Gilbert 2010). Such an approach can rapidly generate data from thousands of subjects and, with appropriate data quality checks, could represent a viable alternative to conventional large population studies.

Given the cost and time involved in properly conducting these human longitudinal studies, a commitment to data sharing in standardized repositories is crucial. The advancements of "big data" can be effectively employed in this field. For instance, until recently, MEG/EEG was lagging MRI in terms of collecting and curating large data repositories of normal variants and disease phenotypes. Reasons for this delay include the lack of a standard file format for raw data and the large volume occupied by high-density recordings. Fortunately, these bottlenecks are gradually, and at least partially, being overcome by the increasing availability and versatility of software readers for most native data formats. Storage capacity, especially in the cloud, has now become ubiquitous and more affordable. The Human Connectome

Project was first to distribute MEG data on a large scale from a subsample of its cohort, along with extensive multimodal MRI, behavioral, and genetic data. With about 150 data volumes available, the Open MEG Archives is the second-largest repository of resting-state MEG data, and it additionally contains T1-weighted MRI volumes of participants (Niso et al. 2016). The recent Cambridge Centre for Ageing and Neuroscience initiative features data from about 650 healthy participants aged 18–88, combined with multimodal MRI and extensive cognitive testing.

Larger volumes of data also enable new research tools. The present renaissance of artificial intelligence methods is boosted by access to such large data resources and augmented access to high-performance computing. Resorting to big data tools and methods is becoming increasingly strategic in systems and clinical neuroscience, especially with neuroimaging; data analysis pipelines have grown in sophistication, and data volumes have inflated concurrently with the augmented spatial and temporal resolution of instruments. We have already put forward the scientific motivation to combine multiple data types (e.g., genotypes, imaging and behavioral phenotypes, clinical data, tissue samples), which transforms every research participant's record in a big data volume. In parallel, community awareness is now growing toward expanding the curated value and lifetime of data collections in public research. The increasing number of open data-sharing initiatives emphasizes and substantiates stronger educational, economical, ethical, and societal values in science.

For the neuroimaging and electrophysiology community, this represents a vital opportunity to validate methods more thoroughly and to overcome the limitations of small-sample, low-powered, and consequently poorly reproducible studies that are eventually detrimental to the credibility of the field. At the same time, it should be kept in mind that the concepts of statistical power and sampling refer not only to the number of participants, but also to the number of time points sampled from a given individual. High-density, in-depth longitudinal data from a relatively small number of individuals transitioning from childhood into adulthood may carry great heuristic value and inform the design of large-scale studies with larger samples of individuals.

Other options for large-scale data acquisition include use of clinical data and new technologies. Clinical institutions often have databases and large repositories of data from individuals with and without diseases that could be repurposed for research. New technologies (e.g., smartphone-based ecological momentary assessment tools or wearable technologies that permit open-field measurements of EEG, electrodermal responses, and eye-tracking) can record ongoing behavior and experiences of a subject in real time, in their typical everyday environment.

Ensuring the reliability of results in these studies is also important. Study design needs to include both confirmatory and exploratory outcome measures in a single cohort, to allow for replication of previous results and validation of study methodology. Statistical tools for longitudinal studies that efficiently combine confirmatory and exploratory approaches are available (Brandmaier et al. 2015). Adolescent longitudinal studies can suffer particularly from biases and hidden variables related to environmental factors. Thus, additional qualitative or quantitative data related to lifestyle that are relevant to the adolescent should be obtained, including interactions with parents and peers, school performance, risk- and sensation-seeking behaviors, romantic/sexual experiences, and substance use/addiction.

Animal Model Studies

Animal models of adolescence should be used in parallel with human studies as they provide the opportunity to interact actively with neural networks; they also help establish indicators of causality, often impossible to obtain in research involving human subjects for ethical reasons. Determining chronological ages that correspond to adolescence and adulthood across species is, however, challenging, especially when the duration of adolescence is radically different between species. In addition, developmental animal studies require that the animals have normal adolescent experiences, including social experiences with animals of the same and opposite sex. Thus it is important to develop more naturalistic ecological environments for lab animal breeding and housing, both for rodents and nonhuman primates.

For reasons similar to those described previously for human studies, longitudinal study design should be used for experiments in animal models, with efforts to look for inflection points and trends across multiple measures that resemble patterns of human adolescence (see Figure 15.1). Behavioral assessments of adolescent animals can also be challenging, as many human behaviors do not have identifiable corresponding behaviors in animals. As such, there is a suggestion in the field that use of social nonhuman primates, such as marmosets, may provide better assessments of cognition and certain interindividual interactions.

Specific benefits of animal models include improved spatiotemporal resolution for electrophysiological data, with the opportunity to record local field potentials, multiunit activity, and even action potentials from individual neurons across multiple brain regions simultaneously. They also provide improved access to deep and mesial structures, such as the hippocampus, medial PFC, and striatum—brain regions that are thought to undergo major modifications in the adolescent period. Furthermore, it is possible to interact with specific cell types and neural circuits in these animals using a combination of viral vectors, RNAi technologies, inducible mouse knockout/transgenic lines, optogenetics, designer receptors exclusively activated by designer drugs, and responsive neurostimulation. Such methods have already been used in adolescent animals to help establish brain processes that are causal to expression of a specific phenotype (Niwa et al. 2010; Cho et al. 2015).

Neural Network Modeling Studies

Currently, an underexplored method in developmental neuroscience is neural network modeling. Artificial neural networks can now attain or surpass human-level performance in various cognitive tasks (Esteva et al. 2017). By investigating how these networks learn, it may therefore become possible to gain insights into how brains mature. For instance, training deep neural networks with specific characteristics, such as reinforcement learning with a transient heightened sensitivity to rewards (discussed above), may serve as a testing ground for exploring forms of adolescent plasticity.

Relatively simple implementation of machine-learning decoding techniques in imaging or multichannel electrophysiology for multidimensional signal classification show impressive applications, such as in identifying early components of visual object categorization and in tracking the temporal organization of spatial patterns of brain activity or that of a mnemonic template in the context of perceptual decisions (Myers et al. 2015). The fact that these methods are, for now, independent of signal models make them an attractive complement to researchers for rapid evaluation of their data, for example, to assess the presence and spatiotemporal topography of effects between experimental conditions or cohorts. Representational similarity analyses were extended to the joint processing of MEG brain data with the outputs of a deep neural network, respectively obtained from and trained on the same visual categorization task (Cichy et al. 2016). This innovative and multimodal approach may allow neuromimetic models[1] to refine, and even discover, new principles of brain function applicable to developmental stages "as the adolescent machine learns."

We can also anticipate that artificial agents may soon be able to capture subtle combinations of behavioral and peripheral markers from psychiatric patients, without the interpersonal challenges that such patients experience with human interventions. We may also extrapolate that these agents, in the form of robots or augmented-reality applications, may become part of the palette for future treatment interventions in neuropsychiatry. Of course, current robotic systems are still technically in their "infant development phase," with robotic engineers able to implement only infant-level capacities and infant learning into embodied systems. As this technology inevitably progresses and society increasingly embraces virtualized forms of interactions, we should be prepared to incorporate artificial intelligence into our tool kit for evaluating human development.

---

[1] Neuromimetic models are those in which computational models or methods apply underlying concepts of neural processes.

Testable Hypotheses of Abnormal Brain
Coordination in Adolescence

Research into both the physiology and pathology of the brain provide complementary views of neural function. Often, features of clinical disorders can shed light onto the underlying physiologic processes that have been deranged. Similarly, mechanisms of normal brain processes can provide a starting point to understanding how neuropsychiatric diseases arise and how to most effectively treat them. This concept is particularly relevant to late adolescence, which is characterized by both the emergence of several disorders and the remittance of others.

During late adolescence, mental disorders such as schizophrenia and affective disorders emerge, thus raising the question of the underlying biological vulnerability and mechanisms that confer risk for psychopathology. One possibility is that the nonlinear maturational changes in neural systems during this age period provide windows of vulnerability that either (a) provide favorable conditions for the emergence of an already existing developmental vulnerability mediated by an earlier developmental insult and/or genetic risk or (b) lead to an expression of psychopathology due to an interaction with environmental events, such as the changing social landscape and increases in social stress.

Among the possible neural mechanisms that undergo profound changes during late adolescence are brain coordination in emotion regulation networks, reward-mediated predictions, and large-scale phase synchronization. On a phenomenological level, there is a close relation between the changes in these networks and disorders involving disturbances in affect (mood disorders, personality disorders), reward (psychosis and substance abuse), and cognition (schizophrenia and bipolar disorder), which tend to emerge during this period. In line with structural evidence (Douaud et al. 2014), it is conceivable that developmental modifications in these circuits may have a causal role in the emergence of specific domains of psychopathology during the transition from adolescence to adulthood.

Adolescence and early adulthood also see the emergence of several genetic or presumed genetic epilepsies, including autosomal dominant nocturnal frontal lobe epilepsy (ADNFLE), autosomal dominant partial epilepsy with auditory features, and familial mesial temporal lobe epilepsy. These epilepsies are localized by seizure semiology and epileptiform electrophysiological patterns to frontal and temporal cortices, regions that mature later and undergo more profound changes during adolescence. Although the mechanisms contributing to this developmental stage-specific expression of epilepsy are mostly unknown, in some cases where human genetic mutations have been identified, progress is being made. Many patients with ADNFLE have mutations in the genes coding for neural nicotinic acetylcholine receptors. Conditional mouse models that can reversibly express similar mutations have demonstrated that expression of the abnormal receptor must occur in the juvenile state for epilepsy to result; expression solely in adulthood is insufficient to cause the clinical phenotype (Douaud et al. 2014).

Different childhood epileptic syndromes have a high rate of remittance by late adolescence, suggesting that there are physiologic or compensatory developmental processes that can facilitate "normalization" of brain function. Epilepsies that are likely to remit include Panayiotopoulos syndrome, Gastaut syndrome, and benign rolandic epilepsy of childhood. The abnormal networks in these syndromes are localized by seizure semiology and epileptiform electrophysiological patterns to the occipital and sensory/motor opercular cortices, regions that are earlier to mature during development and less affected by structural and functional changes during adolescence. The mechanisms underlying remittance of epileptic disorders are unknown, but merit further investigation. Taken together, these observations lend support to the notion that the normal developmental processes of late adolescence determine the patterns of dysfunction and recovery that can be expressed during this phase. Such a hypothesis would need to be supported by multimodal prospective longitudinal investigations of patients who are at high risk for development of neuropsychiatric disease, actively experiencing symptoms of the disease, and after remittance or effective treatment, if applicable. Given the ability to assay dynamic brain coordination noninvasively using electrophysiological techniques, further consideration of how these methods could be applied to neuropsychiatric disorders is warranted.

We propose that electrophysiology geared to monitor dynamic brain coordination is a key methodology to investigate the onset, evolution, and remittance of neuropsychiatric disorders. There is emerging interest in human electrophysiology to study the typical brain rhythms (e.g., theta, alpha, beta, gamma) as coupled and interdependent, rather than separate, expressions of physiological mechanisms. Measures of cross-frequency interactions, originally demonstrated in rodent electrophysiology, such as phase amplitude coupling, can now be obtained in human noninvasive data (MEG or EEG) (Baillet 2017). For instance, there is growing evidence that in the human resting state, ongoing activity is structured by bursts of gamma to fast gamma activity, whose amplitude is modulated by the phase of slower oscillations in the delta to alpha ranges (Florin and Baillet 2015). The slower delta to alpha rhythms mark the net excitability of cell assemblies consisting of slow and fast inhibitory and excitatory cells (Buzsáki 2006). Holistic theoretical frameworks for the organization of brain rhythms, such as the model of synchronized gating and others, consider brain network formation and communication to be enabled by the phase alignment of these cycles between regions (Fries 2005; Florin and Baillet 2015). This can be facilitated by the mechanism of dynamical relaying via the thalamus or cortical hub regions.

While gamma bursts could contribute to bottom-up signaling, beta bursts could manifest top-down modulations generated by upstream regions and thereby contribute to the implementation of contextual predictive inference of input signals. We can anticipate that the later phases of maturation in the adolescent brain, especially concerning the prefrontal areas and associated white fiber tracts, could be evaluated indirectly by evolving expressions of cross-frequency coupling in healthy development and the early onset of syndromes that affect, directly or indirectly, cell excitability. Such a dynamical scaffold, among others possible, helps formulate testable hypotheses inspired by pre-clinical/developmental animal models, using human scalp signals. In short, a global roadmap for MEG/EEG electrophysiology and imaging to build on these recent and still relatively sparse advances would ideally

- clarify further the physiological principles structuring the local-to-global dynamics of neural oscillations,
- define measures of regional activation and inter-areal communication in brain systems that are driven by these biological principles,
- use these measures to survey the dynamical repertoire of the resting brain, which remains largely uncharted, and
- understand how sensory inputs interact with this repertoire, enabling functional integration and eventually behavior.

Approaching future MEG/EEG research with this plan would open considerable perspectives, for instance, by verifying that an aberrant repertoire of brain dynamics phenotypes is expressed in diseases. This, in turn, would enable a new generation of electrophysiological markers of pathology and eventually new forms of intervention.

Brain Network Approaches to Diagnosis, Treatment, and Prevention of Adolescent Brain Disorders

Neuropsychiatric diseases that emerge in adolescence can have profound and long-lasting adverse consequences for the affected individual and their interactions with society. Identifying reliable biomarkers of these diseases is necessary to facilitate early detection and appropriate treatment to mitigate these effects. Equally important is ensuring that these diagnostics and therapeutics can be disseminated broadly to the community to reach all those at risk. Evidence from epidemiologic studies of patients with schizophrenia and those with epilepsy indicate that delayed treatment often results in increased difficulty with later control of the disease symptoms. For instance, duration of untreated psychosis is a consistent predictor of outcome for early psychosis (Harrigan et al. 2003). The concept of "kindling" is recognized in sy across development, wherein the frequent occurrence of seizures can decrease the threshold for further seizures. It may be of clinical relevance to consider whether a similar concept may apply to psychiatric disorders, especially during adolescence when affected networks are likely undergoing modifications that could make them more plastic to repeated abnormal cognitive or behavioral experiences (e.g., hallucinations, panic attacks, or rapid cycling of mood).

The quest for better diagnostics and therapeutics for disorders that emerge in adolescence is complicated by the fact that, as discussed previously, the neural circuits most likely to yield biomarkers of disease are the same circuits that undergo modification during normal adolescence. Therefore, biomarkers in this developmental phase may not be stable, but instead be modulated by the specific time during adolescence in which they are assayed. Coupled with the fact that it is difficult to determine where any individual is on a developmental trajectory by obtaining data at a single time point, the notion of a biomarker may have to be modified to provide an effective application to adolescence. Furthermore, it has been shown in numerous instances that cognitive abilities and symptoms, and likely network properties that underlie them, follow a lognormal rather than bimodal distribution of occurrence in the population. This idea is supported by the typical requirement for functional impairment as part of diagnostic criteria for psychiatric disorders, acknowledging that unless certain symptoms or traits are pervasive enough to impair the individual in their daily life or interaction with society, they may exist within a spectrum of normality. As such, determining the threshold of abnormality for any given biomarker will likely remain challenging.

Adolescence also poses particular issues for therapeutic approaches. Certain behaviors and cognitive states, such as risk-taking and embracing a contrahedonic state, are normative and serve a purpose during adolescence, despite being maladaptive in other life stages. Therefore, it is crucial to use developmental-specific norms and to avoid attempting to overnormalize behaviors that are likely necessary for proper maturation of experience-dependent circuits. Current treatments for many neuropsychiatric disorders carry side effects that can themselves affect brain and body health. For instance, antipsychotics used for schizophrenia can induce a metabolic syndrome that adversely affects cardiovascular health. Anticonvulsants used for both epilepsy and mood disorders can be associated with cognitive dysfunction, among other systemic effects. Use of these agents during adolescence may pose additional, unrecognized hazards, as has been identified with the risk of suicide associated with use of certain antidepressants in adolescents but not adults, or the decrease in IQ associated with prenatal exposure to the anticonvulsant, valproic acid.

Given these issues, we suggest that neuropsychiatric disorders during adolescence may need to be reframed based on different combinations of symptoms (behavioral phenotypes) that can more directly be attributed to dysfunction of specific networks. For example, the Diagnostic and Statistical Manual of Mental Disorders largely describes mental disorders as cross-sectional clusters of symptoms, prioritizing clinical reliability over biological validity. This approach impairs our ability to link pathogenic mechanisms with the disorders. To address this challenge, the National Institutes of Mental Health proposed the Research Domain Criteria (RDoC) (Insel et al. 2010). RDoC dissects mental disorders according to a matrix of dimensions or phenotypes with presumed well-defined biological etiology, and provides the basis for research to understand disease on the level of genes, molecules, synapses, and ultimately dynamic brain coordination (Casey et al. 2014).

In addition, such an approach would allow investigators to look for biomarkers of specific neurologic or psychiatric symptoms in biologically plausible anatomical networks. Rather than requiring that any particular biomarker be sensitive and specific for one clinical disorder, a combination of biomarkers could be used to define a disorder, potentially with the existence of some biomarkers in isolation being within the normal spectrum. Assessment of treatment response could also then be focused on specific symptoms and changes in features of the associated biomarker, with objective and clinically relevant outcome measures.

Moreover, to understand, diagnose, and most effectively treat the atypical neuropsychiatric brain, it is important for future research to target the various neural network dysfunctions identified. General network markers for neuropsychiatric pathologies have been discovered and described in detail for the adult brain (Ribary et al. 2014). In several neurological and neuropsychiatric populations, earlier findings demonstrated that (a) resting-state peak-power oscillatory frequency was persistently slowed from an alpha to a theta rate, (b) theta and gamma power were persistently increased, and (c) persistent cross-frequency coupling was observed among theta and gamma rhythms (Llinas et al. 1999). This is understood to result from a deafferentation of thalamus (i.e., chronic pain) or an excess of inhibition of thalamic activity (i.e., Parkinson disease). In addition, extensive review into the current human and animal literature provides possible clues for the underlying typical and atypical neurophysiological mechanisms (Ribary et al. 2017). To the best of our knowledge, such studies have not been performed during adolescence, but their findings may provide a possible neurophysiological framework for studying such typical or atypical developmental trajectories in network abnormalities.

Once systems-level biomarkers for network dysfunction are identified, we can begin to think about novel approaches to therapeutics (Figure 15.2). Evidence suggests that cognitive or environmental interventions can potentially retrain dysfunctional neural networks. Computerized cognitive training in patients with schizophrenia appears to ameliorate some symptoms of the disease and normalize associated biomarkers (Subramaniam et al. 2012). Similarly, the ventral hippocampal lesioning model of schizophrenia can be rescued by environmental strategies in rodents. Cognitive behavioral and other therapy methods likely also have at least some basis in retraining neural networks that subserve higher cognitive functions. Identifying the networks that are dysfunctional will allow targeted cognitive interventions that are more likely to be successful. It is likely that such interventions will be insufficient in isolation to treat moderate to severe manifestations of neuropsychiatric disorders. However, for certain cognitive disorders, such as attention deficit

Disorder	Distinct clinical features (dimensions)										
Schizophrenia 1	A	B		D							
Schizophrenia 2	A		C	D							
Bipolar disorder 1				D	E	F					
Bipolar disorder 2						F	G				
Major depressive disorder 1				D		F		H			
Major depressive disorder 2						F			J		
Frontal lobe epilepsy				D						X	Y

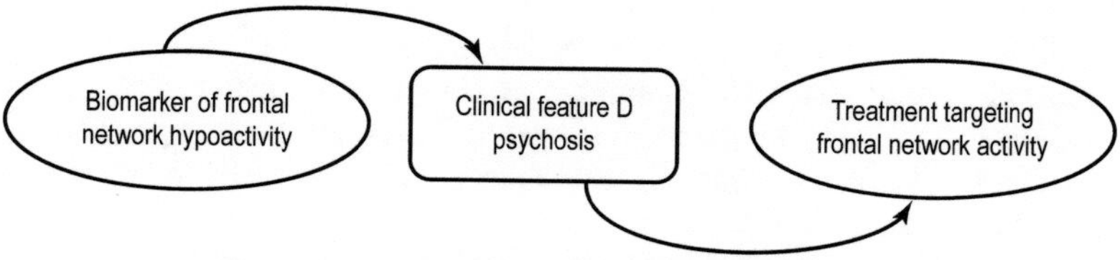

Figure 15.2 Separating neuropsychiatric disorders into dimensions that correspond to functional neural networks may enable insights into novel biomarkers and treatment approaches.

disorder, there is no clear diagnostic line between normal and abnormal function. This uncertainty leaves a "gray area" with a large proportion of adolescents who could certainly benefit from improved cognitive performance, but where the risk-benefit analysis of pharmacological therapy is not positive. One option would be to develop programs that emphasize metacognition (learning to recognize brain mechanisms in one's own behavior) and cognitive strategies (a cognitive "tool kit") to guide adolescents toward more efficient top-down stabilization of their brain dynamics. Of course, any attempt to normalize brain activity according to arbitrary standards and value scales must be avoided; instead, individuals should be provided the means to increase his/her range of options at the behavioral level. The ATOLE program in France, led by Jean-Philippe Lachaux, is an example of such a program, among others currently in operation (e.g., "attentix" in Canada).

If specific networks could be identified as dysfunctional, treatment with direct neural network perturbation could also be employed. Responsive scalp electrical stimulation, deep brain stimulation, and repetitive transcranial magnetic stimulation are currently being used to treat a variety of neurologic diseases, including epilepsy, depression, movement disorders, and stroke. Although mechanisms of benefit and optimal stimulation parameters remain unknown, modest to impressive behavioral and clinical benefit can be observed (Albouy et al. 2017). As our understanding of the pathophysiology of these neural networks progresses, it should be possible to define better, more focused therapeutics. Such approaches could be particularly effective during adolescence, when network plasticity may be more easily invoked.

Outlook

An effective transition from adolescence to adulthood is a fundamental component of a functional society, and better understanding of the neurobiological underpinnings of this change could have far-ranging benefits. We propose that the adolescent brain undergoes numerous nonlinear modifications that set it apart from both the child and adult brain. These changes are characterized by a predilection for specific forms of plasticity that predominantly affect neural networks involved in higher cognitive and emotional processes. There are multiple methods at our disposal to interrogate the adolescent brain, but dedicated and standardized initiatives are required to collect the relevant longitudinal data. A focus on dynamic brain coordination across multiple modalities may allow us to assay more accurately the neurophysiologic processes of typical adolescent development, and identify neural network-level biomarkers and therapeutics for the neuropsychiatric diseases that characteristically emerge during this phase. Perhaps then we will be able to view adolescence as a double positive rather than a double negative: more adventurous, social, and cognitively mature than children, and not yet under the inevitable influences of senescence.

Bibliography

Note: Numbers in square brackets denote the chapter in which an entry is cited.

Abeles, M. 1991. Corticonics. Cambridge: Cambridge Univ. Press. [2]

Abrams, D. A., T. Nicol, S. Zecker, and N. Kraus. 2009. Abnormal Cortical Processing of the Syllable Rate of Speech in Poor Readers. *J. Neurosci.* **29**:7686–7693. [7]

Ackman, J. B., L. Aniksztejn, V. Crépel, et al. 2009. Abnormal Network Activity in a Targeted Genetic Model of Human Double Cortex. *J. Neurosci.* **29**:313–327. [4]

Ahmed, E. I., J. L. Zehr, K. M. Schulz, et al. 2008. Pubertal Hormones Modulate the Addition of New Cells to Sexually Dimorphic Brain Regions. *Nat. Neurosci.* **11**:995–997. [8]

Akbarian, S., and H.-S. Huang. 2006. Molecular and Cellular Mechanisms of Altered GAD1/GAD67 Expression in Schizophrenia and Related Disorders. *Brain Res. Rev.* **52**:293–304. [6]

Akeju, O., K. J. Pavone, J. A. Thum, et al. 2015. Age-Dependency of Sevoflurane-Induced Electroencephalogram Dynamics in Children. *Br. J. Anaesth.* **115(Suppl 1)**:i66–i76. [10]

Akeju, O., K. J. Pavone, M. B. Westover, et al. 2014a. A Comparison of Propofol- and Dexmedetomidine-Induced Electroencephalogram Dynamics Using Spectral and Coherence Analysis. *Anesthesiology* **121**:978–989. [10]

Akeju, O., M. B. Westover, K. J. Pavone, et al. 2014b. Effects of Sevoflurane and Propofol on Frontal Electroencephalogram Power and Coherence. *Anesthesiology* **121**:990–998. [10]

Alavi, M., M. Song, G. L. A. King, et al. 2016. Dscam1 Forms a Complex with Robo1 and the N-Terminal Fragment of Slit to Promote the Growth of Longitudinal Axons. *PLoS Biol.* **14**:e1002560. [3]

Albouy, P., A. Weiss, S. Baillet, and R. J. Zatorre. 2017. Selective Entrainment of Theta Oscillations in the Dorsal Stream Causally Enhances Auditory Working Memory Performance. *Neuron* **94**:193–206. [15]

Alcauter, S., W. Lin, J. K. Smith, et al. 2014. Development of Thalamocortical Connectivity during Infancy and Its Cognitive Correlations. *J. Neurosci.* **34**:9067–9075. [10]

Allen, N. C., S. Bagade, M. B. McQueen, et al. 2008. Systematic Meta-Analyses and Field Synopsis of Genetic Association Studies in Schizophrenia: The SzGene Database. *Nat. Genet.* **40**:827–834. [3]

Allene, C., and R. Cossart. 2010. Early NMDA Receptor-Driven Waves of Activity in the Developing Neocortex: Physiological or Pathological Network Oscillations? *J. Physiol.* **588**:83–91. [4]

Allman, J. 2000. Evolving Brains. Scientific American Library. New York: W. H. Freeman. [4]

Alpár, A., G. Tortoriello, D. Calvigioni, et al. 2014. Endocannabinoids Modulate Cortical Development by Configuring Slit2/Robo1 Signalling. *Nat. Commun.* **5**:4421. [3]

Amedi, A., N. Raz, P. Pianka, R. Malach, and E. Zohary. 2003. Early "Visual" Cortex Activation Correlates with Superior Verbal Memory. *Nat. Neurosci.* **6**:758–766. [11]

An, S., W. Kilb, and H. J. Luhmann. 2014. Sensory-Evoked and Spontaneous Gamma and Spindle Bursts in Neonatal Rat Motor Cortex. *J. Neurosci.* **34**:10870–10883. [4]

Andersen, S. L. 2016. Commentary on the Special Issue on the Adolescent Brain: Adolescence, Trajectories, and the Importance of Prevention. *Neurosci. Biobehav. Rev.* **70**:329–333. [9]

Andersen, S. L., and M. H. Teicher. 2008. Stress, Sensitive Periods and Maturational Events in Adolescent Depression. *Trends Neurosci.* **31**:183–191. [9]

Anderson, S. A., J. D. Classey, F. Conde, J. S. Lund, and D. A. Lewis. 1995. Synchronous Development of Pyramidal Neuron Dendritic Spines and Parvalbumin-Immunoreactive Chandelier Neuron Axon Terminals in Layer III of Monkey Prefrontal Cortex. *Neuroscience* **67**:7–22. [10]

Andreou, C., P. L. Faber, G. Leicht, et al. 2014. Resting-State Connectivity in the Prodromal Phase of Schizophrenia: Insights from EEG Microstates. *Schizophr. Res.* **152**:513–520. [4]

Ango, F., G. Di Cristo, H. Higashiyama, et al. 2004. Ankyrin-Based Subcellular Gradient of Neurofascin, an Immunoglobulin Family Protein, Directs GABAergic Innervation at Purkinje Axon Initial Segment. *Cell* **119**:257–272. [3]

Anitha, A., K. Nakamura, K. Yamada, et al. 2008. Genetic Analyses of Roundabout (ROBO) Axon Guidance Receptors in Autism. *Am. J. Med. Genet. B Neuropsychiatr. Genet.* **147**:1019–1027. [3]

Antonini, A., M. Fagiolini, and M. P. Stryker. 1999. Anatomical Correlates of Functional Plasticity in Mouse Visual Cortex. *J. Neurosci.* **19**:4388–4406. [5]

Antonini, A., and M. P. Stryker. 1993. Rapid Remodeling of Axonal Arbors in the Visual Cortex. *Science* **260**:1819–1821. [5]

Appleton, A. A., D. A. Armstrong, C. Lesseur, et al. 2013. Patterning in Placental 11-B Hydroxysteroid Dehydrogenase Methylation According to Prenatal Socioeconomic Adversity. *PLoS One* **8**:e74691. [6]

Aran, A., N. Rosenfeld, R. Jaron, et al. 2016. Loss of Function of PCDH12 Underlies Recessive Microcephaly Mimicking Intrauterine Infection. *Neurology* **86**:2016–2024. [3]

Arcaro, M. J., P. F. Schade, J. L. Vincent, C. R. Ponce, and M. S. Livingstone. 2017. Seeing Faces Is Necessary for Face-Domain Formation. *Nat. Neurosci.* **20**:1404–1412. [11]

Ariza, J., H. Rogers, E. Hashemi, S. C. Noctor, and V. Martinez-Cerdeno. 2016. The Number of Chandelier and Basket Cells Are Differentially Decreased in Prefrontal Cortex in Autism. *Cereb. Cortex* **28**:411–420. [10]

Artola, A., S. Bröcher, and W. Singer. 1990. Different Voltage-Dependent Thresholds for the Induction of Long-Term Depression and Long-Term Potentiation in Slices of the Rat Visual Cortex. *Nature* **347**:69–72. [2]

Artola, A., and W. Singer. 1987. Long-Term Potentiation and NMDA Receptors in Rat Visual Cortex. *Nature* **330**:649–652. [2]

Ascoli, G. A., L. Alonso-Nanclares, S. A. Anderson, et al. 2008. Petilla Terminology: Nomenclature of Features of GABAergic Interneurons of the Cerebral Cortex. *Nat. Rev. Neurosci.* **9**:557–568. [3]

Assaf, A. A. 2011. Congenital Innervation Dysgenesis Syndrome (CID)/Congenital Cranial Dysinnervation Disorders (CCDDs). *Eye* **25**:1251–1261. [3]

Assali, A., P. Gaspar, and A. Rebsam. 2014. Activity Dependent Mechanisms of Visual Map Formation: From Retinal Waves to Molecular Regulators. *Semin. Cell Dev. Biol.* **35**:136–146. [7]

Athamneh, A. I. M., and D. M. Suter. 2015. Quantifying Mechanical Force in Axonal Growth and Guidance. *Front. Cell. Neurosci.* **9**:359. [3]

Atkinson, J., O. Braddick, M. Nardini, and S. Anker. 2007. Infant Hyperopia: Detection, Distribution, Changes and Correlates: Outcomes from the Cambridge Infant Screening Programs. *Optom. Vis. Sci.* **84**:84–96. [4]

Aton, S. J., C. Broussard, M. Dumoulin, et al. 2013. Visual Experience and Subsequent Sleep Induce Sequential Plastic Changes in Putative Inhibitory and Excitatory Cortical Neurons. *PNAS* **110**:3101–3106. [5]

Attwell, D., and S. B. Laughlin. 2001. An Energy Budget for Signaling in the Grey Matter of the Brain. *J. Cereb. Blood Flow Metab.* **21**:1133–1145. [4]

Atwal, J. K., J. Pinkston-Gosse, J. Syken, et al. 2008. PirB Is a Functional Receptor for Myelin Inhibitors of Axonal Regeneration. *Science* **322**:967–970. [5]

Azevedo, F. A., L. R. Carvalho, L. T. Grinberg, et al. 2009. Equal Numbers of Neuronal and Nonneuronal Cells Make the Human Brain an Isometrically Scaled-up Primate Brain. *J. Comput. Neurosci.* **513**:532–541. [4]

Azmitia, E. C. 2007. Cajal and Brain Plasticity: Insights Relevant to Emerging Concepts of Mind. *Brain Res. Rev.* **55**:395–405. [13]

Badura, A., M. Schonewille, K. Voges, et al. 2013. Climbing Fiber Input Shapes Reciprocity of Purkinje Cell Firing. *Neuron* **78**:700–713. [3]

Baillet, S. 2017. Magnetoencephalography for Brain Electrophysiology and Imaging. *Nat. Neurosci.* **20**:327–339. [15]

Baken, L., I. M. A. van Gruting, E. A. P. Steegers, et al. 2015. Design and Validation of a 3D Virtual Reality Desktop System for Sonographic Length and Volume Measurements in Early Pregnancy Evaluation. *J. Clin. Ultrasound* **43**:164–170. [3]

Baker-Andresen, D., V. S. Ratnu, and T. W. Bredy. 2013. Dynamic DNA Methylation: A Prime Candidate for Genomic Metaplasticity and Behavioral Adaptation. *Trends Neurosci.* **36**:3–13. [6, 7]

Baker, K. D., M. A. Bisby, and R. Richardson. 2016. Impaired Fear Extinction in Adolescent Rodents: Behavioural and Neural Analyses. *Neurosci. Biobehav. Rev.* **70**:59–73. [15]

Baker, R., T. C. Gent, Q. Yang, et al. 2014. Altered Activity in the Central Medial Thalamus Precedes Changes in the Neocortex during Transitions into Both Sleep and Propofol Anesthesia. *J. Neurosci.* **34**:13326–13335. [10]

Bakker, J., S. Honda, N. Harada, and J. Balthazart. 2002. The Aromatase Knock-out Mouse Provides New Evidence That Estradiol Is Required during Development in the Female for the Expression of Sociosexual Behaviors in Adulthood. *J. Neurosci.* **22**:9104–9112. [8]

Balog, J., U. Matthies, L. Naumann, et al. 2014. Social Experience Modulates Ocular Dominance Plasticity Differentially in Adult Male and Female Mice. *NeuroImage* **103**:454–461. [7]

Baltes, P. B., U. Lindenberger, and U. M. Staudinger. 2006. Life Span Theory in Developmental Psychology. In: Handbook of Child Psychology: Theoretical Models of Human Development, ed. W. Damon and R. M. Lerner, pp. 569–664, vol. 1. New York: Wiley. [13]

Bando, Y., T. Hirano, and Y. Tagawa. 2014. Dysfunction of KCNK Potassium Channels Impairs Neuronal Migration in the Developing Mouse Cerebral Cortex. *Cereb. Cortex* **24**:1017–1029. [6]

Banerjee, A., R. V. Rikhye, V. Breton-Provencher, et al. 2016. Jointly Reduced Inhibition and Excitation Underlies Circuit-Wide Changes in Cortical Processing in Rett Syndrome. *PNAS* **113**:E7287–E7729. [7]

Barch, D. M., M. D. Albaugh, S. Avenevoli, et al. 2017. Demographic, Physical and Mental Health Assessments in the Adolescent Brain and Cognitive Development Study: Rationale and Description. *Dev. Cogn. Neurosci.* **pii** :S1878-9293(1817)30068-30063. [4]

Barker, D. 2004. The Developmental Origins of Adult Disease. *J. Am. Coll. Nutr.* **23 (Suppl 6)**:588S–595S. [9]

Baroncelli, L., A. Sale, A. Viegi, et al. 2010. Experience-Dependent Reactivation of Ocular Dominance Plasticity in the Adult Visual Cortex. *Exp. Neurol.* **226**:100–109. [7]

Baroncelli, L., M. Scali, G. Sansevero, et al. 2016. Experience Affects Critical Period Plasticity in the Visual Cortex through an Epigenetic Regulation of Histone Post-Translational Modifications. *J. Neurosci.* **36**:3430–3440. [5]

Bartos, M., I. Vida, and P. Jonas. 2007. Synaptic Mechanisms of Synchronized Gamma Oscillations in Inhibitory Interneuron Networks. *Nat. Rev. Neurosci.* **8**:45–56. [5]

Bashaw, G. J., T. Kidd, D. Murray, T. Pawson, and C. S. Goodman. 2000. Repulsive Axon Guidance: Abelson and Enabled Play Opposing Roles Downstream of the Roundabout Receptor. *Cell* **101**:703–715. [3]

Bates, E., B. O'Connell, and C. Shore. 1987. Language and Communication in Infancy. In: Handbook of Infant Development, ed. J. Osofsky, pp. 149–203. New York: Wiley. [7]

Baum, M. J. 1979. Differentiation of Coital Behavior in Mammals: A Comparative Analysis. *Neurosci. Biobehav. Rev.* **3**:265–284. [8]

Bavamian, S., N. Mellios, J. Lalonde, et al. 2015. Dysregulation of miR-34a Links Neuronal Development to Genetic Risk Factors for Bipolar Disorder. *Mol. Psychiatry* **20**:573–584. [6]

Bavelier, D., D. M. Levi, R. W. Li, Y. Dan, and T. K. Hensch. 2010. Removing Brakes on Adult Brain Plasticity: From Molecular to Behavioral Interventions. *J. Neurosci.* **30**:14964–14971. [5, 13]

Beach, S. R. H., G. H. Brody, A. A. Todorov, T. D. Gunter, and R. A. Philibert. 2011. Methylation at 5HTT Mediates the Impact of Child Sex Abuse on Women's Antisocial Behavior: An Examination of the Iowa Adoptee Sample. *Psychosom. Med.* **73**:83–87. [6]

Beam, C. R., and E. Turkheimer. 2013. Phenotype–Environment Correlations in Longitudinal Twin Models. *Dev. Psychopathol.* **25**:7–16. [13]

Bear, M. F., A. Kleinschmidt, Q. Gu, and W. Singer. 1990. Disruption of Experience-Dependent Synaptic Modifications in Striate Cortex by Infusion of an NMDA Receptor Antagonist. *J. Neurosci.* **10**:909–925. [2]

Bear, M. F., and W. Singer. 1986. Modulation of Visual Cortical Plasticity by Acetylcholine and Noradrenaline. *Nature* **320**:172–176. [2]

Bechdolf, A., A. Ratheesh, S. M. Cotton, et al. 2014. The Predictive Validity of Bipolar At-Risk (Prodromal) Criteria in Help-Seeking Adolescents and Young Adults: A Prospective Study. *Bipolar Disord.* **16**:493–504. [12]

Bedny, M. 2017. Evidence from Blindness for a Cognitively Pluripotent Cortex. *Trends Cogn. Sci.* **21**:637–648. [11]

Bedny, M., A. Pascual-Leone, D. Dodell-Feder, E. Fedorenko, and R. Saxe. 2011. Language Processing in the Occipital Cortex of Congenitally Blind Adults. *PNAS* **108**:4429–4434. [11]

Bedny, M., A. Pascual-Leone, S. Dravida, and R. Saxe. 2012. A Sensitive Period for Language in the Visual Cortex: Distinct Patterns of Plasticity in Congenitally versus Late Blind Adults. *Brain Lang.* **122**:162–170. [11]

Bell, H. C., S. M. Pellis, and B. Kolb. 2010. Juvenile Peer Play Experience and the Development of the Orbitofrontal and Medial Prefrontal Cortex. *Behav. Brain Res.* **207**:7–13. [9, 11]

Bellander, M., L. Bäckman, T. Liu, et al. 2015. Lower Baseline Performance but Greater Plasticity of Working Memory for Carriers of the Val Allele of the COMT Val[158]Met Polymorphism. *Neuropsychology* **29**:247–254. [13]

Belle, M., D. Godefroy, G. Couly, et al. 2017. Tridimensional Visualization and Analysis of Early Human Development. *Cell* **169**:161–173. [3, 4]

Belle, M., D. Godefroy, C. Dominici, et al. 2014. A Simple Method for 3D Analysis of Immunolabeled Axonal Tracts in a Transparent Nervous System. *Cell Rep.* **9**:1191–1201. [4]

Beltz, A. M., and S. A. Berenbaum. 2013. Cognitive Effects of Variations in Pubertal Timing: Is Puberty a Period of Brain Organization for Human Sex-Typed Cognition? *Horm. Behav.* **63**:823–828. [8]

Ben-Ari, Y. 2008. Neuro-Archaeology: Pre-Symptomatic Architecture and Signature of Neurological Disorders. *Trends Neurosci.* **1**:626–636. [4]

———. 2014. The GABA Excitatory/Inhibitory Developmental Sequence: A Personal Journey. *Neuroscience* **279**:187–219. [4]

Ben-Ari, Y., E. Cherubini, R. Corradetti, and J. L. Gaiarsa. 1989. Giant Synaptic Potentials in Immature Rat CA3 Hippocampal Neurones. *J. Physiol.* **416**:303–325. [4]

Ben-Ari, Y., and N. C. Spitzer. 2010. Phenotypic Checkpoints Regulate Neuronal Development. *Trends Neurosci.* **33**:485–492. [4]

Benasich, A. A., N. A. Choudhury, T. Realpe-Bonilla, and C. P. Roesler. 2014. Plasticity in Developing Brain: Active Auditory Exposure Impacts Prelinguistic Acoustic Mapping. *J. Neurosci.* **34**:13349–13363. [7]

Benasich, A. A., and P. Tallal. 2002. Infant Discrimination of Rapid Auditory Cues Predicts Later Language Impairment. *Behav. Brain Res.* **136**:31–49. [7]

Benders, M. J., K. Palmu, C. Menache, et al. 2015. Early Brain Activity Relates to Subsequent Brain Growth in Premature Infants. *Cereb. Cortex* **5**:3014–3024. [4]

Bennett, C., L. J. Voss, J. P. Barnard, and J. W. Sleigh. 2009. Practical Use of the Raw Electroencephalogram Waveform during General Anesthesia: The Art and Science. *Anesth. Analg.* **109**:539–550. [10]

Berardi, N., T. Pizzorusso, and L. Maffei. 2000. Critical Periods during Sensory Development. *Curr. Opin. Neurobiol.* **10**:138–145. [11]

Berking, M., and P. Wupperman. 2012. Emotion Regulation and Mental Health: Recent Findings, Current Challenges, and Future Directions. *Curr. Opin. Psychiatry* **25**:128–134. [12]

Beurdeley, M., J. Spatazza, H. H. Lee, et al. 2012. Otx2 Binding to Perineuronal Nets Persistently Regulates Plasticity in the Mature Visual Cortex. *J. Neurosci.* **32**:9429–9437. [5]

Bi, G. Q., and M. M. Poo. 1998. Synaptic Modifications in Cultured Hippocampal Neurons: Dependence on Spike Timing, Synaptic Strength, and Postsynaptic Cell Type. *J. Neurosci.* **18**:10464–10472. [2]

Bielle, F., P. Marcos-Mondéjar, M. Keita, et al. 2011a. Slit2 Activity in the Migration of Guidepost Neurons Shapes Thalamic Projections during Development and Evolution. *Neuron* **69**:1085–1098. [3]

Bielle, F., P. Marcos-Mondéjar, E. Leyva-Díaz, et al. 2011b. Emergent Growth Cone Responses to Combinations of Slit1 and Netrin 1 in Thalamocortical Axon Topography. *Curr. Biol.* **21**:1748–1755. [3]

Bifari, F., I. Decimo, A. Pino, et al. 2017. Neurogenic Radial Glia-Like Cells in Meninges Migrate and Differentiate into Functionally Integrated Neurons in the Neonatal Cortex. *Cell Stem Cell* **20**:360–373. [3]

Bird, A. 2007. Perceptions of Epigenetics. *Nature* **447**:396–398. [6]

Biswal, B., F. Z. Yetkin, V. M. Haughton, and J. S. Hyde. 1995. Functional Connectivity in the Motor Cortex of Resting Human Brain Using Echo-Planar MRI. *Magn. Reson. Med.* **34**:537–541. [11]

Bitzenhofer, S. H., J. Ahlbeck, A. Wolff, et al. 2017. Layer-Specific Optogenetic Activation of Pyramidal Neurons Causes Beta-Gamma Entrainment of Neonatal Networks. *Nat. Commun.* **8**:14563. [4]

Bitzenhofer, S. H., K. Sieben, K. D. Siebert, M. Spehr, and I. L. Hanganu-Opatz. 2015. Oscillatory Activity in Developing Prefrontal Networks Results from Theta-Gamma-Modulated Synaptic Inputs. *Cell Rep.* **11**:486–497. [4]

Blakemore, C., L. J. Garey, and F. Vital-Durand. 1978. The Physiological Effects of Monocular Deprivation and Their Reversal in the Monkey's Visual Cortex. *J. Physiol. (Lond)* **283**:223–262. [5]

Blakemore, S.-J. 2008. The Social Brain in Adolescence. *Nat. Rev. Neurosci.* **9**:267–277. [9]

Blakemore, S.-J. and K. Mills. 2014. Is Adolescence a Sensitive Period for Sociocultural Processing? *Annu. Rev. Psychol.* **65**:187–207. [14]

Bliss, T. V. P., and T. Lomo. 1973. Long-Lasting Potentiation of Synaptic Transmission in the Dentate Area of the Anaesthetized Rabbit Following Stimulation of the Perforant Path. *J. Physiol.* **232**:331–356. [2]

Blockus, H., and A. Chédotal. 2012. Dystroglycan Adds More Sugars to the Midline Cocktail. *Neuron* **76**:864–867. [3]

———. 2015. Disorders of Axon Guidance. In: The Genetics of Neurodevelopmental Disorders, ed. K. J. Mitchell, pp. 155–195. Hoboken: Wiley. [3]

Blumberg, M. S., H. G. Marques, and F. Iida. 2013. Twitching in Sensorimotor Development from Sleeping Rats to Robots. *Curr. Biol.* **23**:R532–R537. [4]

Bochner, D. N., R. W. Sapp, J. D. Adelson, et al. 2014. Blocking PirB up-Regulates Spines and Functional Synapses to Unlock Visual Cortical Plasticity and Facilitate Recovery from Amblyopia. *Sci. Transl. Med.* **6**:258ra140. [5]

Bolhuis, J. J., and M. Gar. 2006. Neuronal Mechanisms of Bird Song Memory. *Nat. Rev. Neurosci.* **7**:347–357. [7]

Bonhoeffer, F., and A. Gierer. 1984. How Do Retinal Axons Find Their Targets on the Tectum? *Trends Neurosci.* **7**:378–381. [2]

Börgers, C., S. Epstein, and N. J. Kopell. 2005. Background Gamma Rhythmicity and Attention in Cortical Local Circuits: A Computational Study. *PNAS* **102**:7002–7007. [10]

Börgers, C., and N. J. Kopell. 2008. Gamma Oscillations and Stimulus Selection. *Neural Comput.* **20**:383–414. [2]

Borghol, N., M. Suderman, W. McArdle, et al. 2012. Associations with Early-Life Socio-Economic Position in Adult DNA Methylation. *Int. J. Epidemiol.* **41**:62–74. [6]

Bornstein, M. H., and M. D. Sigman. 1986. Continuity in Mental Development from Infancy. *Child Dev.* **57**:251–274. [7]

Borodinsky, L. N., C. M. Root, J. A. Cronin, et al. 2004. Activity-Dependent Homeostatic Specification of Transmitter Expression in Embryonic Neurons. *Nature* **429**:523–530. [4]

Borrell, V., and O. Marín. 2006. Meninges Control Tangential Migration of Hem-Derived Cajal-Retzius Cells via CXCL12/CXCR4 Signaling. *Nat. Neurosci.* **9**:1284–1293. [3]

Bouvier, J., M. Thoby-Brisson, N. Renier, et al. 2010. Hindbrain Interneurons and Axon Guidance Signaling Critical for Breathing. *Nat. Neurosci.* **13**:1066–1074. [4]

Boyce, W. T., and M. S. Kobor. 2015. Development and the Epigenome: The Synapse of Gene-Environment Interplay. *Dev. Sci.* **18**:1–23. [6]

Braams, B. R., A. C. van Duijvenvoorde, J. S. Peper, and E. A. Crone. 2015. Longitudinal Changes in Adolescent Risk-Taking: A Comprehensive Study of Neural Responses to Rewards, Pubertal Development, and Risk-Taking Behavior. *J. Neurosci.* **35**:7226–7238. [15]

Brainard, M. S., and A. J. Doupe. 2002. What Songbirds Teach Us About Learning. *Nature* **417**:351–358. [7]

Brandmaier, A. M., T. von Oertzen, P. Ghisletta, C. Hertzog, and U. Lindenberger. 2015. Lifespan: A Tool for the Computer-Aided Design of Longitudinal Studies. *Front. Psychol.* **6**:272. [15]

Bredy, T. W., Q. Lin, W. Wei, D. Baker-Andresen, and J. S. Mattick. 2011. MicroRNA Regulation of Neural Plasticity and Memory. *Neurobiol. Learn. Mem.* **96**:89–94. [6]

Brehmer, Y., S.-C. Li, V. Müller, T. von Oertzen, and U. Lindenberger. 2007. Memory Plasticity across the Life Span: Uncovering Children's Latent Potential. *Dev. Psychol.* **43**:465–478. [13]

Brehmer, Y., S.-C. Li, B. Straube, et al. 2008. Comparing Memory Skill Maintenance across the Lifespan: Preservation in Adults, Increase in Children. *Psychol. Aging* **23**:227–238. [13]

Breiner, K., A. Li, A. O. Cohen, et al. 2018. Combined Effects of Peer Presence, Social Cues and Rewards on Cognitive Control in Adolescents. *Dev. Psychobiol.* doi: 10.1002/dev.21599. [14]

Brenhouse, H., and S. Andersen. 2008. Delayed Extinction and Stronger Reinstatement of Cocaine Conditioned Place Preference in Adolescent Rats, Compared to Adults. *Behav. Neurosci.* **122**:460–465. [14]

Brenhouse, H. C., and J. M. Schwarz. 2016. Immunoadolescence: Neuroimmune Development and Adolescent Behavior. *Neurosci. Biobehav. Rev.* **70**:288–299. [9]

Brennand, K. J., A. Simone, J. Jou, et al. 2011. Modelling Schizophrenia Using Human Induced Pluripotent Stem Cells. *Nature* **473**:221. [3]

Brickley, S. G., M. I. Aller, C. Sandu, et al. 2007. TASK-3 Two-Pore Domain Potassium Channels Enable Sustained High-Frequency Firing in Cerebellar Granule Neurons. *J. Neurosci.* **27**:9329–9340. [6]

Bridi, M. S., and T. Abel. 2013. The NR4A Orphan Nuclear Receptors Mediate Transcription-Dependent Hippocampal Synaptic Plasticity. *Neurobiol. Learn. Mem.* **105**:151–158. [6]

Bröcher, S., A. Artola, and W. Singer. 1992a. Agonists of Cholinergic and Noradrenergic Receptors Facilitate Synergistically the Induction of Long-Term Potentiation in Slices of Rat Visual Cortex. *Brain Res. Cogn. Brain Res.* **573**:27–36. [2]

———. 1992b. Intracellular Injection of Ca^{++} Chelators Blocks Induction of Long-Term Depression in Rat Visual Cortex. *PNAS* **89**:123–127. [2]

Brock, O., M. J. Baum, and J. Bakker. 2011. The Development of Female Sexual Behavior Requires Prepubertal Estradiol. *J. Neurosci.* **31**:5574–5578. [8]

Brockmann, M. D., B. Pöschel, N. Cichon, and I. L. Hanganu-Opatz. 2011. Coupled Oscillations Mediate Directed Interactions between Prefrontal Cortex and Hippocampus of the Neonatal Rat. *Neuron* **71**:332–347. [4]

Brookes, M. J., M. Woolrich, H. Luckhoo, et al. 2011. Investigating the Electrophysiological Basis of Resting State Networks Using Magnetoencephalography. *PNAS* **108**:16783–16788. [11]

Brown, B. B., N. Mounts, S. D. Lamborn, and L. Steinberg. 1993. Parenting Practices and Peer Group Affiliation in Adolescence. *Child Dev.* **642**:467–482. [14]

Brown, E. N., R. Lydic, and N. D. Schiff. 2010. General Anesthesia, Sleep, and Coma. *New Engl. J. Med.* **363**:2638–2650. [10]

Brown, E. N., P. L. Purdon, and C. J. Van Dort. 2011. General Anesthesia and Altered States of Arousal: A Systems Neuroscience Analysis. *Annu. Rev. Neurosci.* **34**:601–628. [10]

Brückner, G., D. Hausen, W. Härtig, et al. 1999. Cortical Areas Abundant in Extracellular Matrix Chondroitin Sulphate Proteoglycans Are Less Affected by Cytoskeletal Changes in Alzheimer's Disease. *Neuroscience* **92**:791–805. [7]

Bruner, J. 1972. Nature and Uses of Immaturity. *Am. Psychol.* **27**:687–708. [14]

Brunet, I., E. Gordon, J. Han, et al. 2014. Netrin-1 Controls Sympathetic Arterial Innervation. *J. Clin. Investig.* **124**:3230–3240. [3]

Brunet, I., C. Weinl, M. Piper, et al. 2005. The Transcription Factor Engrailed-2 Guides Retinal Axons. *Nature* **438**:94–98. [3]

Budday, S., C. Raybaud, and E. Kuhl. 2014. A Mechanical Model Predicts Morphological Abnormalities in the Developing Human Brain. *Sci. Rep.* **4**:5644. [3]

Buisseret, P., E. Gary-Bobo, and M. Imbert. 1978. Ocular Motility and Recovery of Orientational Properties of Visual Cortical Neurones in Dark-Reared Kittens. *Nature* **272**:816–817. [2]

Buisseret, P., and W. Singer. 1983. Proprioceptive Signals from Extraocular Muscles Gate Experience Dependent Modifications of Receptive Fields in the Kitten Visual Cortex. *Exp. Brain Res.* **51**:443–450. [2]

Buracas, G., A. Zador, M. Deweese, and T. Albright. 1998. Efficient Discrimination of Temporal Patterns by Motion-Sensitive Neurons in Primate Visual Cortex. *Neuron* **20**:959–969. [2]

Burleson, C. A., R. W. Pedersen, S. Seddighi, et al. 2016. Social Play in Juvenile Hamsters Alters Dendritic Morphology in the Medial Prefrontal Cortex and Attenuates Effects of Social Stress in Adulthood. *Behav. Neurosci.* **130**:437–447. [9, 11]

Butler, S. J., and J. Dodd. 2003. A Role for Bmp Heterodimers in Roof Plate-Mediated Repulsion of Commissural Axons. *Neuron* **38**:389–401. [3]

Buzsáki, G. 2006. Rhythms of the Brain. Oxford: Oxford Univ. Press. [2, 7, 15]

Buzsáki, G., C. A. Anastassiou, and C. Koch. 2012. The Origin of Extracellular Fields and Currents: EEG, ECoG, LFP and Spikes. *Nat. Rev. Neurosci.* **13**:407–420. [10]

Buzsáki, G., K. Kaila, and M. E. Raichle. 2007. Inhibition and Brain Work. *Neuron* **56**:771–783. [4]

Buzsáki, G., N. Logothetis, and W. Singer. 2013. Scaling Brain Size, Keeping Timing: Evolutionary Preservation of Brain Rhythms. *Neuron* **80**:751–764. [2]

Buzsáki, G., and K. Mizuseki. 2014. The Log-Dynamic Brain: How Skewed Distributions Affect Network Operations. *Nat. Rev. Neurosci.* **15**:264–278. [7]

Buzsáki, G., and E. W. Schomburg. 2015. What Does Gamma Coherence Tell Us About Inter-Regional Neural Communication? *Nat. Neurosci.* **18**:484–489. [7]

Byrne, E. M., T. Carrillo-Roa, A. K. Henders, et al. 2013. Monozygotic Twins Affected with Major Depressive Disorder Have Greater Variance in Methylation Than Their Unaffected Co-Twin. *Transl. Psychiatry* **3**:e269. [6]

Bystron, I., C. Blakemore, and P. Rakic. 2008. Development of the Human Cerebral Cortex: Boulder Committee Revisited. *Nat. Rev. Neurosci.* **9**:110–122. [3]

Bystron, I., P. Rakic, Z. Molnár, and C. Blakemore. 2006. The First Neurons of the Human Cerebral Cortex. *Nat. Neurosci.* **9**:880–886. [3]

Cao-Lei, L., D. P. Laplante, and S. King. 2016. Prenatal Maternal Stress and Epigenetics: Review of the Human Research. *Curr. Mol. Biol. Rep.* **2**:16–25. [6]

Cao-Lei, L., R. Massart, M. J. Suderman, et al. 2014. DNA Methylation Signatures Triggered by Prenatal Maternal Stress Exposure to a Natural Disaster: Project Ice Storm. *PLoS One* **9**:e107653. [6]

Cardin, J. A., M. Carlen, K. Meletis, et al. 2009. Driving Fast-Spiking Cells Induces Gamma Rhythm and Controls Sensory Responses. *Nature* **459**:663–667. [10]

Cariboni, A., W. D. Andrews, F. Memi, et al. 2012. Slit2 and Robo3 Modulate the Migration of Gnrh-Secreting Neurons. *Dev. Change* **139**:3326–3331. [3]

Carskadon, M. 2011. Sleep in Adolescents: The Perfect Storm. *Pediat. Clin. North Am.* **58**:637–647. [14]

Carulli, D., T. Pizzorusso, J. C. Kwok, et al. 2010. Animals Lacking Link Protein Have Attenuated Perineuronal Nets and Persistent Plasticity. *Brain* **133**:2331–2347. [5]

Casey, B. J. 2013. The Teenage Brain: An Overview. *Curr. Dir. Psychol.* **22**:80–81. [9]

Casey, B. J., A. Galván, and L. H. Somerville. 2015. Beyond Simple Models of Adolescence to an Integrated Circuit-Based Account: A Commentary. *Dev. Cogn. Neurosci.* **17**:128–130. [9]

Casey, B. J., S. Getz, and A. Galván. 2008. The Adolescent Brain. *Dev. Rev.* **28**:62–77. [14]

Casey, B. J., M. E. Oliveri, and T. Insel. 2014. A Neurodevelopmental Perspective on the Research Domain Criteria (RDoC) Framework. *Biol. Psychiatry* **76**:350–353. [15]

Casoni, F., S. A. Malone, M. Belle, et al. 2016. Development of the Neurons Controlling Fertility in Humans: New Insights from 3D Imaging and Transparent Fetal Brains. *Development* **143**:3969–3981. [3]

Caspi, A., T. Vishne, A. Reichenberg, et al. 2009. Refractive Errors and Schizophrenia. *Schizophr. Res.* **107**:238–241. [5]

Castellani, V., A. Chédotal, M. Schachner, C. Faivre-Sarrailh, and G. Rougon. 2000. Analysis of the L1-Deficient Mouse Phenotype Reveals Cross-Talk between Sema3a and L1 Signaling Pathways in Axonal Guidance. *Neuron* **27**:237–249. [3]

Chai, G., L. Zhou, M. Manto, et al. 2014. Celsr3 Is Required in Motor Neurons to Steer Their Axons in the Hindlimb. *Nat. Neurosci.* **17**:1171–1179. [3]

Chanen, A. M., L. K. McCutcheon, M. Jovev, H. J. Jackson, and P. D. McGorry. 2007. Prevention and Early Intervention for Borderline Personality Disorder. *Med. J. Aust.* **187**:S18–21. [12]

Changeux, J.-P., and S. Dehaene. 1989. Neuronal Models of Cognitive Functions. *Cognition* **33**:63–109. [13]

Charron, F., E. Stein, J. Jeong, A. P. McMahon, and M. Tessier-Lavigne. 2003. The Morphogen Sonic Hedgehog Is an Axonal Chemoattractant That Collaborates with Netrin-1 in Midline Axon Guidance. *Cell* **113**:11–23. [3]

Chauvet, S., S. Cohen, Y. Yoshida, et al. 2007. Gating of Sema3E/PlexinD1 Signaling by Neuropilin-1 Switches Axonal Repulsion to Attraction during Brain Development. *Neuron* **56**:807–822. [3]

Chédotal, A. 2011. Further Tales of the Midline. *Curr. Opin. Neurobiol.* **21**:68–75. [3]

———. 2014. Development and Plasticity of Commissural Circuits: From Locomotion to Brain Repair. *Trends Neurosci.* **37**:551–562. [3]

Chédotal, A., and L. J. Richards. 2010. Wiring the Brain: The Biology of Neuronal Guidance. *Cold Spring Harb. Pers. Biol.* **2**:1–17. [3]

Chein, J., D. Albert, L. O'Brien, K. Uckert, and L. Steinberg. 2011. Peers Increase Adolescent Risk Taking by Enhancing Activity in the Brain's Reward Circuitry. *Dev. Sci.* **14**:F1–10. [14, 15]

Chein, J. M., and W. Schneider. 2012. The Brain's Learning and Control Architecture. *Curr. Dir. Psychol.* **21**:78–84. [13]

Chen, L., H. Pan, T. A. Tuan, et al. 2015. Brain-Derived Neurotrophic Factor (BDNF) Val66Met Polymorphism Influences the Association of the Methylome with Maternal Anxiety and Neonatal Brain Volumes. *Dev. Psychopathol.* **27**:137–150. [6]

Chen, Z., B. B. Gore, H. Long, L. Ma, and M. Tessier-Lavigne. 2008. Alternative Splicing of the Robo3 Axon Guidance Receptor Governs the Midline Switch from Attraction to Repulsion. *Neuron* **58**:325–332. [3]

Cheour, M., R. Ceponiene, A. Lehtokoski, et al. 1998. Development of Language-Specific Phoneme Representations in the Infant Brain. *Nat. Neurosci.* **1**:351–353. [7]

Ching, S., A. Cimenser, P. L. Purdon, E. N. Brown, and N. J. Kopell. 2010. Thalamocortical Model for a Propofol-Induced Alpha-Rhythm Associated with Loss of Consciousness. *PNAS* **107**:22665–22670. [10]

Cho, K.-H., H.-J. Jang, Y.-H. Jo, W. Singer, and D.-J. Rhie. 2012. Cholinergic Induction of Input-Specific Late-Phase LTP via Localized Ca^{2+} Release in the Visual Cortex. *J. Neurosci.* **32**:4520–4530. [2]

Cho, K. K., R. Hoch, A. T. Lee, et al. 2015. Gamma Rhythms Link Prefrontal Interneuron Dysfunction with Cognitive Inflexibility in Dlx5/6(+/–) Mice. *Neuron* **85**:1332–1343. [10, 15]

Cho, R. Y., C. P. Walker, N. R. Polizzotto, et al. 2013. Development of Sensory Gamma Oscillations and Cross-Frequency Coupling from Childhood to Early Adulthood. *Cereb. Cortex* **25**:1509–1518. [12]

Chugani, H. T. 1993. Positron Emission Tomography Scanning: Applications in Newborns. *Clin. Perinatol.* **20**:395–409. [7]

Chugani, H. T., M. E. Phelps, and J. C. Mazziotta. 1987. Positron Emission Tomography Study of Human Brain Functional Development. *Ann. Neurol.* **22**:487–497. [15]

Chung, S.-H., H. Marzban, K. Aldinger, et al. 2011. Zac1 Plays a Key Role in the Development of Specific Neuronal Subsets in the Mouse Cerebellum. *Neural Dev.* **6**:25. [6]

Ciccarelli, A., and M. Giustetto. 2014. Role of ERK Signaling in Activity-Dependent Modifications of Histone Proteins. *Neuropharmacology* **80**:34–44. [6]

Cichon, N. B., M. Denker, S. Grün, and I. L. Hanganu-Opatz. 2014. Unsupervised Classification of Neocortical Activity Patterns in Neonatal and Pre-Juvenile Rodents. *Front. Neur. Circuits* **8**:50. [4]

Cichy, R. M., A. Khosla, D. Pantazis, A. Torralba, and A. Oliva. 2016. Comparison of Deep Neural Networks to Spatio-Temporal Cortical Dynamics of Human Visual Object Recognition Reveals Hierarchical Correspondence. *Sci. Rep.* **6**:27755. [15]

Clarke, A. R., R. J. Barry, R. McCarthy, and M. Selikowitz. 2001. Age and Sex Effects in the EEG: Development of the Normal Child. *Clin. Neurophysiol.* **112**:806–814. [7]

Cohen, J. R., R. F. Asarnow, F. W. Sabb, et al. 2010. A Unique Adolescent Response to Reward Prediction Errors. *Nat. Neurosci.* **13**:669–671. [15]

Cohen, L. G., P. Celnik, A. Pascual-Leone, et al. 1997. Functional Relevance of Cross-Modal Plasticity in Blind Humans. *Nature* **389**:180–183. [11]

Colak, D., S.-J. Ji, B. T. Porse, and S. R. Jaffrey. 2013. Regulation of Axon Guidance by Compartmentalized Nonsense-Mediated mRNA Decay. *Cell* **153**:1252–1265. [3]

Coleman, J. E., M. Nahmani, J. P. Gavornik, et al. 2010. Rapid Structural Remodeling of Synapses Parallels Experience-Dependent Functional Plasticity in Mouse Primary Visual Cortex. *J. Neurosci.* **30**:9670–9682. [5]

Collignon, O., G. Dormal, G. Albouy, et al. 2013. Impact of Blindness Onset on the Functional Organization and the Connectivity of the Occipital Cortex. *Brain* **136**:2769–2783. [11]

Collingridge, G., and W. Singer. 1990. Excitatory Amino Acid Receptors and Synaptic Plasticity. *Trends Pharmacol. Sci.* **11**:290–296. [2]

Collins, W. A. 2003. More Than Myth: The Developmental Significance of Romantic Relationships during Adolescence. *J. Res. Adolesc.* **1**:1–24. [14]

Colonnese, M. T., A. Kaminska, M. Minlebaev, et al. 2010. A Conserved Switch in Sensory Processing Prepares Developing Neocortex for Vision. *Neuron* **67**:480–484. [4]

Condé, F., J. S. Lund, and D. A. Lewis. 1996. The Hierarchical Development of Monkey Visual Cortical Regions as Revealed by the Maturation of Parvalbumin-Immunoreactive Neurons. *Brain Res. Dev. Brain Res.* **96**:261–276. [7]

Cong, L., F. A. Ran, D. Cox, et al. 2013. Multiplex Genome Engineering Using CRISPR/Cas Systems. *Science* **339**:819–823. [3]

Conway, C. D., K. M. Howe, N. K. Nettleton, et al. 2011. Heparan Sulfate Sugar Modifications Mediate the Functions of Slits and Other Factors Needed for Mouse Forebrain Commissure Development. *J. Neurosci.* **31**:1955–1970. [3]

Cook, C., N. D. Goodman, and L. E. Schulz. 2011. Where Science Starts: Spontaneous Experiments in Preschoolers Exploratory Play. *Cognition* **120**:341–349. [11]

Cooper, J. B., R. S. Newbower, and R. J. Kitz. 1984. An Analysis of Major Errors and Equipment Failures in Anesthesia Management: Considerations for Prevention and Detection. *Anesthesiology* **60**:34–42. [10]

Cooper, L. N., and M. F. Bear. 2012. The Bcm Theory of Synapse Modification at 30: Interaction of Theory with Experiment. *Nat. Rev. Neurosci.* **13**:798–810. [5]

Cornelissen, L., S. E. Kim, P. L. Purdon, E. N. Brown, and C. B. Berde. 2015. Age-Dependent Electroencephalogram (EEG) Patterns during Sevoflurane General Anesthesia in Infants. *eLife* **4**:e06513. [10]

Crair, M. C., D. C. Gillespie, and M. P. Stryker. 1998. The Role of Visual Experience in the Development of Columns in Cat Visual Cortex. *Science* **279**:566–570. [7]

Crone, E. A., and R. E. Dahl. 2012. Understanding Adolescence as a Period of Social-Affective Engagement and Goal Flexibility. *Nat. Rev. Neurosci.* **13**:636–650. [14, 15]

Cruceanu, C., M. Alda, C. Nagy, et al. 2013. H3K4 Tri-Methylation in Synapsin Genes Leads to Different Expression Patterns in Bipolar Disorder and Major Depression. *Int. J. Neuropsychopharm.* **16**:289–299. [6]

Cruz, D. A., S. M. Eggan, and D. A. Lewis. 2003. Postnatal Development of Pre- and Postsynaptic GABA Markers at Chandelier Cell Connections with Pyramidal Neurons in Monkey Prefrontal Cortex. *J. Comp. Neurol.* **465**:385–400. [10]

Cunningham, M. G., S. Bhattacharyya, and F. M. Benes. 2002. Amygdalo-Cortical Sprouting Continues into Early Adulthood: Implications for the Development of Normal and Abnormal Function during Adolescence. *J. Comp. Neurol.* **453**:116–130. [12, 15]

Dahlin, E., A. Stigsdotter Neely, A. Larsson, L. Bäckman, and L. Nyberg. 2008. Transfer of Learning after Updating Training Mediated by the Striatum. *Science* **320**:1510–1512. [13]

Dascenco, D., M.-L. Erfurth, A. Izadifar, et al. 2015. Slit and Receptor Tyrosine Phosphatase 69D Confer Spatial Specificity to Axon Branching via Dscam1. *Cell* **162**:1140–1154. [3]

Davidow, J. Y., K. Foerde, A. Galván, and D. Shohamy. 2016. An Upside to Reward Sensitivity: The Hippocampus Supports Enhanced Reinforcement Learning in Adolescence. *Neuron* **92**:93–99. [14, 15]

Davidson, A. J., G. H. Huang, C. S. Rebmann, and C. Ellery. 2005. Performance of Entropy and Bispectral Index as Measures of Anaesthesia Effect in Children of Different Ages. *Br. J. Anaesth.* **95**:674–679. [10]

Davis, E. C., J. E. Shryne, and R. A. Gorski. 1996. Structural Sexual Dimorphisms in the Anteroventral Periventricular Nucleus of the Rat Hypothalamus Are Sensitive to Gonadal Steroids Perinatally, but Develop Peripubertally. *Neuroendocrinology* **63**:142–148. [8]

Daw, N. W. 1998. Critical Periods and Amblyopia. *Arch Ophthalmol* **116**:502–505. [5]
———. 2013. Visual Development. New York: Springer Science and Business Media. [5]
Daw, N. W., and H. J. Wyatt. 1976. Kittens Reared in a Unidirectional Environment: Evidence for a Critical Period. *J. Physiol.* **257**:155–170. [5]
Day, J. J., and J. D. Sweatt. 2011. Cognitive Neuroepigenetics: A Role for Epigenetic Mechanisms in Learning and Memory. *Neurobiol. Learn. Mem.* **96**:2–12. [6]
Dayan, E., and L. G. Cohen. 2011. Neuroplasticity Subserving Motor Skill Learning. *Neuron* **72**:443–454. [13]
Deck, M., L. Lokmane, S. Chauvet, et al. 2013. Pathfinding of Corticothalamic Axons Relies on a Rendezvous with Thalamic Projections. *Neuron* **77**:472–484. [3]
Deen, B., K. Koldewyn, N. Kanwisher, and R. Saxe. 2015. Functional Organization of Social Perception and Cognition in the Superior Temporal Sulcus. *Cereb. Cortex* **25**:4596–4609. [11]
de Heering, A., and D. Maurer. 2012. Face Memory Deficits in Patients Deprived of Early Visual Input by Bilateral Congenital Cataracts. *Dev. Psychobiol.* **56**:96–108. [11]
De Lorme, K., M. R. Bell, and C. L. Sisk. 2013. The Teenage Brain: Social Reorientation and the Adolescent Brain-the Role of Gonadal Hormones in the Male Syrian Hamster. *Curr. Dir. Psychol.* **22**:128–133. [8, 11]
De Lorme, K. C., and C. L. Sisk. 2013. Pubertal Testosterone Programs Context-Appropriate Agonistic Behavior and Associated Neural Activation Patterns in Male Syrian Hamsters. *Physiol. Behav.* **112-113**:1–7. [8, 11]
———. 2016. The Organizational Effects of Pubertal Testosterone on Sexual Proficiency in Adult Male Syrian Hamsters. *Physiol. Behav.* **165**:273–277. [8, 11]
Dennison, M., S. Whittle, M. Yücel, et al. 2013. Mapping Subcortical Brain Maturation during Adolescence: Evidence of Hemisphere- and Sex-Specific Longitudinal Changes. *Dev. Sci.* **5**:772–791. [14]
Depienne, C., D. Bouteiller, A. Méneret, et al. 2012. RAD51 Haploinsufficiency Causes Congenital Mirror Movements in Humans. *Am. J. Hum. Genet.* **90**:301–307. [3]
Depienne, C., M. Cincotta, S. Billot, et al. 2011. A Novel DCC Mutation and Genetic Heterogeneity in Congenital Mirror Movements. *Neurology* **79**:260–264. [3]
Diamond, L. M., and J. A. Dickenson. 2012. The Neuroimaging of Love and Desire: Review and Future Directions. *Clinical Neuropsychiatry* **9**:39–46. [14]
Diaz Heijtz, R., S. Wang, F. Anuar, et al. 2011. Normal Gut Microbiota Modulates Brain Development and Behavior. *PNAS* **108**:3047–3052. [9]
Dickendesher, T. L., K. T. Baldwin, Y. A. Mironova, et al. 2012. NgR1 and NgR3 Are Receptors for Chondroitin Sulfate Proteoglycans. *Nat. Neurosci.* **15**:703–712. [5]
Dickson, B. J. 2002. Molecular Mechanisms of Axon Guidance. *Science* **298**:1959–1964. [3]
Di Cristo, G., B. Chattopadhyaya, S. J. Kuhlman, et al. 2007. Activity-Dependent PSA Expression Regulates Inhibitory Maturation and Onset of Critical Period Plasticity. *Nat. Neurosci.* **10**:1569–1577. [5]
Diesmann, M., M.-O. Gewaltig, and A. Aertsen. 1999. Stable Propagation of Synchronous Spiking in Cortical Neural Networks. *Nature* **402**:529–533. [2]
Di Meglio, T., C. F. Kratochwil, N. Vilain, et al. 2013. Ezh2 Orchestrates Topographic Migration and Connectivity of Mouse Precerebellar Neurons. *Science* **339**:204–207. [6]
Djurisic, M., G. S. Vidal, M. Mann, et al. 2013. PirB Regulates a Structural Substrate for Cortical Plasticity. *PNAS* **110**:20771–20776. [5]
Do, K. Q., M. Cuenod, and T. K. Hensch. 2015. Targeting Oxidative Stress and Aberrant Critical Period Plasticity in the Developmental Trajectory to Schizophrenia. *Schizophr. Bull.* **41**:835–446. [5]

Doesburg, S. M., K. Tingling, M. J. MacDonald, and E. W. Pang. 2016. Development of Network Synchronization Predicts Language Abilities. *J. Cogn. Neurosci.* **28**:55–68. [1, 11]

Donato, F., R. I. Jacobsen, M.-B. Moser, and E. I. Moser. 2017. Stellate Cells Drive Maturation of the Entorhinal-Hippocampal Circuit. *Science* **355**:1172–1172. [2, 4]

Donato, F., S. B. Rompani, and P. Caroni. 2013. Parvalbumin-Expressing Basket-Cell Network Plasticity Induced by Experience Regulates Adult Learning. *Nature* **504**:272–276. [5]

Dosenbach, N. U., B. Nardos, A. L. Cohen, et al. 2010. Prediction of Individual Brain Maturity Using fMRI. *Science* **329**:1358–1361. [11, 15]

Douaud, G., A. R. Groves, C. K. Tamnes, et al. 2014. A Common Brain Network Links Development, Aging, and Vulnerability to Disease. *PNAS* **111**:17648–17653. [15]

Doudna, J. A., and E. Charpentier. 2014. The New Frontier of Genome Engineering with CRISPR-Cas9. *Science* **346**:1258096. [3, 7]

Draganski, B., C. Gaser, V. Busch, et al. 2004. Changes in Grey Matter Induced by Training. *Nature* **427**:311–312. [13]

Drescher, U., C. Kremoser, C. Handwerker, et al. 1995. *In Vitro* Guidance of Retinal Ganglion Cell Axons by RAGS, a 25 kDa Tectal Protein Related to Ligands for Eph Receptor Tyrosine Kinases. *Cell* **82**:359–370. [3]

Ducharme, S., M. D. Albaugh, T. V. Nguyen, et al. 2016. Trajectories of Cortical Thickness Maturation in Normal Brain Development: The Importance of Quality Control Procedures. *NeuroImage* **125**:267–279. [8]

Duffy, K. R., A. J. Lingley, K. D. Holman, and D. E. Mitchell. 2016. Susceptibility to Monocular Deprivation Following Immersion in Darkness Either Late into or Beyond the Critical Period. *J. Comp. Neurol.* **524**:2643–2653. [5]

Duffy, K. R., and M. S. Livingstone. 2005. Loss of Neurofilament Labeling in the Primary Visual Cortex of Monocularly Deprived Monkeys. *Cereb. Cortex* **15**:1146–1154. [5]

Duffy, K. R., and D. E. Mitchell. 2013. Darkness Alters Maturation of Visual Cortex and Promotes Fast Recovery from Monocular Deprivation. *Curr. Biol.* **23**:382–386. [5]

Duffy, K. R., K. M. Murphy, M. P. Frosch, and M. S. Livingstone. 2007. Cytochrome Oxidase and Neurofilament Reactivity in Monocularly Deprived Human Primary Visual Cortex. *Cereb. Cortex* **17**:1283–1291. [5]

Dulcis, D., P. Jamshidi, S. Leutgeb, and N. C. Spitzer. 2013. Neurotransmitter Switching in the Adult Brain Regulates Behavior. *Science* **340**:449–453. [4]

Dumontheil, I. 2014. Development of Abstract Thinking during Childhood and Adolescence: The Role of Rostrolateral Prefrontal Cortex. *Dev. Cogn. Neurosci.* **10**:57–76. [11]

Dupont, E., I. L. Hanganu, W. Kilb, S. Hirsch, and H. J. Luhmann. 2006. Rapid Developmental Switch in the Mechanisms Driving Early Cortical Columnar Networks. *Nature* **439**:79–83. [4]

Durand, S., A. Patrizi, K. B. Quast, et al. 2012. NMDA Receptor Regulation Prevents Regression of Visual Cortical Function in the Absence of Mecp2. *Neuron* **76**:1078–1090. [5]

Dwyer, A. A., R. Quinton, N. Pitteloud, and D. Morin. 2015. Psychosexual Development in Men with Congenital Hypogonadotropic Hypogonadism on Long-Term Treatment: A Mixed Methods Study. *Sex Med.* **3**:32–41. [8]

Eaton, N. C., H. M. Sheehan, and E. M. Quinlan. 2016. Optimization of Visual Training for Full Recovery from Severe Amblyopia in Adults. *Learn. Memory* **23**:99–103. [5]

Edelman, G. M. 1987. Neural Darwinism: The Theory of Neuronal Group Selection. New York: Basic Books. [13]

Edwards, T. J., E. H. Sherr, A. J. Barkovich, and L. J. Richards. 2014. Clinical, Genetic and Imaging Findings Identify New Causes for Corpus Callosum Development Syndromes. *Brain* **137**:1579–1613. [3]

Eiland, L., J. Ramroop, M. N. Hill, J. Manley, and B. S. McEwen. 2012. Chronic Juvenile Stress Produces Corticolimbic Dendritic Architectural Remodeling and Modulates Emotional Behavior in Male and Female Rats. *Psychoneuroendocrinology* **37**:39–47. [9]

Eiraku, M., N. Takata, H. Ishibashi, et al. 2011. Self-Organizing Optic-Cup Morphogenesis in Three-Dimensional Culture. *Nature* **472**:51–56. [3]

Elbaz, B. 2016. Activity-Dependent Myelination Shapes Conduction Velocity. *J. Neurosci.* **36**:11585–11586. [2]

Elbert, T., C. Pantev, C. Wienbruch, B. Rockstroh, and E. Taub. 1995. Increased Cortical Representation of the Fingers of the Left Hand in String Players. *Science* **270**:305–307. [13]

Elman, J. L., E. A. Bates, M. H. Johnson, et al. 1996. Rethinking Innateness: A Connectionist Perspective on Development. Cambridge, MA: MIT Press. [13]

Emberson, L. L., B. D. Zinszer, R. D. S. Raizada, and R. N. Aslin. 2017. Decoding the Infant Mind: Multivariate Pattern Analysis (MVPA) Using fNIRS. *PLoS One* **12**:e0172500. [7]

Engel, A. K., and W. Singer. 2001. Temporal Binding and the Neural Correlates of Sensory Awareness. *Trends Cogn. Sci.* **5**:16–25. [7]

Enwright, J. F., S. Sanapala, A. Foglio, et al. 2016. Reduced Labeling of Parvalbumin Neurons and Perineuronal Nets in the Dorsolateral Prefrontal Cortex of Subjects with Schizophrenia. *Neuropsychopharmacology* **41**:2206–2214. [10]

Ermentrout, G. B., and D. Kleinfeld. 2001. Traveling Electrical Waves in Cortex: Insights from Phase Dynamics and Speculation on a Computational Role. *Neuron* **29**:33–44. [2]

Ernst, M., E. E. Nelson, S. Jazbec, et al. 2005. Amygdala and Nucleus Accumbens in Responses to Receipt and Omission of Gains in Adults and Adolescents. *NeuroImage* **25**:1279–1291. [15]

Ernst, M., D. Pine, and M. Hardin. 2006. Triadic Model of the Neurobiology of Motivated Behavior in Adolescence. *Psychol. Med.* **36**:299–312. [14]

Erskine, L., S. Reijntjes, T. Pratt, et al. 2011. Vegf Signaling through Neuropilin 1 Guides Commissural Axon Crossing at the Optic Chiasm. *Neuron* **70**:951–965. [3]

Esposito, E. A., M. J. Jones, J. R. Doom, et al. 2016. Differential DNA Methylation in Peripheral Blood Mononuclear Cells in Adolescents Exposed to Significant Early but Not Later Childhood Adversity. *Dev. Psychopathol.* **28**:1385–1399. [6]

Essex, M. J., T. B. W., C. Hertzman, et al. 2013. Epigenetic Vestiges of Early Developmental Adversity: Childhood Stress Exposure and DNA Methylation in Adolescence. *Child Dev.* **84**:58–75. [6]

Esteva, A., B. Kuprel, R. A. Novoa, et al. 2017. Dermatologist-Level Classification of Skin Cancer with Deep Neural Networks. *Nature* **542**:115–118. [15]

Eyre, J. A. 2000. Functional Corticospinal Projections Are Established Prenatally in the Human Foetus Permitting Involvement in the Development of Spinal Motor Centres. *Brain* **123**:51–64. [3]

———. 2003. Development and Plasticity of the Corticospinal System in Man. *Neural Plast.* **10**:93–106. [3]

Fagiolini, M., J. M. Fritschy, K. Löw, et al. 2004. Specific GABAA Circuits for Visual Cortical Plasticity. *Science* **303**:1681–1683. [5]

Fagiolini, M., and T. K. Hensch. 2000. Inhibitory Threshold for Critical Period Activation in Primary Visual Cortex. *Nature* **404**:183–186. [5, 10]

Fagiolini, M., C. L. Jensen, and F. A. Champagne. 2009. Epigenetic Influences on Brain Development and Plasticity. *Curr. Opin. Neurobiol.* **19**:207–212. [6]

Fagiolini, M., H. Katagiri, H. Miyamoto, et al. 2003. Separable Features of Visual Cortical Plasticity Revealed by N-Methyl-D-Aspartate Receptor 2a Signaling. *PNAS* **100**:2854–2859. [5]

Fagiolini, M., T. Pizzorusso, N. Berardi, L. Domenici, and L. Maffei. 1994. Functional Postnatal Development of the Rat Primary Visual Cortex and the Role of Visual Experience: Dark Rearing and Monocular Lid Suture. *Vision Res.* **34**:709–720. [5]

Fair, D. A., A. L. Cohen, N. U. Dosenbach, et al. 2008. The Maturing Architecture of the Brain's Default Network. *PNAS* **105**:4028–4032. [11, 15]

Fair, D. A., A. L. Cohen, J. D. Power, et al. 2009. Functional Brain Networks Develop from a "Local to Distributed" Organization. *PLoS Comput. Biol.* **5**:e1000381. [11]

Fan, J., J. Byrne, M. S. Worden, et al. 2007. The Relation of Brain Oscillations to Attentional Networks. *J. Neurosci.* **27**:6197–6206. [7]

Farré, P., M. J. Jones, M. J. Meaney, et al. 2015. Concordant and Discordant DNA Methylation Signatures of Aging in Human Blood and Brain. *Epigenetics Chromatin* **8**:19. [6]

Feinberg, I., and I. G. Campbell. 2010. Sleep EEG Changes during Adolescence: An Index of a Fundamental Brain Reorganization. *Brain Cogn.* **72**:56–65. [10]

Fell, J., P. Klaver, H. Elfadil, et al. 2003. Rhinal-Hippocampal Theta Coherence during Declarative Memory Formation: Interaction with Gamma Synchronization? *Eur. J. Neurosci.* **17**:1082–1088. [2]

Fell, J., P. Klaver, K. Lehnertz, et al. 2001. Human Memory Formation Is Accompanied by Rhinal-Hippocampal Coupling and Decoupling. *Nat. Neurosci.* **4**:1259–1264. [2]

Fell, J., E. Ludowig, B. P. Staresina, et al. 2011. Medial Temporal Theta/Alpha Power Enhancement Precedes Successful Memory Encoding: Evidence Based on Intracranial EEG. *J. Neurosci.* **31**:5392–5397. [2]

Field, E. F., I. Q. Whishaw, M. L. Forgie, and S. M. Pellis. 2004. Neonatal and Pubertal, but Not Adult, Ovarian Steroids Are Necessary for the Development of Female-Typical Patterns of Dodging to Protect a Food Item. *Behav. Neurosci.* **118**:1293–1304. [8]

Fields, R. D. 2008. White Matter in Learning, Cognition, and Psychiatric Disorders. *Trends Neurosci.* **31**:361–370. [9]

Fleming, J. J., and P. M. England. 2010. AMPA Receptors and Synaptic Plasticity: A Chemist's Perspective. *Nat. Chem. Biol.* **6**:89–97. [6]

Flores, F. J., K. E. Hartnack, A. B. Fath, et al. 2017. Thalamocortical Synchronization during Induction and Emergence from Propofol-Induced Unconsciousness. *PNAS* **114**:E6660–E6668. [10]

Florin, E., and S. Baillet. 2015. The Brain's Resting-State Activity Is Shaped by Synchronized Cross-Frequency Coupling of Neural Oscillations. *NeuroImage* **111**:26–35. [11, 15]

Fortenberry, J. D. 2014. Sexual Learning, Sexual Experience, and Healthy Adolescent Sex. *New Dir. Child Adolesc. Dev.* **144**:71–86. [14]

Fox, S. E., P. Levitt, and C. A. Nelson. 2010. How the Timing and Quality of Early Experiences Influence the Development of Brain Architecture. *Child Dev.* **81**:28–40. [6]

Fransson, P., B. Skiold, S. Horsch, et al. 2007. Resting-State Networks in the Infant Brain. *PNAS* **104**:15531–15536. [15]

Franze, K. 2013. The Mechanical Control of Nervous System Development. *Development* **140**:3069–3077. [3]

Fredriksen, K., J. Rhodes, R. Reddby, and N. Way. 2004. Sleepless in Chicago: Tracking the Effects of Adolescent Sleep Loss during the Middle School Years. *Soc. Res. Child Dev.* **75**:84–95. [14]

Freeman, W. J. 1994. Characterization of State Transitions in Spatially Distributed, Chaotic, Nonlinear, Dynamical Systems in Cerebral Cortex. *Integr. Physiol. Behav. Sci.* **29**:294–306. [1]

Frenkel, M. Y., and M. F. Bear. 2004. How Monocular Deprivation Shifts Ocular Dominance in Visual Cortex of Young Mice. *Neuron* **44**:917–923. [5, 7]

Freund, J., A. M. Brandmaier, L. Lewejohann, et al. 2013. Emergence of Individuality in Genetically Identical Mice. *Science* **340**:756–759. [13]

Frey, U., H. Schroeder, and H. Matthies. 1990. Dopaminergic Antagonists Prevent Long-Term Maintenance of Posttetanic LTP in the CA1 Region of Rat Hippocampal Slices. *Brain Res.* **522**:69–75. [6]

Friedmann, N., and D. Rusou. 2015. Critical Period for First Language: The Crucial Role of Language Input during the First Year of Life. *Curr. Opin. Neurobiol.* **35**:27–34. [11]

Fries, P. 2005. A Mechanism for Cognitive Dynamics: Neuronal Communication through Neuronal Coherence. *Trends Cogn. Sci.* **9**:474–480. [15]

———. 2009. Neuronal Gamma-Band Synchronization as a Fundamental Process in Cortical Computation. *Annu. Rev. Neurosci.* **32**:209–224. [2]

Fries, P., D. Nikolić, and W. Singer. 2007. The Gamma Cycle. *Trends Neurosci.* **30**:309–316. [7]

Fries, P., J. H. Reynolds, A. E. Rorie, and R. Desimone. 2001. Modulation of Oscillatory Neuronal Synchronization by Selective Visual Attention. *Science* **291**:1560–1563. [2]

Friston, K. 2010. The Free-Energy Principle: A Unified Brain Theory? *Nat. Rev. Neurosci.* **11**:127–138. [15]

Frodl, T., M. Szyf, A. Carballedo, et al. 2015. DNA Methylation of the Serotonin Transporter Gene (SLC6A4) Is Associated with Brain Function Involved in Processing Emotional Stimuli. *J. Psych. Neurosci. (Japan)* **40**:296–305. [6]

Fu, Y., M. Kaneko, Y. Tang, A. Alvarez-Buylla, and M. P. Stryker. 2015. A Cortical Disinhibitory Circuit for Enhancing Adult Plasticity. *eLife* **4**:e05558. [5]

Fu, Y., J. M. Tucciarone, J. S. Espinosa, et al. 2014. A Cortical Circuit for Gain Control by Behavioral State. *Cell* **156**:1139–1152. [5]

Fuhrmann, D., L. J. Knoll, and S.-J. Blakemore. 2015. Adolescence as a Sensitive Period of Brain Development. *Trends Cogn. Sci.* **19**:558–566. [9, 11]

Fujii, T., Y. Iijima, H. Kondo, et al. 2007. Failure to Confirm an Association between the PLXNA2 Gene and Schizophrenia in a Japanese Population. *Prog. Neuropsychopharmacol. Biol. Psychiatry* **31**:873–877. [3]

Fuligni, A. J., V. Tseng, and M. Lam. 1999. Attitudes toward Family Obligations among American Adolescents from Asian, Latin American, and European Backgrounds. *Soc. Res. Child Dev.* **70**:1030–1044. [14]

Furman, W., and L. Shaffer. 2003. The Role of Romantic Relationships in Adolescent Development. In: Adolescent Romantic Relations and Sexual Behavior: Theory, Research, and Practical Implications, ed. P. Florsheim, pp. 3–22. Mahwah, NJ: Lawrence Erlbaum. [14]

Fusar-Poli, P., S. Borgwardt, A. Bechdolf, et al. 2013. The Psychosis High-Risk State: A Comprehensive State-of-the-Art Review. *JAMA Psychiatry* **70**:107–120. [12]

Galván, A. 2014. Insights About Adolescent Behavior, Plasticity and Policy from Neuroscience Research. *Neuron* **83**:262–265. [14]

Galván, A., T. A. Hare, C. E. Parra, et al. 2006. Earlier Development of the Accumbens Relative to Orbitofrontal Cortex Might Underlie Risk-Taking Behavior in Adolescents. *J. Neurosci.* **26**:6885–6892. [12, 15]

Galván, A., and K. M. McGlennen. 2013. Enhanced Striatal Sensitivity to Aversive Reinforcement in Adolescents versus Adults. *J. Cogn. Neurosci.* **25**:284–296. [15]

Gaser, C., and G. Schlaug. 2003. Brain Structures Differ between Musicians and Non-Musicians. *J. Neurosci.* **23**:9240–9245. [13]

Gaudreau, H., J. Carrier, and J. Montplaisir. 2001. Age-Related Modifications of Nrem Sleep EEG: From Childhood to Middle Age. *J. Sleep Res.* **10**:165–172. [10]

Gaughwin, P., M. Ciesla, H. Yang, B. Lim, and P. Brundin. 2011. Stage-Specific Modulation of Cortical Neuronal Development by Mmu-miR-134. *Cereb. Cortex* **21**:1857–1869. [6]

Gdalyahu, A., E. Tring, P.-O. Polack, et al. 2012. Associative Fear Learning Enhances Sparse Network Coding in Primary Sensory Cortex. *Neuron* **75**:121–132. [13]

Ge, W.-P., A. Miyawaki, F. H. Gage, Y. N. Jan, and L. Y. Ja. 2012. Local Generation of Glia Is a Major Astrocyte Source in Postnatal Cortex. *Nature* **484**:376–380. [9]

Gee, D. G., K. L. Humphreys, J. Flannery, et al. 2013. A Developmental Shift from Positive to Negative Connectivity in Human Amygdala-Prefrontal Circuitry. *J. Neurosci.* **33**:4584–4593. [12, 15]

Geier, C. F., R. Terwilliger, T. Teslovich, K. Velanova, and B. Luna. 2010. Immaturities in Reward Processing and Its Influence on Inhibitory Control in Adolescence. *Cereb. Cortex* **20**:1613–1629. [15]

German, M., N. A. Gonzales, and L. Dumka. 2009. Familism Values as a Protective Factor for Mexican-Origin Adolescents Exposed to Deviant Peers. *J. Early Adolesc.* **29**:16–42. [14]

Gervain, J., B. W. Vines, L. M. Chen, et al. 2013. Valproate Reopens Critical-Period Learning of Absolute Pitch. *Front. Syst. Neurosci.* **7**:102. [5, 7]

Ghisletta, P., K. M. Kennedy, K. M. Rodrigue, U. Lindenberger, and N. Raz. 2010. Adult Age Differences and the Role of Cognitive Resources in Perceptual-Motor Skill Acquisition: Application of a Multilevel Negative Exponential Model. *J. Gerontol. B* **65B**:163–173. [13]

Ghosh, A., A. Antonini, S. K. McConnell, and C. J. Shatz. 1990. Requirements of Subplate Neurons in the Formation of Thalamocortical Connections. *Nature* **347**:179–181. [4]

Ghosh, A., and C. J. Shatz. 1992. Involvement of Subplate Neurons in the Formation of Ocular Dominance Columns. *Science* **255**:1441–1443. [4]

Giedd, J. N., J. Blumenthal, N. O. Jeffries, et al. 1999. Brain Development during Childhood and Adolescence: A Longitudinal MRI Study. *Nat. Neurosci.* **2**:861–863. [8]

Giedd, J. N., and J. L. Rapoport. 2010. Structural MRI of Pediatric Brain Development: What Have We Learned and Where Are We Going? . *Neuron* **67**:728–734. [11]

Gilbert, C. D., and T. N. Wiesel. 1989. Columnar Specificity of Intrinsic Horizontal and Corticocortical Connections in Cat Visual Cortex. *J. Neurosci.* **9**:2432–2442. [2]

Gilmour, D., H. Knaut, H.-M. Maischein, and C. Nüsslein-Volhard. 2004. Towing of Sensory Axons by Their Migrating Target Cells *in Vivo*. *Nat. Neurosci.* **7**:491–492. [3]

Giraud, A. L., and D. Poeppel. 2012. Cortical Oscillations and Speech Processing: Emerging Computational Principles and Operations. *Nat. Neurosci.* **15**:511–517. [7]

Glausier, J. R., K. N. Fish, and D. A. Lewis. 2014. Altered Parvalbumin Basket Cell Inputs in the Dorsolateral Prefrontal Cortex of Schizophrenia Subjects. *Mol. Psychiatry* **19**:30–36. [10]

Goel, A., and H. K. Lee. 2007. Persistence of Experience-Induced Homeostatic Synaptic Plasticity through Adulthood in Superficial Layers of Mouse Visual Cortex. *J. Neurosci.* **27**:6692–6700. [5]

Gogolla, N., A. E. Takesian, G. Feng, M. Fagiolini, and T. K. Hensch. 2014. Sensory Integration in Mouse Insular Cortex Reflects GABA Circuit Maturation. *Neuron* **83**:894–905. [5]

Gogtay, N., J. N. Giedd, L. Lusk, et al. 2004. Dynamic Mapping of Human Cortical Development during Childhood through Early Adulthood. *PNAS* **101**:8174–8179. [8, 11]

Gordon, J. A., and M. P. Stryker. 1996. Experience-Dependent Plasticity of Binocular Responses in the Primary Visual Cortex of the Mouse. *J. Neurosci.* **16**:3274–3286. [5]

Gou, Z., N. Choudhury, and A. A. Benasich. 2011. Resting Frontal Gamma Power at 16, 24 and 36 Months Predicts Individual Differences in Language and Cognition at 4 and 5 Years. *Behav. Brain Res.* **220**:263–270. [7]

Gougoux, F., R. J. Zatorre, M. Lassonde, P. Voss, and F. Lepore. 2005. A Functional Neuroimaging Study of Sound Localization: Visual Cortex Activity Predicts Performance in Early-Blind Individuals. *PLoS Biol.* **3**:e27. [11]

Grant, A., F. Fathalli, G. Rouleau, R. Joober, and C. Flores. 2012. Association between Schizophrenia and Genetic Variation in DCC: A Case-Control Study. *Schizophr. Res.* **137**:26–31. [3]

Gray, C. M., and D. A. McCormick. 1996. Chattering Cells: Superficial Pyramidal Neurons Contributing to the Generation of Synchronous Oscillations in the Visual Cortex. *Science* **274**:109–113. [2]

Graziano, M. S. A. 2016. Ethological Action Maps: A Paradigm Shift for the Motor Cortex. *Trends Cogn. Sci.* **20**:121–132. [13]

Greenough, W. T., J. E. Black, and C. S. Wallace. 1987. Experience and Brain Development. *Child Dev.* **58**:539–559. [8, 11]

Greifzu, F., E. Kalogeraki, and S. Löwel. 2016. Environmental Enrichment Preserved Lifelong Ocular Dominance Plasticity, but Did Not Improve Visual Abilities. *Neurobiol. Aging* **May**:130–137. [5]

Greifzu, F., J. Pielecka-Fortuna, E. Kalogeraki, et al. 2013. Environmental Enrichment Extends Ocular Dominance Plasticity into Adulthood and Protects from Stroke-Induced Impairments of Plasticity. *PNAS* **111**:1150–1155. [5]

Grienberger, C., X. Chen, and A. Konnerth. 2015. Dendritic Function *in Vivo*. *Trends Neurosci.* **38**:45–54. [2]

Grillner, S. 2006. Biological Pattern Generation: The Cellular and Computational Logic of Networks in Motion. *Neuron* **52**:751–766. [2]

Grosmark, A. D., and G. Buzsáki. 2016. Diversity in Neural Firing Dynamics Supports Both Rigid and Learned Hippocampal Sequences. *Science* **351**:1440–1443. [7]

Gu, Q., and W. Singer. 1995. Involvement of Serotonin in Developmental Plasticity of Kitten Visual Cortex. *Eur. J. Neurosci.* **7**:1146–1153. [2]

Gu, X., E. C. Olson, and N. C. Spitzer. 1994. Spontaneous Neuronal Calcium Spikes and Waves during Early Differentiation. *J. Neurosci.* **14**:6325–6335. [4]

Gu, X., and N. C. Spitzer. 1995. Distinct Aspects of Neuronal Differentiation Encoded by Frequency of Spontaneous Ca^{++} Transients. *Nature* **375**:784–787. [4]

Gu, Y., S. Huang, M. C. Chang, et al. 2013. Obligatory Role for the Immediate Early Gene NARP in Critical Period Plasticity. *Neuron* **79**:335–346. [5]

Gu, Y., T. Tran, S. Murase, et al. 2016. Neuregulin-Dependent Regulation of Fast-Spiking Interneuron Excitability Controls the Timing of the Critical Period. *J. Neurosci.* **36**:10285–10295. [5]

Guibert, S., and M. Weber. 2013. Functions of DNA Methylation and Hydroxymethylation in Mammalian Development. *Curr. Top. Dev. Biol.* **104**:47–83. [6]

Guillery, R. W., and J. H. Kaas. 1973. Genetic Abnormality of the Visual Pathways in a "White" Tiger. *Science* **180**:1287–1289. [3]

Guilmatre, A., G. Huguet, R. Delorme, and T. Bourgeron. 2014. The Emerging Role of Shank Genes in Neuropsychiatric Disorders. *Dev. Neurobiol.* **74**:113–122. [4]

Guirado, R., M. Perez-Rando, D. Sanchez-Matarredona, E. Castrén, and J. Nacher. 2014. Chronic Fluoxetine Treatment Alters the Structure, Connectivity and Plasticity of Cortical Interneurons. *Int J Neuropsychopharmacol* **17**:1635–1646. [5]

Guo, J. U., D. K. Ma, H. Mo, et al. 2011. Neuronal Activity Modifies the DNA Methylation Landscape in the Adult Brain. *Nat. Neurosci.* **14**:1345–1351. [6, 13]

Guo, J. U., Y. Su, J. H. Shin, et al. 2013. Distribution, Recognition and Regulation of Non-CpG Methylation in the Adult Mammalian Brain. *Nat. Neurosci.* **17**:215–222. [6]

Guo, Y., S. Huang, R. De Pasquale, et al. 2012. Dark Exposure Extends the Integration Window for Spike-Timing-Dependent Plasticity. *J. Neurosci.* **32**:15027–15035. [5]

Gurnot, C., I. Martin-Subero, S. M. Mah, et al. 2015. Prenatal Antidepressant Exposure Associated with CYP2E1 DNA Methylation Change in Neonates. *Epigenetics* **10**:361–372. [6]

Gutierrez-Arcelus, M., T. Lappalainen, S. B. Montgomery, et al. 2013. Passive and Active DNA Methylation and the Interplay with Genetic Variation in Gene Regulation. *eLife* **2**: [6]

Guy, A. T., Y. Nagatsuka, N. Ooashi, et al. 2015. Glycerophospholipid Regulation of Modality-Specific Sensory Axon Guidance in the Spinal Cord. *Science* **349**:974–977. [3]

Gweon, H., D. Dodell-Feder, M. Bedny, and R. Saxe. 2012. Theory of Mind Performance in Children Correlates with Functional Specialization of a Brain Region for Thinking About Thoughts. *Child Dev.* **83**:1853–1868. [11]

Gweon, H., H. Pelton, J. A. Konopka, and L. E. Schulz. 2014. Sins of Omission: Children Selectively Explore When Teachers Are under-Informative. *Cognition* **132**:335–341. [11]

Haaf, T. 2006. Methylation Dynamics in the Early Mammalian Embryo: Implications of Genome Reprogramming Defects for Development. *Curr. Top. Microbiol. Immunol.* **310**:13–22. [6]

Hagenauer, M., and T. Lee. 2012. The Neuroendocrine Control of the Circadian System. *Front. Neuroendocrin.* **33**:211–229. [14]

Hanganu, I. L., Y. Ben-Ari, and R. Khazipov. 2006. Retinal Waves Trigger Spindle Bursts in the Neonatal Rat Visual Cortex. *J. Neurosci.* **26**:6728–6736. [4]

Hanganu, I. L., W. Kilb, and H. J. Luhmann. 2002. Functional Synaptic Projections onto Subplate Neurons in Neonatal Rat Somatosensory Cortex. *J. Neurosci.* **22**:7165–7176. [4]

Hanganu, I. L., A. Okabe, V. Lessmann, and H. J. Luhmann. 2009. Cellular Mechanisms of Subplate-Driven and Cholinergic Input-Dependent Network Activity in the Neonatal Rat Somatosensory Cortex. *Cereb. Cortex* **19**:89–105. [4]

Hanganu-Opatz, I. L. 2010. Between Molecules and Experience: Role of Early Patterns of Coordinated Activity for the Development of Cortical Maps and Sensory Abilities. *Brain Res. Rev.* **64**:160–176. [4]

Hannula-Jouppi, K., N. Kaminen-Ahola, M. Taipale, et al. 2005. The Axon Guidance Receptor Gene ROBO1 Is a Candidate Gene for Developmental Dyslexia. *PLoS Genet.* **1**:e50. [3]

Hanover, J. L., Z. J. Huang, S. Tonegawa, and M. P. Stryker. 1999. Brain-Derived Neurotrophic Factor Overexpression Induces Precocious Critical Period in Mouse Visual Cortex. *J. Neurosci.* **19**:RC40. [5]

Hansel, C., A. Artola, and W. Singer. 1996. Different Threshold Levels of Postsynaptic $[Ca^{2+}]_i$ Have to Be Reached to Induce LTP and LTD in Neocortical Pyramidal Cells. *J. Physiol. (Paris)* **90**:317–319. [2]

————. 1997. Relation between Dendritic Ca^{2+} Levels and the Polarity of Synaptic Long-Term Modifications in Rat Visual Cortex Neurons. *Eur. J. Neurosci.* **9**:2309–2322. [2]

Harauzov, A., M. Spolidoro, G. Dicristo, et al. 2010. Reducing Intracortical Inhibition in the Adult Visual Cortex Promotes Ocular Dominance Plasticity. *J. Neurosci.* **30**:361–371. [5]

Hare, T. A., N. Tottenham, A. Galván, et al. 2008. Biological Substrates of Emotional Reactivity and Regulation in Adolescence during an Emotional Go-Nogo Task. *Biol. Psychiatry* **63**:927–934. [12, 15]

Hargus, G., O. Cooper, M. Deleidi, et al. 2010. Differentiated Parkinson Patient-Derived Induced Pluripotent Stem Cells Grow in the Adult Rodent Brain and Reduce Motor Asymmetry in Parkinsonian Rats. *PNAS* **107**:15921–15926. [3]

Harrigan, S. M., P. D. McGorry, and H. Krstev. 2003. Does Treatment Delay in First-Episode Psychosis Really Matter? *Psychol. Med.* **33**:97–110. [15]

Hartung, H., M. D. Brockmann, B. Pöschel, V. De Feo, and I. L. Hanganu-Opatz. 2016a. Thalamic and Entorhinal Network Activity Differently Modulates the Functional Development of Prefrontal-Hippocampal Interactions. *J. Neurosci.* **36**:3676–36790. [4]

Hartung, H., N. Cichon, V. De Feo, et al. 2016b. From Shortage to Surge: A Developmental Switch in Hippocampal-Prefrontal Coupling in a Gene-Environment Model of Neuropsychiatric Disorders. *Cereb. Cortex* **26**:4265–4281. [4]

Harwell, C. C., L. C. Fuentealba, A. Gonzalez-Cerrillo, et al. 2015. Wide Dispersion and Diversity of Clonally Related Inhibitory Interneurons. *Neuron* **87**:999–1007. [3]

Harwerth, R. S., E. L. Smith, G. C. Duncan, M. L. Crawford, and G. K. von Noorden. 1986. Multiple Sensitive Periods in the Development of the Primate Visual System. *Science* **232**:235–238. [5]

Hasenstaub, A., S. Otte, E. Callaway, and T. J. Sejnowski. 2010. Metabolic Cost as a Unifying Principle Governing Neuronal Biophysics. *PNAS* **107**:12329–12334. [4]

Hashemi, E., J. Ariza, H. Rogers, S. C. Noctor, and V. Martinez-Cerdeno. 2017. The Number of Parvalbumin-Expressing Interneurons Is Decreased in the Medial Prefrontal Cortex in Autism. *Cereb. Cortex* **27**:1931–1943. [10]

Hashemirad, F., M. Zoghi, P. B. Fitzgerald, and S. Jaberzadeh. 2016. The Effect of Anodal Transcranial Direct Current Stimulation on Motor Sequence Learning in Healthy Individuals: A Systematic Review and Meta-Analysis. *Brain Cogn.* **102**:1–12. [13]

Hashimoto, K., and M. Kano. 2005. Postnatal Development and Synapse Elimination of Climbing Fiber to Purkinje Cell Projection in the Cerebellum. *Neurosci. Res.* **53**:221–228. [7]

Hashimoto, T., Q. L. Nguyen, D. Rotaru, et al. 2009. Protracted Developmental Trajectories of GABAA Receptor Alpha1 and Alpha2 Subunit Expression in Primate Prefrontal Cortex. *Biol. Psychiatry* **65**:1015–1023. [12]

Hassan, B. A., and P. R. Hiesinger. 2015. Beyond Molecular Codes: Simple Rules to Wire Complex Brains. *Cell* **163**:285–291. [3]

Hayashi, S., Y. Inoue, H. Kiyonari, et al. 2016. Protocadherin-17 Mediates Collective Axon Extension by Recruiting Actin Regulator Complexes to Interaxonal Contacts. *Dev. Cell* **38**:331. [3]

He, H.-Y., W. Hodos, and E. M. Quinlan. 2006. Visual Deprivation Reactivates Rapid Ocular Dominance Plasticity in Adult Visual Cortex. *J. Neurosci.* **26**:2951–2955. [5, 7]

He, H. Y., B. Ray, K. Dennis, and E. M. Quinlan. 2007. Experience-Dependent Recovery of Vision Following Chronic Deprivation Amblyopia. *Nat. Neurosci.* **10**:1134–1136. [5]

He, S., Z. Li, S. Ge, Y.-C. Yu, and S.-H. Shi. 2015. Inside-out Radial Migration Facilitates Lineage-Dependent Neocortical Microcircuit Assembly. *Neuron* **86**:1159–1166. [3]

Hebb, D. O. 1949. The Organization of Behavior. New York: Wiley. [2]

Hebbard, P. C., R. R. King, C. W. Malsbury, and C. W. Harley. 2003. Two Organizational Effects of Pubertal Testosterone in Male Rats: Transient Social Memory and a Shift Away from Long-Term Potentiation Following a Tetanus in Hippocampal CA1. *Exp. Neurol.* **182**:470–475. [8]

Heidemann, S. R., and D. Bray. 2015. Tension-Driven Axon Assembly: A Possible Mechanism. *Front. Cell. Neurosci.* **9**:1–4. [3]

Heidenreich, M., and F. Zhang. 2016. Applications of CRISPR-Cas Systems in Neuroscience. *Nat Rev Neurosci.* **17**:36–44. [7]

Held, R., and A. Hein. 1963. Movement-Produced Stimulation in the Development of Visually Guided Behavior. *J. Comp. Physiol. Psych.* **56**:872–876. [2, 7]

Helfer, J. L., L. H. Calizo, W. K. Dong, et al. 2009. Binge-Like Postnatal Alcohol Exposure Triggers Cortical Gliogenesis in Adolescent Rats. *J. Comp. Neurol.* **514**:259–271. [9]

Heller, A. S., and B. J. Casey. 2016. The Neurodynamics of Emotion: Delineating Typical and Atypical Emotional Processes during Adolescence. *Dev. Sci.* **19**:3–18. [12]

Hemmings, H. C., Jr., M. H. Akabas, P. A. Goldstein, et al. 2005. Emerging Molecular Mechanisms of General Anesthetic Action. *Trends Pharmacol. Sci.* **26**:503–510. [10]

Henikoff, S., and J. M. Greally. 2016. Epigenetics, Cellular Memory and Gene Regulation. *Curr. Biol.* **26**:R644–R648. [6]

Henikoff, S., and M. M. Smith. 2015. Histone Variants and Epigenetics. *Cold Spring Harb. Perspect. Biol.* **7**:a019364. [6]

Hennou, S., I. Khalilov, D. Diabira, Y. Ben-Ari, and H. Gozlan. 2002. Early Sequential Formation of Functional GABA(a) and Glutamatergic Synapses on CA1 Interneurons of the Rat Foetal Hippocampus. *Eur. J. Neurosci.* **16**:197–208. [4]

Hensch, T. K. 2005. Critical Period Plasticity in Local Cortical Circuits. *Nat. Rev. Neurosci.* **6**:877–888. [5, 11, 13]

Hensch, T. K., M. Fagiolini, N. Mataga, et al. 1998. Local GABA Circuit Control of Experience-Dependent Plasticity in Developing Visual Cortex. *Science* **282**:1504–1508. [5]

Herculano-Houzel, S. 2009. The Human Brain in Numbers: A Linearly Scaled-up Primate Brain. *Front. Hum. Neurosci.* **3**:31. [4]

———. 2011. Scaling of Brain Metabolism with a Fixed Energy Budget Per Neuron: Implications for Neuronal Activity, Plasticity and Evolution. *PLoS One* **6**:e17514. [4]

Herculano-Houzel, S., and R. Lent. 2005. Isotropic Fractionator: A Simple, Rapid Method for the Quantification of Total Cell and Neuron Numbers in the Brain. *J. Neurosci.* **25**:2518–2521. [4]

Herculano-Houzel, S., M. H. J. Munk, S. Neuenschwander, and W. Singer. 1999. Precisely Synchronized Oscillatory Firing Patterns Require Electroencephalographic Activation. *J. Neurosci.* **19**:3992–4010. [2]

Herculano-Houzel, S., C. Watson, and G. Paxinos. 2013. Distribution of Neurons in Functional Areas of the Mouse Cerebral Cortex Reveals Quantitatively Different Cortical Zones. *Front. Neuroanat.* **7**:35. [4]

Herrmann, C. S., M. H. Munk, and A. K. Engel. 2004. Cognitive Functions of Gamma Band Activity: Memory Match and Utilization. *Trends Cogn. Sci.* **8**:347–355. [7]

Herting, M. M., and E. R. Sowell. 2017. Puberty and Structural Brain Development in Humans. *Front. Neuroendocrin.* **44**:122–137. [8]

Hess, R. F., and B. Thompson. 2015. Amblyopia and the Binocular Approach to Its Therapy. *Vis Res* **114**:4–16. [5]

Heyer, C. B., and H. D. Lux. 1976. Properties of a Facilitating Calcium Current in Pacemaker Neurons of the Snail Helix Pomatia. *J. Physiol. (Lond.)* **262**:319–348. [2]

Hier, D. B., and W. F. Crowley, Jr. 1982. Spatial Ability in Androgen-Deficient Men. *New Engl. J. Med.* **306**:1202–1205. [8]

Himmler, B. T., S. M. Pellis, and B. Kolb. 2013. Juvenile Play Experience Primes Neurons in the Medial Prefrontal Cortex to Be More Responsive to Later Experiences. *Neurosci. Lett.* **556**:42–45. [9]

Hirsch, S., and H. J. Luhmann. 2008. Pathway-Specificity in N-Methyl-D-Aspartate Receptor-Mediated Synaptic Inputs onto Subplate Neurons. *Neuroscience* **153**:1092–1102. [4]

Hochedlinger, K., and R. Jaenisch. 2015. Induced Pluripotency and Epigenetic Reprogramming. *Cold Spring Harb. Perspect. Biol.* **7**:a019448. [4]

Hoekstra, E. J., L. von Oerthel, A. J. A. van der Linden, et al. 2013. Lmx1a Is an Activator of Rgs4 and Grb10 and Is Responsible for the Correct Specification of Rostral and Medial mdDA Neurons. *Eur. J. Neurosci.* **37**:23–32. [6]

Hofer, S. B., and T. Bonhoeffer. 2010. Dendritic Spines: The Stuff That Memories Are Made of? *Curr. Biol.* **20**:R157–R159. [13]

Hofer, S. B., T. D. Mrsic-Flogel, T. Bonhoeffer, and M. Hübener. 2006. Prior Experience Enhances Plasticity in Adult Visual Cortex. *Nat. Neurosci.* **9**:127–132. [7]

———. 2009. Experience Leaves a Lasting Structural Trace in Cortical Circuits. *Nature* **457**:313–317. [5, 7]

Hoffmann, A., G. Daniel, U. Schmidt-Edelkraut, and D. Spengler. 2014. Roles of Imprinted Genes in Neural Stem Cells. *Epigenomics* **6**:515–532. [6]

Hoffmann, M. B., F. R. Kaule, N. Levin, et al. 2012. Plasticity and Stability of the Visual System in Human Achiasma. *Neuron* **75**:393–401. [3]

Hofstetter, S., N. Friedmann, and Y. Assaf. 2017. Rapid Language-Related Plasticity: Microstructural Changes in the Cortex after a Short Session of New Word Learning. *Brain Struct. Funct.* **222**:1231–1241. [9]

Hofstetter, S., I. Tavor, S. Tzur Moryosef, and Y. Assaf. 2013. Short-Term Learning Induces White Matter Plasticity in the Fornix. *J. Neurosci.* **33**:12844–12850. [9]

Hoftman, G. D., and D. A. Lewis. 2011. Postnatal Developmental Trajectories of Neural Circuits in the Primate Prefrontal Cortex: Identifying Sensitive Periods for Vulnerability to Schizophrenia. *Schizophr. Bull.* **37**:493–503. [12]

Hompes, T., B. Izzi, E. Gellens, et al. 2013. Investigating the Influence of Maternal Cortisol and Emotional State during Pregnancy on the DNA Methylation Status of the Glucocorticoid Receptor Gene (NR3C1) Promoter Region in Cord Blood. *J. Psychiatr. Res.* **47**:880–891. [6]

Hopfield, J. J. 1982. Neural Networks and Physical Systems with Emergent Collective Computational Abilities. *PNAS* **79**:2554–2558. [4]

Hsu, P. D., E. S. Lander, and F. Zhang. 2014. Development and Applications of CRISPR-Cas9 for Genome Engineering. *Cell* **157**:1262–1278. [3]

Huang, S., Y. Gu, E. M. Quinlan, and A. Kirkwood. 2010. A Refractory Period for Rejuvenating GABAergic Synaptic Transmission and Ocular Dominance Plasticity with Dark Exposure. *J. Neurosci.* **30**:16636–16642. [5]

Huang, X., S. K. Stodieck, B. Goetze, et al. 2015. Progressive Maturation of Silent Synapses Governs the Duration of a Critical Period. *PNAS* **112**:E3131–3140. [5]

Huang, Z. J., G. Di Cristo, and F. Ango. 2007. Development of GABA Innervation in the Cerebral and Cerebellar Cortices. *Nat. Rev. Neurosci.* **8**:673–686. [3]

Huang, Z. J., A. Kirkwood, T. Pizzorusso, et al. 1999. BDNF Regulates the Maturation of Inhibition and the Critical Period of Plasticity in Mouse Visual Cortex. *Cell* **98**:7965–7980. [5]

Hubel, D. H., and T. N. Wiesel. 1970. The Period of Susceptibility to the Physiological Effects of Unilateral Eye Closure in Kittens. *J. Physiol.* **206**:419–436. [11]

Hubel, D. H., T. N. Wiesel, and S. LeVay. 1977. Plasticity of Ocular Dominance Columns in Monkey Striate Cortex. *Philos Trans R Soc Lond Biol Sci.* **278**:377–409. [5]

Hübener, M., and T. Bonhoeffer. 2010. Searching for Engrams. *Neuron* **67**:363–371. [13]

———. 2014. Neuronal Plasticity: Beyond the Critical Period. *Cell* **159**:727–737. [13, 15]

Huerta, P. T., and J. E. Lisman. 1995. Bidirectional Synaptic Plasticity Induced by a Single Burst during Cholinergic Theta Oscillation in CA1 *in Vitro*. *Neuron* **15**:1053–1063. [2]

Humphrey, T. 1944. Primitive Neurons in the Embryonic Human Central Nervous System. *J. Comp. Neurol.* **81**:1–45. [3]

Hutson, L. D., and C. B. Chien. 2002. Pathfinding and Error Correction by Retinal Axons: The Role of Astray/Robo2. *Neuron* **33**:205–217. [3]

Huttenlocher, P. R. 1990. Morphometric Study of Human Cerebral Cortex Development. *Neuropsychologia* **28**:517–527. [7]

Iaccarino, H. F., A. C. Singer, A. J. Martorell, et al. 2016. Gamma Frequency Entrainment Attenuates Amyloid Load and Modifies Microglia. *Nature* **540**:230–235. [2]

Igarashi, K. M., L. Lu, L. L. Colgin, M.-B. Moser, and E. I. Moser. 2014. Coordination of Entorhinal-Hippocampal Ensemble Activity during Associative Learning. *Nature* **510**:143–147. [2]

Illingworth, R. S., and A. P. Bird. 2009. CpG Islands: A Rough Guide. *FEBS Lett.* **583**:1713–1720. [6]

Inacio, A. R., A. Nasretdinov, J. Lebedeva, and R. Khazipov. 2016. Sensory Feedback Synchronizes Motor and Sensory Neuronal Networks in the Neonatal Rat Spinal Cord. *Nat. Commun.* **7**:13060. [4]

Insel, T. R. 2010. Rethinking Schizophrenia. *Nature* **468**:187–193. [12]

Insel, T. R., B. Cuthbert, M. Garvey, et al. 2010. Research Domain Criteria (RDoC): Toward a New Classification Framework for Research on Mental Disorders. *Am. J. Psychiatry* **167**:748–751. [15]

Ishida, M., and G. E. Moore. 2013. The Role of Imprinted Genes in Humans. *Mol. Aspects Med.* **34**:826–840. [6]

Islam, S. M., Y. Shinmyo, T. Okafuji, et al. 2009. Draxin, a Repulsive Guidance Protein for Spinal Cord and Forebrain Commissures. *Science* **323**:388–393. [3]

Iwai, Y., M. Fagiolini, K. Obata, and T. K. Hensch. 2003. Rapid Critical Period Induction by Tonic Inhibition in Visual Cortex. *J. Neurosci.* **23**:6695–6702. [5]

Iyer, K. K., J. A. Roberts, L. Hellstrom-Westas, et al. 2015. Cortical Burst Dynamics Predict Clinical Outcome Early in Extremely Preterm Infants. *Brain* **138**:2206–2218. [4]

Izzi, L., and F. Charron. 2011. Midline Axon Guidance and Human Genetic Disorders. *Clin. Genet.* **80**:226–234. [3]

Jaffe, A. E., and R. A. Irizarry. 2014. Accounting for Cellular Heterogeneity Is Critical in Epigenome-Wide Association Studies. *Genome Biol.* **15**:R31. [6]

Jakab, A., G. Kasprian, E. Schwartz, et al. 2015. Disrupted Developmental Organization of the Structural Connectome in Fetuses with Corpus Callosum Agenesis. *NeuroImage* **111**:277–288. [3]

Janiesch, P. C., H. S. Krüger, B. Pöschel, and I. L. Hanganu-Opatz. 2011. Cholinergic Control in Developing Prefrontal-Hippocampal Networks. *J. Neurosci.* **31**:17955–17970. [4]

Jarcho, J. M., B. E. Benson, R. C. Plate, et al. 2012. Developmental Effects of Decision-Making on Sensitivity to Reward: An fMRI Study. *Dev. Cogn. Neurosci.* **2**:437–447. [15]

Jeanmonod, D., F. L. Rice, and H. Van der Loos. 1981. Mouse Somatosensory Cortex: Alterations in the Barrelfield Following Receptor Injury at Different Early Postnatal Ages. *Neuroscience* **6**:1503–1535. [2]

Jen, J. C., W.-M. M. Chan, T. M. Bosley, et al. 2004. Mutations in a Human ROBO Gene Disrupt Hindbrain Axon Pathway Crossing and Morphogenesis. *Science* **304**:1509–1513. [3, 4]

Jenkins, W. M., M. M. Merzenich, M. T. Ochs, T. Allard, and E. Guic-Robles. 1990. Functional Reorganization of Primary Somatosensory Cortex in Adult Owl Monkeys after Behaviorally Controlled Tactile Stimulation. *J. Neurophysiol.* **63**:82–104. [2]

Jevtovic-Todorovic, V., A. R. Absalom, K. Blomgren, et al. 2013. Anaesthetic Neurotoxicity and Neuroplasticity: An Expert Group Report and Statement Based on the BJA Salzburg Seminar. *Br. J. Anaesth.* **111**:143–151. [10]

Jinek, M., K. Chylinski, I. Fonfara, et al. 2012. A Programmable Dual-RNA-Guided DNA Endonuclease in Adaptive Bacterial Immunity. *Science* **337**:816–821. [3]

John, E. R., H. Ahn, L. Prichep, et al. 1980. Developmental Equations for the Electroencephalogram. *Science* **210**:1255–1258. [7]

Johnson, J. S., and E. L. Newport. 1989. Critical Period Effects in Second Language Learning: The Influence of Maturational State on the Acquisition of English as a Second Language. *Cogn. Psychol.* **21**:60–99. [11]

Johnson, M. H., and J. Shrager. 1996. Dynamic Plasticity Influences the Emergence of Function in a Simple Cortical Array. *Neural Netw.* **9**:1119–1129. [15]

Johnson, R. T., S. M. Breedlove, and C. L. Jordan. 2013. Androgen Receptors Mediate Masculinization of Astrocytes in the Rat Posterodorsal Medial Amygdala during Puberty. *J. Comp. Neurol.* **521**:2298–2309. [8]

Jones, M. J., A. P. Fejes, and M. S. Kobor. 2013. DNA Methylation, Genotype and Gene Expression: Who Is Driving and Who Is Along for the Ride? *Genome Biol.* **14**:126. [6]

Jones, M. J., S. R. Moore, and M. Kobor. 2018. Principles and Challenges of Applying Epigenetic Epidemiology to Psychology. *Annu. Rev. Psychol.* **69**:459–485. [6]

Jones, P. A. 2012. Functions of DNA Methylation: Islands, Start Sites, Gene Bodies and Beyond. *Nat. Rev. Genet.* **13**:484–492. [6]

Jordan, C. J., and S. L. Andersen. 2017. Sensitive Periods of Substance Abuse: Early Risk for the Transition to Dependence. *Dev. Cogn. Neurosci.* **25**:29–44. [9]

Juraska, J. M., C. L. Sisk, and L. L. DonCarlos. 2013. Sexual Differentiation of the Adolescent Rodent Brain: Hormonal Influences and Developmental Mechanisms. *Horm. Behav.* **64**:203–210. [8]

Kameyama, K., K. Sohya, T. Ebina, et al. 2010. Difference in Binocularity and Ocular Dominance Plasticity between GABAergic and Excitatory Cortical Neurons. *J. Neurosci.* **30**:1551–1559. [5]

Kaminsky, Z., M. Tochigi, P. Jia, et al. 2012. A Multi-Tissue Analysis Identifies HLA Complex Group 9 Gene Methylation Differences in Bipolar Disorder. *Mol. Psychiatry* **17**:728–740. [6]

Kaneko, M., D. Stellwagen, R. C. Malenka, and M. P. Stryker. 2008. Tumor Necrosis Factor-Alpha Mediates One Component of Competitive, Experience-Dependent Plasticity in Developing Visual Cortex. *Neuron* **58**:673–680. [5]

Kaneko, M., and M. P. Stryker. 2014. Sensory Experience during Locomotion Promotes Recovery of Function in Adult Visual Cortex. *eLife* **3**:e02798. [5, 7]

Kang, H.-J., J.-M. Kim, R. Stewart, et al. 2013. Association of SLC6A4 Methylation with Early Adversity, Characteristics and Outcomes in Depression. *Progress in Neuro-Psychopharmacology and Biological Psychiatry* **44**:23–28. [6]

Kanherkar, R. R., N. Bhatia-Dey, and A. B. Csoka. 2014. Epigenetics across the Lifespan. *Front. Cell Dev. Biol.* **2**:49. [9]

Kanjlia, S., C. Lane, L. Feigenson, and M. Bedny. 2016. Absence of Visual Experience Modifies the Neural Basis of Numerical Thinking. *PNAS* **113**:11172–11177. [11]

Kanold, P. O., P. Kara, R. C. Reid, and C. J. Shatz. 2003. Role of Subplate Neurons in Functional Maturation of Visual Cortical Columns. *Science* **301**:521–525. [4]

Kanold, P. O., and C. J. Shatz. 2006. Subplate Neurons Regulate Maturation of Cortical Inhibition and Outcome of Ocular Dominance Plasticity. *Neuron* **51**:627–638. [4]

Kaplan, B. J., J. E. Fisher, S. G. Crawfor, C. J. Field, and B. Kolb. 2004. Improved Mood and Behavior during Treatment with a Mineral-Vitamin Supplement: An Open-Label Case Series of Children. *J. Child Adolesc. Psychopharmacol.* **14**:115–122. [9]

Katagiri, H., M. Fagiolini, and T. K. Hensch. 2007. Optimization of Somatic Inhibition at Critical Period Onset in Mouse Visual Cortex. *Neuron* **53**:805–812. [5]

Kayser, C., R. F. Salazar, and P. König. 2003. Responses to Natural Scenes in Cat V1. *J. Neurophysiol.* **90**:1910–1920. [2]

Kelly, A. M., A. Di Martino, L. Q. Uddin, et al. 2009. Development of Anterior Cingulate Functional Connectivity from Late Childhood to Early Adulthood. *Cereb. Cortex* **19**:640–657. [15]

Kempermann, G. 2011. Seven Principles in the Regulation of Adult Neurogenesis. *Eur. J. Neurosci.* **33**:1018–1024. [13]

Kessler, R. C., P. Berglund, O. Demler, et al. 2005. Lifetime Prevalence and Age-of-Onset Distributions of DSM-IV Disorders in the National Comorbidity Survey Replication. *Arch. Gen. Psychiatry* **62**:593–602. [12]

Khazipov, R., M. Esclapez, O. Caillard, et al. 2001. Early Development of Neuronal Activity in the Primate Hippocampus *in Utero. J. Neurosci.* **21**:9770–9781. [4]

Khazipov, R., and H. J. Luhmann. 2006. Early Patterns of Electrical Activity in the Developing Cerebral Cortex of Humans and Rodents. *Trends Neurosci.* **29**:414–418. [4]

Khazipov, R., A. Sirota, X. Leinekugel, et al. 2004. Early Motor Activity Drives Spindle Bursts in the Developing Somatosensory Cortex. *Nature* **432**:758–761. [2, 4]

Khundrakpam, B. S., A. Reid, J. Brauer, et al. 2013. Developmental Changes in Organization of Structural Brain Networks. *Cereb. Cortex* **23**:2072–2085. [9]

Kiessling, M. C., A. Büttner, C. Butti, et al. 2014. Cerebellar Granule Cells Are Generated Postnatally in Humans. *Brain Struct. Funct.* **219**:1271–1286. [3]

Kilgard, M. P. 2012. Harnessing Plasticity to Understand Learning and Treat Disease. *Trends Neurosci.* **36**:715–722. [13]

Killingsworth, M. A., and D. T. Gilbert. 2010. A Wandering Mind Is an Unhappy Mind. *Science* **330**:932. [15]

Kinnally, E. L., C. Feinberg, D. Kim, et al. 2011. DNA Methylation as a Risk Factor in the Effects of Early Life Stress. *Brain Behav. Immun.* **25**:1548–1553. [6]

Kinsbourne, M., and R. E. Hicks. 1978. Functional Cerebral Space: A Model for Overflow, Transfer and Interference Effects in Human Performance: A Tutorial Review. In: Attention and Performance, ed. J. Requin, pp. 345–362, vol. VII. Hillsdale, NJ: Erlbaum. [13]

Kiorpes, L. 2015. Visual Development in Primates: Neural Mechanisms and Critical Periods. *Dev. Neurobiol.* **75**:1080–1090. [5]

Kiorpes, L., C. Tang, and J. A. Movshon. 2006. Sensitivity to Visual Motion in Amblyopic Macaque Monkeys. *Vis Neurosci* **23**:247–256. [5]

Kitzbichler, M. G., S. Khan, S. Ganesan, et al. 2015. Altered Development and Multifaceted Band-Specific Abnormalities of Resting State Networks in Autism. *Biol. Psychiatry* **77**:794–804. [10]

Kleim, J. A., S. Barbay, N. R. Cooper, et al. 2002. Motor Learning-Dependent Synaptogenesis Is Localized to Functionally Reorganized Motor Cortex. *Neurobiol. Learn. Mem.* **77**:63–77. [13]

Kleinschmidt, A., M. F. Bear, and W. Singer. 1987. Blockade of "NMDA" Receptors Disrupts Experience-Dependent Plasticity of Kitten Striate Cortex. *Science* **238**:355–358. [2]

Klengel, T., D. Mehta, C. Anacker, et al. 2012. Allele-Specific FKBP5 DNA Demethylation Mediates Gene–Childhood Trauma Interactions. *Nat. Neurosci.* **16**:33–41. [6]

Klengel, T., J. Pape, E. B. Binder, and D. Mehta. 2014. The Role of DNA Methylation in Stress-Related Psychiatric Disorders. *Neuropharmacology* **80**:115–132. [6]

Knudsen, E. I. 2002. Instructed Learning in the Auditory Localization Pathway of the Barn Owl. *Nature* **417**:322–328. [7]

———. 2004. Sensitive Periods in the Development of the Brain and Behavior. *J. Cogn. Neurosci.* **16**:1412–1425. [9]

Knudsen, E. I., and P. F. Knudsen. 1989. Visuomotor Adaptation to Displacing Prisms by Adult and Baby Barn Owls. *J. Neurosci.* **9**:3297–3305. [2, 7]

Kobayashi, A., R. L. Parker, A. P. Wright, et al. 2014. Lynx1 Supports Neuronal Health in the Mouse Dorsal Striatum during Aging: An Ultrastructural Investigation. *J. Mol. Neurosci.* **53**:525–536. [7]

Kobayashi, Y., Z. Ye, and T. K. Hensch. 2015. Clock Genes Control Cortical Critical Period Timing. *Neuron* **86**:264–275. [4]

Kohonen, T. 1982. Self-Organized Formation of Topologically Correct Feature Maps. *Biol. Cybern.* **43**:59–69. [2]

Kolb, B. 1995. Brain Plasticity and Behavior. Hillsdale, NJ: Lawrence Erlbaum. [9]

Kolb, B., and R. Gibb. 1991. Sparing of Function after Neonatal Frontal Lesions Correlates with Increased Cortical Dendritic Branching: A Possible Mechanism for the Kennard Effect. *Behav. Brain Res.* **43**:51–56. [9]

———. 2010. Tactile Stimulation Facilitates Functional Recovery and Dendritic Change after Neonatal Medial Frontal or Posterior Parietal Lesions in Rats. *Behav. Brain Res.* **214**:115–120. [9]

Kolb, B., R. Gibb, and G. Gorny. 2003. Experience-Dependent Changes in Dendritic Arbor and Spine Density in Neocortex Vary with Age and Sex. *Neurobiol. Learn. Mem.* **79**:1–10. [9]

Kolb, B., R. Gibb, G. Gorny, and I. Q. Whishaw. 1998. Possible Brain Regrowth after Cortical Lesions in Rats. *Behav. Brain Res.* **91**:127–141. [9]

Kolb, B., L. Harker, S. R. de Melo, and R. Gibb. 2017. Stress and Prefrontal Cortical Plasticity in the Developing Brain. *Cogn. Dev.* **42**:15–26. [9]

Kolb, B., C. Morshead, C. Gonzalez, et al. 2007. Growth Factor-Stimulated Generation of New Cortical Tissue and Functional Recovery after Stroke Damage to the Motor Cortex of Rats. *J. Cereb. Blood Flow Metab.* **27**:983–997. [9]

Kolb, B., R. Mychasiuk, A. Muhammad, and R. Gibb. 2013. Brain Plasticity in the Developing Brain. *Prog. Brain Res.* **207**:35–64. [9]

Koleske, A. J. 2013. Molecular Mechanisms of Dendrite Stability. *Nat. Rev. Neurosci.* **14**:536–550. [9]

Kopell, N., G. B. Ermentrout, M. A. Whittington, and R. D. Traub. 2000. Gamma Rhythms and Beta Rhythms Have Different Synchronization Properties. *PNAS* **97**:1867–1872. [2]

Korecka, J. A., S. Levy, and O. Isacson. 2016. *In Vivo* Modeling of Neuronal Function, Axonal Impairment and Connectivity in Neurodegenerative and Neuropsychiatric Disorders Using Induced Pluripotent Stem Cells. *Mol. Cell. Neurosci.* **73**:3–12. [3]

Koser, D. E., A. J. Thompson, S. K. Foster, et al. 2016. Mechanosensing Is Critical for Axon Growth in the Developing Brain. *Nat. Neurosci.* **19**:1592–1598. [3]

Kosik, K. S., T. J. Sejnowski, M. E. Raichle, A. Ciechanover, and D. Baltimore. 2016. War and Cancer: A Path toward Understanding Neurodegeneration. *Science* **353**:872–873. [4]

Koss, W. A., C. E. Belden, A. D. Hristov, and J. M. Juraska. 2014. Dendritic Remodeling in the Adolescent Medial Prefrontal Cortex and the Basolateral Amygdala of Male and Female Rats. *Synapse* **68**:61–72. [8]

Koss, W. A., M. M. Lloyd, R. N. Sadowski, L. M. Wise, and J. M. Juraska. 2015. Gonadectomy before Puberty Increases the Number of Neurons and Glia in the Medial Prefrontal Cortex of Female, but Not Male, Rats. *Dev. Psychobiol.* **57**:305–312. [8]

Krishnan, K., B. S. Wang, J. Lu, et al. 2015. MeCP2 Regulates the Timing of Critical Period Plasticity That Shapes Functional Connectivity in Primary Visual Cortex. *PNAS* **112**:E4782–4791. [5]

Kronenberg, G., A. Bick-Sander, E. Bunk, et al. 2006. Physical Exercise Prevents Age-Related Decline in Precursor Cell Activity in the Mouse Dentate Gyrus. *Neurobiol. Aging* **27**:1505–1513. [13]

Kuhl, P. K. 2004. Early Language Acquisition: Cracking the Speech Code. *Nat. Rev. Neurosci.* **5**:831–843. [6]

Kuhl, P. K., E. Stevens, A. Hayashi, et al. 2006. Infants Show a Facilitation Effect for Native Language Phonetic Perception between 6 and 12 Months. *Dev. Sci.* **9**:F13–F21. [7]

Kuhl, P. K., F.-M. Tsao, and H.-M. Liu. 2003. Foreign-Language Experience in Infancy: Efects of Short-Term Exposure and Social Interaction on Phonetic Learning. *PNAS* **100**:9096–9101. [7]

Kuhl, P. K., K. A. Williams, F. Lacerda, K. N. Stevens, and B. Lindblom. 1992. Linguistic Experience Alters Phonetic Perception in Infants by 6 Months of Age. *Science* **255**:606–608. [7]

Kuhlman, S. J., N. D. Olivas, E. Tring, et al. 2013. A Disinhibitory Microcircuit Initiates Critical-Period Plasticity in the Visual Cortex. *Nature* **56**:908–923. [5]

Kühn, S., T. Gleich, R. C. Lorenz, U. Lindenberger, and J. Gallinat. 2014. Playing Super Mario Induces Structural Brain Plasticity: Grey Matter Changes Resulting from Training with a Commercial Video Game. *Mol. Psychiatry* **19**:265–271. [13]

Kühn, S., and U. Lindenberger. 2016. Research on Human Plasticity in Adulthood: A Lifespan Agenda. In: Handbook of the Psychology of Aging, ed. K. W. Schaie and S. L. Willis, pp. 105–123. San Diego: Elsevier. [13]

Kundakovic, M., and F. A. Champagne. 2015. Early-Life Experience, Epigenetics, and the Developing Brain. *Neuropsychopharmacology* **40**:141–153. [6]

Kurjak, A., R. K. Pooh, L. T. Merce, et al. 2005. Structural and Functional Early Human Development Assessed by Three-Dimensional and Four-Dimensional Sonography. *Fertil. Steril.* **84**:1285–1299. [3]

Kuwajima, T., Y. Yoshida, N. Takegahara, et al. 2012. Optic Chiasm Presentation of Semaphorin6d in the Context of Plexin-A1 and Nr-CAM Promotes Retinal Axon Midline Crossing. *Neuron* **74**:676–690. [3]

Kuzawa, C. W., H. T. Chugani, L. I. Grossman, et al. 2014. Metabolic Costs and Evolutionary Implications of Human Brain Development. *PNAS* **111**:13010–13015. [13]

Lam, L. L., E. Emberly, H. B. Fraser, et al. 2012. Factors Underlying Variable DNA Methylation in a Human Community Cohort. *PNAS* **109**:17253–17260. [6]

Lamminmaki, S., S. Massinen, J. Nopola-Hemmi, J. Kere, and R. Hari. 2012. Human ROBO1 Regulates Interaural Interaction in Auditory Pathways. *J. Neurosci.* **32**:966–971. [3]

Lander, S. S., D. Linder-Shacham, and S. Gaisler, I. 2017. Differential Effects of Social Isolation in Adolescent and Adult Mice on Behavior and Cortical Gene Expression. *Behav. Brain Res.* **316**:245–254. [15]

Lane, C., S. Kanjlia, A. Omaki, and M. Bedny. 2015. "Visual" Cortex of Congenitally Blind Adults Responds to Syntactic Movement. *J. Neurosci. Methods* **35**:12859–12868. [11]

Langers, D. R. M., R. M. Sanchez-Panchuelo, S. T. Francis, K. Krumbholz, and D. A. Hall. 2014. Neuroimaging Paradigms for Tonotopic Mapping (II): The Influence of Acquisition Protocol. *NeuroImage* **100**:663–675. [13]

Larroche, J. C. 1981. The Marginal Layer in the Neocortex of a 7 Week-Old Human Embryo: A Light and Electron Microscopic Study. *Anat. Embryol.B160* **162**:301–312. [3]

Laughlin, S. B., and T. J. Sejnowski. 2003. Communication in Neuronal Networks. *Science* **301**:1870–1874. [4]

Laumonnerie, C., Y. G. Tong, H. Alstermark, and S. I. Wilson. 2015. Commissural Axonal Corridors Instruct Neuronal Migration in the Mouse Spinal Cord. *Nat. Commun.* **6**:7028. [3]

Lawrence, P. 1973. The Development of Spatial Patterns in the Integument of Insects. In: Developmental Systems: Insects, ed. S. Counce and C. Waddington, pp. 157–209. New York: Academic Press. [4]

LeBlanc, J. J., G. DeGregorio, E. Centofante, et al. 2015. Visual Evoked Potential Detect Cortical Processing Deficits in Rett Syndrome. *Ann. Neurol.* **78**:775–786. [7, 10]

LeCun, Y., Y. Bengio, and G. Hinton. 2015. Deep Learning. *Nature* **521**:436–444. [13]

Lee, C. W., R. J. Cooper, and T. Austin. 2017a. Diffuse Optical Tomography to Investigate the Newborn Brain. *Pediatr. Res.* **82**:376–386. [7]

Lee, F. S., H. Heimer, J. N. Giedd, et al. 2014. Adolescent Mental Health: Opportunity and Obligation. *Science* **346**:547–549. [7, 12]

Lee, J. M., O. Akeju, K. Terzakis, et al. 2017b. A Prospective Study of Age-Dependent Changes in Propofol-Induced Electroencephalogram Oscillations in Children. *Anesthesiology* **127**:293–306. [10]

Lee, W. C., H. Huang, G. Feng, et al. 2006. Dynamic Remodeling of Dendritic Arbors in GABAergic Interneurons of Adult Visual Cortex. *PLoS Biol.* **4**:e29. [5]

Lefebvre, J. L., D. Kostadinov, W. V. Chen, T. Maniatis, and J. R. Sanes. 2012. Proto-cadherins Mediate Dendritic Self-Avoidance in the Mammalian Nervous System. *Nature* **488**:517. [3]

Lehmann, K., and S. Löwel. 2008. Age-Dependent Ocular Dominance Plasticity in Adult Mice. *PLoS One* **3**:e3120. [7]

Leicht, G., S. Vauth, N. Polomac, et al. 2016. EEG-Informed fMRI Reveals a Disturbed Gamma-Band-Specific Network in Subjects at High Risk for Psychosis. *Schizophr. Bull.* **42**:239–249. [4]

Le Magueresse, C., and H. Monyer. 2013. GABAergic Interneurons Shape the Func-tional Maturation of the Cortex. *Neuron* **77**:388–405. [10]

Lemonnier, E., N. Villeneuve, S. Sonie, et al. 2017. Effects of Bumetanide on Neurobe-havioral Function in Children and Adolescents with Autism Spectrum Disorders. *Transl. Psychiatry* **7**:e1056. [4]

Lennartsson, A., E. Arner, M. Fagiolini, et al. 2015. Remodeling of Retrotransposon Elements during Epigenetic Induction of Adult Visual Cortical Plasticity by HDAC Inhibitors. *Epigenetics Chromatin* **8**:55. [5–7]

Lenroot, R. K., and J. N. Giedd. 2006. Brain Development in Children and Adolescents: Insights from Anatomical Magnetic Resonance Imaging. *Neurosci. Biobehav. Rev.* **30**:718–729. [8]

Lerch, J. P., A. J. W. van der Kouwe, A. Raznahan, et al. 2017. Studying Neuroanatomy Using MRI. *Nat. Neurosci.* **20**:314–326. [13]

Lesnick, T. G., S. Papapetropoulos, D. C. Mash, et al. 2007. A Genomic Pathway Ap-proach to a Complex Disease: Axon Guidance and Parkinson Disease. *PLoS Genet.* **3**:e98. [3]

Letzkus, J. J., S. B. Wolff, E. M. Meyer, et al. 2011. A Disinhibitory Microcircuit for Associative Fear Learning in the Auditory Cortex. *Nature* **480**:331. [2, 5]

Leung, L. C., V. Urbančič, M.-L. Baudet, et al. 2013. Coupling of NF-Protocadherin Signaling to Axon Guidance by Cue-Induced Translation. *Nat. Neurosci.* **16**:166–173. [3]

LeVay, S., M. P. Stryker, and C. J. Shatz. 1978. Ocular Dominance Columns and Their Development in Layer IV of the Cat's Visual Cortex: A Quantitative Study. *J. Comp. Neurol.* **179**:223–244. [5]

Levenstein, D., B. O. Watson, J. Rinzel, and G. Buzsáki. 2017. Sleep Regulation of the Distribution of Cortical Firing Rates. *Curr. Opin. Neurobiol.* **44**:34–42. [4]

Levi, D. M., D. C. Knill, and D. Bavelier. 2015. Stereopsis and Amblyopia: A Mini-Review. *Vis Res* **114**:17–30. [5]

Lewis, L. D., V. S. Weiner, E. A. Mukamel, et al. 2012. Rapid Fragmentation of Neuronal Networks at the Onset of Propofol-Induced Unconsciousness. *PNAS* **109**:E3377–3386. [10]

Li, Y., H. Lu, P. Cheng, et al. 2012. Clonally Related Visual Cortical Neurons Show Similar Stimulus Feature Selectivity. *Nature* **486**:118–121. [3]

Li, Y., J. Muffat, A. Omer, et al. 2017. Induction of Expansion and Folding in Human Cerebral Organoids. *Cell Stem Cell* **20**:385–396. [3]

Lieberam, I., D. Agalliu, T. Nagasawa, J. Ericson, and T. M. Jessell. 2005. A Cxcl12-Cxcr4 Chemokine Signaling Pathway Defines the Initial Trajectory of Mammalian Motor Axons. *Neuron* **47**:667–679. [4]

Lima, B., W. Singer, and S. Neuenschwander. 2011. Gamma Responses Correlate with Temporal Expectation in Monkey Primary Visual Cortex. *J. Neurosci.* **31**:15919–15931. [2]

Lin, A., S. J. Wood, B. Nelson, et al. 2015. Outcomes of Nontransitioned Cases in a Sample at Ultra-High Risk for Psychosis. *Am. J. Psychiatry* **172**:249–258. [12]

Lin, L., T. G. Lesnick, D. M. Maraganore, and O. Isacson. 2009. Axon Guidance and Synaptic Maintenance: Preclinical Markers for Neurodegenerative Disease and Therapeutics. *Trends Neurosci* **32**:142–149. [3]

Lindenberger, U. 2014. Human Cognitive Aging: Corriger la Fortune? *Science* **346**:572–578. [13]

Lindenberger, U., E. Wenger, and M. Lövdén. 2017. Towards a Stronger Science of Human Plasticity. *Nat. Rev. Neurosci.* **18**:261–262. [13]

Lister, R., E. A. Mukamel, J. R. Nery, et al. 2013. Global Epigenomic Reconfiguration During Mammalian Brain Development. *Science* **341**:1237905–1237905. [6]

Liu, S.-C., T. Delbruck, G. Indiveri, A. Whatley, and R. Douglas, eds. 2015. Event-Based Neuromorphic Systems. New York: Wiley. [4]

Liu, X., S. Ramirez, P. T. Pang, et al. 2012a. Optogenetic Stimulation of a Hippocampal Engram Activates Fear Memory Recall. *Nature* **484**:381–385. [2]

Liu, Y., S. K. Murphy, A. P. Murtha, et al. 2012b. Depression in Pregnancy, Infant Birth Weight and DNA Methylation of Imprint Regulatory Elements. *Epigenetics* **7**:735–746. [6]

Liu, Y., J. Shi, C.-C. Lu, et al. 2005. Ryk-Mediated Wnt Repulsion Regulates Posterior-Directed Growth of Corticospinal Tract. *Nat. Neurosci.* **8**:1151–1159. [3]

Llinas, R. R., U. Ribary, D. Jeanmonod, E. Kronberg, and P. P. Mitra. 1999. Thalamo-cortical Dysrhythmia: A Neurological and Neuropsychiatric Syndrome Characterized by Magnetoencephalography. *PNAS* **96**:15222–15227. [15]

Logue, S., J. Chein, T. Gould, E. Holliday, and L. Steinberg. 2014 Adolescent Mice, Unlike Adults, Consume More Alcohol in the Presence of Peers Than Alone. *Dev. Sci.* **17**:79–85. [14]

Lokmane, L., R. Proville, N. Narboux-Nême, et al. 2013. Sensory Map Transfer to the Neocortex Relies on Pretarget Ordering of Thalamic Axons. *Curr. Biol.* **23**:810–816. [3]

López-Bendito, G., A. Cautinat, J. A. Sánchez, et al. 2006. Tangential Neuronal Migration Controls Axon Guidance: A Role for Neuregulin-1 in Thalamocortical Axon Navigation. *Cell* **125**:127–142. [3]

LoTurco, J. J., D. F. Owens, M. J. S. Heath, M. B. E. Davis, and A. R. Kriegstein. 1995. GABA and Glutamate Depolarize Cortical Progenitor Cells and Inhibit DNA Synthesis. *Neuron* **15**:1287–1298. [4]

Lövdén, M., L. Bäckman, U. Lindenberger, S. Schaefer, and F. Schmiedek. 2010. A Theoretical Framework for the Study of Adult Cognitive Plasticity. *Psychol. Bull.* **136**:659–676. [13, 15]

Lövdén, M., S. Schaefer, H. Noack, et al. 2011. Performance-Related Increases in Hippocampal N-Acetylaspartate (NAA) Induced by Spatial Navigation Training Are Restricted to BDNF Val Homozygotes. *Cereb. Cortex* **21**:1435–1442. [13]

———. 2012. Spatial Navigation Training Protects the Hippocampus against Age-Related Changes during Early and Late Adulthood. *Neurobiol. Aging* **33**:620. e629–620.e622. [13]

Lövdén, M., E. Wenger, J. Mårtensson, U. Lindenberger, and L. Bäckman. 2013. Structural Brain Plasticity in Adult Learning and Development. *Neurosci. Biobehav. Rev.* **37**:2296–2310. [13]

Löwel, S., and W. Singer. 1992. Selection of Intrinsic Horizontal Connections in the Visual Cortex by Correlated Neuronal Activity. *Science* **255**:209–212. [2]

Lubin, F. D., T. L. Roth, and J. D. Sweatt. 2008. Epigenetic Regulation of BDNF Gene Transcription in the Consolidation of Fear Memory. *J. Neurosci.* **28**:10576–10586. [6]

Luders, E., K. L. Narr, P. M. Thompson, and A. W. Toga. 2009. Neuroanatomical Correlates of Intelligence. *Intelligence* **37**:156–163. [13]

Luhmann, H. J. 2017. Review of Imaging Network Activities in Developing Rodent Cerebral Cortex *in Vivo*. *Neurophotonics* **4**:031202. [4]

Luhmann, H. J., W. Kilb, and I. L. Hanganu-Opatz. 2009. Subplate Cells: Amplifiers of Neuronal Activity in the Developing Cerebral Cortex. *Front. Neuroanat.* **3**:19. [4]

Luhmann, H. J., R. A. Reiprich, I. L. Hanganu, and W. Kilb. 2000. Cellular Physiology of the Neonatal Rat Cerebral Cortex: Intrinsic Membrane Properties, Sodium and Calcium Currents. *J. Neurosci. Res.* **62**:574–584. [4]

Luhmann, H. J., A. Sinning, J. W. Yang, et al. 2016. Spontaneous Neuronal Activity in Developing Neocortical Networks: From Single Cells to Large-Scale Interactions. *Front. Neur. Circuits* **10**:40. [4]

Lumsden, A. G., and A. M. Davies. 1983. Earliest Sensory Nerve Fibres Are Guided to Peripheral Targets by Attractants Other Than Nerve Growth Factor. *Nature* **306**:786–788. [3]

Luo, H., and D. Poeppel. 2012. Cortical Oscillations in Auditory Perception and Speech: Evidence for Two Temporal Windows in Human Auditory Cortex. *Front. Psychol.* **3**:170. [7]

Lv, L., X. Han, Y. Sun, X. Wang, and Q. Dong. 2012. Valproic Acid Improves Locomotion *in Vivo* after Sci and Axonal Growth of Neurons *in Vitro*. *Exp. Neurol.* **233**:783–790. [3]

Lyuksyutova, A. I., C.-C. Lu, N. Milanesio, et al. 2003. Anterior-Posterior Guidance of Commissural Axons by Wnt-Frizzled Signaling. *Science* **302**:1984–1988. [3]

Mah, S., M. R. Nelson, L. E. DeLisi, et al. 2006. Identification of the Semaphorin Receptor PLXNA2 as a Candidate for Susceptibility to Schizophrenia. *Mol. Psychiatry* **11**:471–478. [3]

Mahmoudzadeh, M., G. Dehaene-Lambertz, M. Fournier, et al. 2013. Syllabic Discrimination in Premature Human Infants Prior to Complete Formation of Cortical Layers. *PNAS* **110**:4846–4851. [7]

Mainen, Z. F., and T. J. Sejnowski. 1995. Reliability of Spike Timing in Neocortical Neurons. *Science* **268**:1503–1506. [2]

Maitre, N. L., W. E. Lambert, J. L. Aschner, and A. P. Key. 2013. Cortical Speech Sound Differentiation in the Neonatal Intensive Care Unit Predicts Cognitive and Language Development in the First 2 Years of Life. *Dev. Med. Child Neurol.* **55**:834–839. [7]

Majdan, M., and C. J. Shatz. 2006. Effects of Visual Experience on Activity-Dependent Gene Regulation in Cortex. *Nat Neurosci.* **9**:650–659. [7]

Makita, T., H. M. Sucov, C. E. Gariepy, M. Yanagisawa, and D. D. Ginty. 2008. Endothelins Are Vascular-Derived Axonal Guidance Cues for Developing Sympathetic Neurons. *Nature* **452**:759–763. [3]

Mandai, K., T. Guo, C. St. Hillaire, et al. 2009. Lig Family Receptor Tyrosine Kinase-Associated Proteins Modulate Growth Factor Signals during Neural Development. *Neuron* **63**:614–627. [3]

Manent, J. B., I. Jorquera, I. Mazzucchelli, et al. 2007. Fetal Exposure to GABA-Acting Antiepileptic Drugs Generates Hippocampal and Cortical Dysplasias. *Epilepsia* **48**:684–693. [4]

Manent, J. B., and J. J. LoTurco. 2009. Dcx Reexpression Reduces Subcortical Band Heterotopia and Seizure Threshold in an Animal Model of Neuronal Migration Disorder. *Nat. Med.* **15**:84–90. [4]

Manitt, C., C. Eng, M. Pokinko, et al. 2013. DCC Orchestrates the Development of the Prefrontal Cortex During Adolescence and Is Altered in Psychiatric Patients. *Transl. Psychiatry* **3**:e338. [3]

Maret, S., U. Faraguna, A. B. Nelson, C. Cirelli, and G. Tononi. 2012. Sleep and Waking Modulate Spine Turnover in the Adolescent Mouse Cortex. *Nat. Neurosci.* **14**:1418–1420. [14]

Margolis, K. G., Z. Li, K. Stevanovic, et al. 2016. Serotonin Transporter Variant Drives Preventable Gastrointestinal Abnormalities in Development and Function. *J. Clin. Invest.* **126**:2221–2235. [3]

Marillat, V., C. Sabatier, V. Failli, et al. 2004. The Slit Receptor Rig-1/Robo3 Controls Midline Crossing by Hindbrain Precerebellar Neurons and Axons. *Neuron* **43**:69–79. [3]

Markham, J. A., J. R. Morris, and J. M. Juraska. 2007. Neuron Number Decreases in the Rat Ventral, but Not Dorsal, Medial Prefrontal Cortex between Adolescence and Adulthood. *Neuroscience* **144**:961–968. [8]

Markram, H., J. Lübke, M. Frotscher, and B. Sakmann. 1997. Regulation of Synaptic Efficacy by Coincidence of Postsynaptic APS and EPSPs. *Science* **275**:213–215. [2]

Marshall, L., H. Helgadóttir, M. Mölle, and J. Born. 2006. Boosting Slow Oscillations during Sleep Potentiates Memory. *Nature* **44**:610–613. [2]

Mårtensson, J., J. Eriksson, N. C. Bodammer, et al. 2012. Growth of Language-Related Brain Areas after Foreign Language Learning. *NeuroImage* **63**:240–244. [13]

Martinez, M. E., M. Charalambous, A. Saferali, et al. 2014. Genomic Imprinting Variations in the Mouse Type 3 Deiodinase Gene between Tissues and Brain Regions. *Mol. Endocrinol.* **28**:1875–1886. [6]

Mash, C., M. H. Bornstein, and M. E. Arterberry. 2013. Brain Dynamics in Young Infants' Recognition of Faces. *Neuroreport* **24**:359–363. [7]

Mastick, G. S., and S. S. Easter. 1996. Initial Organization of Neurons and Tracts in the Embryonic Mouse Fore- and Midbrain. *Dev. Biol.* **173**:79–94. [3]

Masuda, Y., S. O. Dumoulin, S. Nakadomari, and B. A. Wandell. 2008. V1 Projection Zone Signals in Human Macular Degeneration Depend on Task, Not Stimulus. *Cereb. Cortex* **18**:2483–2493. [5]

Mataga, N., Y. Mizuguchi, and T. K. Hensch. 2004. Experience-Dependent Pruning of Dendritic Spines in Visual Cortex by Tissue Plasminogen Activator. *Neuron* **44**:1031–1041. [5, 7]

Mataga, N., N. Nagai, and T. K. Hensch. 2002. Permissive Proteolytic Activity for Visual Cortical Plasticity. *PNAS* **99**:7717–7721. [5]

Matthies, U., J. Balog, and K. Lehmann. 2013. Temporally Coherent Visual Stimuli Boost Ocular Dominance Plasticity. *J. Neurosci.* **33**:11774–11778. [7]

Mattick, J. S., and I. V. Makunin. 2006. Non-Coding RNA. *Hum. Mol. Genet.* R17–29. [6]

Maurer, D., and T. K. Hensch. 2012. Amblyopia: Background to the Special Issue on Stroke Recovery. *Dev. Psychobiol.* **54**:224–238. [11]

Maya Vetencourt, J. F., A. Sale, A. Viegi, et al. 2008. The Antidepressant Fluoxetine Restores Plasticity in the Adult Visual Cortex. *Science* **18**:385–388. [5]

Mayberry, R. 1998. The Critical Period for Language Acquisition and the Deaf Child's Language Comprehension: A Psycholinguistic Approach. *Ann. Sci. Univ. Franche-Comté: Bull. d'Audiophonologie* **15**:349–358. [11]

Mayer, C., R. C. Bandler, and G. Fishell. 2016. Lineage Is a Poor Predictor of Interneuron Positioning within the Forebrain. *Neuron* **92**:45–51. [3]

McCall, R. B., and M. S. Carriger. 1993. A Meta-Analysis of Infant Habituation and Recognition Memory Performance as Predictors of Later IQ. *Child Dev.* **64**:57–79. [7]

McCarthy, M. M., E. N. Brown, and N. Kopell. 2008. Potential Network Mechanisms Mediating Electroencephalographic Beta Rhythm Changes during Propofol-Induced Paradoxical Excitation. *J. Neurosci.* **28**:13488–13504. [10]

McClelland, J. L. 1996. Integration of Information: Reflections on the Theme of Attention and Performance XVI. In: Attention and Performance XVI, ed. T. Inui and J. L. McClelland, pp. 633–656. Cambridge, MA: MIT Press. [13]

McConnell, S. K., A. Ghosh, and C. J. Shatz. 1989. Subplate Neurons Pioneer the First Axon Pathway from the Cerebral Cortex. *Science* **245**:978–982. [4]

McCurry, C. L., J. D. Shepherd, D. Tropea, et al. 2010. Loss of Arc Renders the Visual Cortex Impervious to the Effects of Sensory Deprivation or Experience. *Nat. Neurosci.* **13**:450–457. [5]

McGee, A. W., Y. Yang, Q. S. Fischer, N. W. Daw, and S. M. Strittmatter. 2005. Experience-Driven Plasticity of Visual Cortex Limited by Myelin and Nogo Receptor. *Science* **309**:2222–2226. [5]

McGorry, P. D. 2010. Risk Syndromes, Clinical Staging and DSM V: New Diagnostic Infrastructure for Early Intervention in Psychiatry. *Schizophr. Res.* **120**:49–53. [12]

McGowan, P. O., A. Sasaki, A. C. D'Alessio, et al. 2009. Epigenetic Regulation of the Glucocorticoid Receptor in Human Brain Associates with Childhood Abuse. *Nat. Neurosci.* **12**:342–348. [6]

Meaney, M. J. 2010. Epigenetics and the Biological Definition of Gene X Environment Interactions. *Child Dev.* **81**:41–79. [6]

Mehler, J., P. Jusczyk, G. Lambertz, et al. 1988. A Precursor of Language Acquisition in Young Infants. *Cognition* **29**:143–178. [7]

Meister, M., R. O. L. Wong, D. A. Baylor, and C. J. Shatz. 1991. Synchronous Bursts of Action Potentials in Ganglion Cells of the Developing Mammalian Retina. *Science* **252**:939–943. [2]

Melin, M., B. Carlsson, H. Anckarsater, et al. 2006. Constitutional Downregulation of Sema5a Expression in Autism. *Neuropsychobiology* **54**:64–69. [3]

Mellios, N., H. Sugihara, J. Castro, et al. 2011. miR-132, an Experience-Dependent microRNA, Is Essential for Visual Cortex Plasticity. *Nat. Neurosci.* **14**:1240–1242. [7]

Mellios, N., and M. Sur. 2012. The Emerging Role of microRNAs in Schizophrenia and Autism Spectrum Disorders. *Front. Psychiatry* **3**:39. [7]

Meneret, A., Q. Welniarz, O. Trouillard, and E. Roze. 2015. Congenital Mirror Movements: From Piano Player to Opera Singer. *Neurology* **84**:860–860. [3]

Merikangas, K. R., J. P. He, M. Burstein, et al. 2010. Lifetime Prevalence of Mental Disorders in U.S. Adolescents: Results from the National Comorbidity Survey Replication: Adolescent Supplement (NCS-A). *J. Am. Acad. Child Adolesc. Psychiatry* **49**:980–989. [8]

Mersfelder, E. L., and M. R. Parthun. 2006. The Tale Beyond the Tail: Histone Core Domain Modifications and the Regulation of Chromatin Structure. *Nucleic Acids Res.* **34**:2653–2662. [6]

Messina, A., N. Ferraris, S. Wray, et al. 2011. Dysregulation of Semaphorin7a/B1-Integrin Signaling Leads to Defective GnRH-1 Cell Migration, Abnormal Gonadal Development and Altered Fertility. *Hum. Mol. Genet.* **20**:4759–4774. [3]

Meyer, D., T. Bonhoeffer, and V. Scheuss. 2014. Balance and Stability of Synaptic Structures during Synaptic Plasticity. *Neuron* **82**:430–443. [13]

Michelsen, K. A., S. Acosta-Verdugo, M. Benoit-Marand, et al. 2015. Area-Specific Reestablishment of Damaged Circuits in the Adult Cerebral Cortex by Cortical Neurons Derived from Mouse Embryonic Stem Cells. *Neuron* **85**:982–997. [3]

Milh, M., A. Kaminska, C. Huon, et al. 2007. Rapid Cortical Oscillations and Early Motor Activity in Premature Human Neonate. *Cereb. Cortex* **17**:1582–1594. [4]

Mill, J., E. L. Dempster, A. Caspi, et al. 2006. Evidence for Monozygotic Twin (MZ) Discordance in Methylation Level at Two CpG Sites in the Promoter Region of the Catechol-O-Methyltransferase (COMT) Gene. *Am. J. Med. Genet. B Neuropsychiatr. Genet.* **141B**:421–425. [6]

Mill, J., T. Tang, Z. Kaminsky, et al. 2008. Epigenomic Profiling Reveals DNA-Methylation Changes Associated with Major Psychosis. *Am. J. Hum. Genet.* **82**:696–711. [6]

Milnor, J. 1985. On the Concept of Attractor. *Commun. Math. Physics* **99**:177–195. [4]

Ming, G.-L., and H. Song. 2011. Adult Neurogenesis in the Mammalian Brain: Significant Answers and Significant Questions. *Neuron* **70**:687–702. [6]

Minlebaev, M., M. Colonnese, T. Tsintsadze, A. Sirota, and R. Khazipov. 2011. Early Gamma Oscillations Synchronize Developing Thalamus and Cortex. *Science* **334**:226–229. [2]

Mire, E., C. Mezzera, E. Leyva-Díaz, et al. 2012. Spontaneous Activity Regulates Robo1 Transcription to Mediate a Switch in Thalamocortical Axon Growth. *Nat. Neurosci.* **15**:1134–1143. [3]

Miskovic, V., X. Ma, C. A. Chou, et al. 2015. Developmental Changes in Spontaneous Electrocortical Activity and Network Organization from Early to Late Childhood. *NeuroImage* **118**:237–247. [10, 11]

Mitchell, D. E., and K. R. Duffy. 2014. The Case from Animal Studies for Balanced Binocular Treatment Strategies for Human Amblyopia. *Opthalmic Physiol Opt* **34**:129–145. [5]

Mitchell, D. E., and S. MacKinnon. 2002. The Present and Potential Impact of Research on Animal Models for Clinical Treatment of Stimulus Deprivation Amblyopia. *Clin Exp Optom* **85**:5–18. [5]

Mitchell, D. E., K. MacNeill, N. A. Crowder, K. Holman, and K. R. Duffy. 2016. Recovery of Visual Functions in Amblyopic Animals Following Brief Exposure to Total Darkness. *J. Physiol.* **594**:149–167. [5]

Miwa, J. M., I. Ibanez-Tallon, G. W. Crabtree, et al. 1999. Lynx1, an Endogenous Toxin-Like Modulator of Nicotinic Acetylcholine Receptors in the Mammalian CNS. *Neuron* **23**:105–114. [5]

Mnih, V., K. Kavukcuoglu, D. Silver, et al. 2015. Human-Level Control through Deep Reinforcement Learning. *Nature* **518**:529–533. [13]

Mohr, M. A., L. L. DonCarlos, and C. L. Sisk. 2017. Inhibiting Production of New Brain Cells during Puberty or Adulthood Blunts the Hormonally Induced Surge of Luteinizing Hormone in Female Rats. *eNeuro* **4**:e0133–0117. [8]

Mohr, M. A., F. L. Garcia, L. L. DonCarlos, and C. L. Sisk. 2016. Neurons and Glial Cells Are Added to the Female Rat Anteroventral Periventricular Nucleus During Puberty. *Endocrinology* **157**:2393–2402. [8]

Mohr, M. A., and C. L. Sisk. 2013. Pubertally Born Neurons and Glia Are Functionally Integrated into Limbic and Hypothalamic Circuits of the Male Syrian Hamster. *PNAS* **110**:4792–4797. [8]

Molina-Luna, K., B. Hertler, M. M. Buitrago, and A. R. Luft. 2008. Motor Learning Transiently Changes Cortical Somatotopy. *NeuroImage* **40**:1748–1754. [13]

Molnár, Z., R. Adams, and C. Blakemore. 1998. Mechanisms Underlying the Early Establishment of Thalamocortical Connections in the Rat. *J. Neurosci.* **18**:5723–5745. [3]

Monfils, M.-H., I. Driscoll, H. Kamitakahara, et al. 2006. FGF-2-Induced Cell Proliferation Stimulates Anatomical, Neurophysiological, and Functional Recovery from Neonatal Motor Cortex Injury. *Eur. J. Neurosci.* **24**:739–749. [9]

Monnier, P. P., A. Sierra, P. Macchi, et al. 2002. Rgm Is a Repulsive Guidance Molecule for Retinal Axons. *Nature* **419**:392–395. [3]

Montey, K. L., N. C. Eaton, and E. M. Quinlan. 2013. Repetitive Visual Stimulation Enhances Recovery from Severe Amblyopia. *Learn Mem* **20**:311–317. [5]

Montey, K. L., and E. M. Quinlan. 2011. Recovery from Chronic Monocular Deprivation Following Reactivation of Thalamocortical Plasticity by Dark Exposure. *Nat. Commun.* **2**:317. [5]

Moon, C., H. Lagercrantz, and P. K. Kuhl. 2013. Language Experienced *in Utero* Affects Vowel Perception after Birth: A Two-Country Study. *Acta Paediatr.* **102**:156–160. [7]

Morales, D., and A. Kania. 2016. Cooperation and Crosstalk in Axon Guidance Cue Integration: Additivity, Synergy, and Fine-Tuning in Combinatorial Signaling. *Dev. Neurobiol.* **77**:891–904. [3]

Morishita, H., and T. K. Hensch. 2008. Critical Period Revisited: Impact on Vision. *Curr. Opin. Neurobiol.* **18**:101–107. [5, 7, 11]

Morishita, H., J. M. Miwa, N. Heintz, and T. K. Hensch. 2010. Lynx1, a Cholinergic Brake, Limits Plasticity in Adult Visual Cortex. *Science* **330**:1238–1240. [5, 7]

Morishita, W., H. Marie, and R. C. Malenka. 2005. Distinct Triggering and Expression Mechanisms Underlie LTD of AMPA and NMDA Synaptic Responses. *Nat. Neurosci.* **8**:1043–1050. [2]

Morrison, K. E., A. B. Rodgers, C. P. Morgan, and T. L. Bale. 2014. Epigenetic Mechanisms in Pubertal Brain Maturation. *Neuroscience* **264**:17–24. [9]

Morrison, P. D., J. Nottage, J. M. Stone, et al. 2011. Disruption of Frontal Theta Coherence by Delta9-Tetrahydrocannabinol Is Associated with Positive Psychotic Symptoms. *Neuropsychopharmacology* **36**:827–836. [12]

Mosing, M. A., G. Madison, N. L. Pedersen, R. Kuja-Halkola, and F. Ullén. 2014. Practice Does Not Make Perfect: No Causal Effect of Music Practice on Music Ability. *Psychol. Sci.* **25**:1795–1803. [13]

Movshon, J. A., and M. R. Dürsteler. 1977. Effects of Brief Periods of Unilateral Eye Closure on the Kitten's Visual System. *J. Neurophysiol.* **40**:1255–1265. [5]

Mrsic-Flogel, T. D., S. B. Hofer, K. Ohki, et al. 2007. Homeostatic Regulation of Eye-Specific Responses in Visual Cortex during Ocular Dominance Plasticity. *Neuron* **54**:961–972. [7]

Muckli, L., M. J. Naumer, and W. Singer. 2009. Bilateral Visual Field Maps in a Patient with Only One Hemisphere. *PNAS* **106**:13034–13039. [2, 3]

Mueller, S. C., V. Temple, E. Oh, et al. 2008. Early Androgen Exposure Modulates Spatial Cognition in Congenital Adrenal Hyperplasia (CAH). *Psychoneuroendocrinology* **33**:973–980. [8]

Muguruma, K., A. Nishiyama, H. Kawakami, K. Hashimoto, and Y. Sasai. 2015. Self-Organization of Polarized Cerebellar Tissue in 3D Culture of Human Pluripotent Stem Cells. *Cell Rep.* **10**:537–550. [3]

Müller, V., W. Gruber, W. Klimesch, and U. Lindenberger. 2009. Lifespan Differences in Cortical Dynamics of Auditory Perception. *Dev. Sci.* **12**:839–853. [15]

Munk, M. H. J., P. R. Roelfsema, P. König, A. K. Engel, and W. Singer. 1996. Role of Reticular Activation in the Modulation of Intracortical Synchronization. *Science* **272**:271–274. [2]

Murphy, K. M., G. Roumeliotis, K. Williams, B. R. Beston, and D. G. Jones. 2015. Binocular Visual Training to Promote Recovery from Monocular Deprivation. *Vis Res* **114**:68–78. [5]

Bibliography

Musacchia, G., N. A. Choudhury, S. Ortiz-Mantilla, et al. 2013. Oscillatory Support for Rapid Frequency Change Processing in Infants. *Neuropsychologia* **51**:2812–2824. [7]

Musacchia, G., S. Ortiz-Mantilla, N. A. Choudhury, et al. 2017. Active Auditory Experience in Infancy Promotes Brain Plasticity in Theta and Gamma Oscillations. *Dev. Cogn. Neurosci.* **26**:9–19. [7]

Musset, B., S. G. Meuth, G. X. Liu, et al. 2006. Effects of Divalent Cations and Spermine on the K+ Channel TASK-3 and on the Outward Current in Thalamic Neurons. *J. Physiol.* **572**:639–657. [6]

Mychasiuk, R., and G. A. S. Metz. 2016. Epigenetic and Gene Expression Changes in the Adolescent Brain: What Have We Learned from Animal Models? *Neurosci. Biobehav. Rev.* **70**:189–197. [9]

Mychasiuk, R., A. Muhammad, and B. Kolb. 2014. Environmental Enrichment Alters Structural Plasticity of the Adolescent Brain but Does Not Remediate the Effects of Prenatal Nicotine Exposure. *Synapse* **68**:293–305. [9]

Myers, N. E., G. Rohenkohl, V. Wyart, et al. 2015. Testing Sensory Evidence against Mnemonic Templates. *eLife* **4**:e09000. [15]

Nabbout, R., C. Depienne, C. Chiron, and O. Dulac. 2011. Protocadherin 19 Mutations in Girls with Infantile-Onset Epilepsy. *Neurology* **76**:1193–1194. [3]

Nagakura, I., A. Van Wart, J. Petravicz, D. Tropea, and M. Sur. 2014. Stat1 Regulates the Homeostatic Component of Visual Cortical Plasticity via an AMPA Receptor-Mediated Mechanism. *J. Neurosci.* **34**:10256–10263. [5]

Narayanan, R. T., T. Seidenbecher, C. Kluge, et al. 2007. Dissociated Theta Synchronization in Amygdalo-Hippocampal Circuits during Various Stages of Fear Memory. *Eur. J. Neurosci.* **25**:1823–1831. [2]

National Sleep Foundation. 2014. Sleep in the Modern Family Poll Results. Sleep in America Poll. https://sleepfoundation.org/sleep-polls-data/sleep-in-america-poll/2014-sleep-and-family (accessed Aug. 2, 2017). [14]

Naumova, O. Y., M. Lee, R. Koposov, et al. 2012. Differential Patterns of Whole-Genome DNA Methylation in Institutionalized Children and Children Raised by Their Biological Parents. *Dev. Psychopathol.* **24**:143–155. [6]

Nava, E., and B. Röder. 2011. Adaption and Maladaption: Insights from Brain Plasticity. *Prog. Brain Res.* **191**:177–194. [13]

Navlakha, S., A. L. Barth, and Z. Bar-Joseph. 2015. Decreasing-Rate Pruning Optimizes the Construction of Efficient and Robust Distributed Networks. *PLoS Comput. Biol.* **11**:e1004347. [4]

Nemati, F., and B. Kolb. 2010. Motor Cortex Injury Has Different Behavioral and Anatomical Effects in Juvenile and Adolescent Rats. *Behav. Neurosci.* **24**:612–622. [9]

———. 2011. FGF-2 Induces Behavioral Recovery after Early Adolescent Injury to the Motor Cortex of Rats. *Behav. Brain Res.* **225**:184–191. [9]

Nestler, E. J., C. J. Pena, M. Kundakovic, A. Mitchell, and S. Akbarian. 2016. Epigenetic Basis of Mental Illness. *Neuroscientist* **22**:447–463. [6]

Neuhaus-Follini, A., and G. J. Bashaw. 2015. Crossing the Embryonic Midline: Molecular Mechanisms Regulating Axon Responsiveness at an Intermediate Target. *Wiley Interdisc. Rev. Dev. Biol.* **4**:377–389. [3]

Neveu, D., and R. S. Zucker. 1996. Postsynaptic Levels of $[Ca^{2+}]_i$ Needed to Trigger LTD and LTP. *Neuron* **16**:619–629. [2]

Neveu, M. M., and G. Jeffery. 2007. Chiasm Formation in Man Is Fundamentally Different from That in the Mouse. *Eye* **21**:1264–1270. [3]

Niell, C. M., and M. P. Stryker. 2010. Modulation of Visual Responses by Behavioral State in Mouse Visual Cortex. *Neuron* **65**:472–479. [5]

Niessing, J., B. Ebisch, K. E. Schmidt, et al. 2005. Hemodynamic Signals Correlate Tightly with Synchronized Gamma Oscillations. *Science* **309**:948–951. [2]

Nikolova, Y. S., and A. R. Hariri. 2015. Can We Observe Epigenetic Effects on Human Brain Function? *Trends Cogn. Sci.* **19**:366–373. [6]

Nikolova, Y. S., K. C. Koenen, S. Galea, et al. 2014. Beyond Genotype: Serotonin Transporter Epigenetic Modification Predicts Human Brain Function. *Nat. Neurosci.* **17**:1153–1155. [6]

Nimmervoll, B., R. White, J. W. Yang, et al. 2013. LPS-Induced Microglial Secretion of TNF-Alpha Increases Activity-Dependent Neuronal Apoptosis in Neonatal Cerebral Cortex. *Cereb. Cortex* **23**:1742–1755. [4]

Niquille, M., S. Garel, F. Mann, et al. 2009. Transient Neuronal Populations Are Required to Guide Callosal Axons: A Role for Semaphorin 3C. *PLoS Biol.* **7**:e1000230. [3]

Niso, G., C. Rogers, J. T. Moreau, et al. 2016. Omega: The Open MEG Archive. *NeuroImage* **124**:1182–1187. [15]

Niwa, M., H. Jaaro-Peled, S. Tankou, et al. 2013. Adolescent Stress-Induced Epigenetic Control of Dopaminergic Neurons via Glucocorticoids. *Science* **339**:335–339. [12]

Niwa, M., A. Kamiya, R. Murai, et al. 2010. Knockdown of DISC1 by *in Utero* Gene Transfer Disturbs Postnatal Dopaminergic Maturation in the Frontal Cortex and Leads to Adult Behavioral Deficits. *Neuron* **65**:480–489. [15]

Non, A. L., A. M. Binder, L. D. Kubzansky, and K. B. Michels. 2014. Genome-Wide DNA Methylation in Neonates Exposed to Maternal Depression, Anxiety, or SSRI Medication during Pregnancy. *Epigenetics* **9**:964–972. [6]

Nordt, M., S. Hoehl, and S. Weigelt. 2016. The Use of Repetition Suppression Paradigms in Developmental Cognitive Neuroscience. *Cortex* **80**:61–75. [7]

Nowak, L., P. Bregestovski, P. Ascher, A. Herbet, and A. Prochiantz. 1984. Magnesium Gates Glutamate-Activated Channels in Mouse Central Neurones. *Nature* **307**:462–465. [2]

Nugent, A. A., A. L. Kolpak, and E. C. Engle. 2012. Human Disorders of Axon Guidance. *Curr. Opin. Neurobiol.* **22**:837–843. [3]

Nutsch, V. L., R. G. Will, T. Hattori, D. J. Tobiansky, and J. M. Dominguez. 2014. Sexual Experience Influences Mating-Induced Activity in Nitric Oxide Synthase-Containing Neurons in the Medial Preoptic Area. *Neurosci. Lett.* **579**:92–96. [14]

O'Connor, R., and M. Tessier-Lavigne. 1999. Identification of Maxillary Factor, a Maxillary Process-Derived Chemoattractant for Developing Trigeminal Sensory Axons. *Neuron* **44**:165–178. [3]

Oberlander, T. F., J. Weinberg, M. Papsdorf, et al. 2008. Prenatal Exposure to Maternal Depression, Neonatal Methylation of Human Glucocorticoid Receptor Gene (NR3C1) and Infant Cortisol Stress Responses. *Epigenetics* **3**:97–106. [6]

Okada, A., F. Charron, S. Morin, et al. 2006. Boc Is a Receptor for Sonic Hedgehog in the Guidance of Commissural Axons. *Nature* **444**:369–373. [3]

O'Leary, T. P., M. R. Kutcher, D. E. Mitchell, and K. R. Duffy. 2012. Recovery of Neurofilament Following Early Monocular Deprivation. *Front Syst Neurosci* **6**:22. [5]

Olson, C. R., and R. D. Freeman. 1975. Progressive Changes in Kitten Striate Cortex during Monocular Vision. *J. Neurophysiol.* **38**:26–32. [5]

O'Rahilly, R., and F. Müller. 1987. Developmental Stages in Human Embryos, vol. 637. Washington, D.C.: Carnegie Institution. [3]

———. 2008. Significant Features in the Early Prenatal Development of the Human Brain. *Ann. Anat.* **190**:105–118. [3]

Oray, S., A. Majewska, and M. Sur. 2004. Dendritic Spine Dynamics Are Regulated by Monocular Deprivation and Extracellular Matrix Degradation. *Neuron* **44**:1021–1030. [5]

Ortigue, S., F. Bianchi-Demicheli, N. Patel, C. Frum, and J. W. Lewis. 2010. Neuroimaging of Love: Fmri Meta-Analysis Evidence toward New Perspectives in Sexual Medicine. *Journal of Sexual Medicine* **11**:3541–3552. [14]

Ortiz-Mantilla, S., J. A. Hämäläinen, G. Musacchia, and A. A. Benasich. 2013. Enhancement of Gamma Oscillations Signals Preferential Processing of Native over Foreign Phonemic Contrasts in Infants. *J. Neurosci.* **33**:18746–18754. [7]

Ortiz-Mantilla, S., J. A. Hämäläinen, T. Realpe-Bonilla, and A. A. Benasich. 2016. Oscillatory Dynamics Underlying Perceptual Narrowing of Native Phoneme Mapping from 6 to 12 Months of Age. *J. Neurosci.* **36**:12095–12105. [2, 7, 10]

Ouellet-Morin, I., C. C. Y. Wong, A. Danese, et al. 2013. Increased Serotonin Transporter Gene (Sert) DNA Methylation Is Associated with Bullying Victimization and Blunted Cortisol Response to Stress in Childhood: A Longitudinal Study of Discordant Monozygotic Twins. *Psychol. Med.* **43**:1813–1823. [6]

Owens, J. 2014. Insufficient Sleep in Adolescents and Young Adults: An Update on Causes and Consequences. *Pediatrics* **134**:e921–e932. [14]

Owens, M. T., D. A. Feldheim, M. P. Stryker, and J. W. Triplett. 2015. Stochastic Interaction between Neural Activity and Molecular Cues in the Formation of Topographic Maps. *Neuron* **87**:1261–1273. [4]

Pan, P., A. S. Fleming, D. Lawson, J. M. Jenkins, and P. O. McGowan. 2014. Within- and between-Litter Maternal Care Alter Behavior and Gene Regulation in Female Offspring. *Behav. Neurosci.* **128**:736–748. [6]

Paolicelli, R. C., G. Bolasco, F. Pagani, et al. 2011. Synaptic Pruning by Microglia Is Necessary for Normal Brain Development. *Science* **333**:1456–1458. [7]

Papenberg, G., U. Lindenberger, and L. Bäckman. 2015. Aging-Related Magnification of Genetic Effects on Cognitive and Brain Integrity. *Trends Cogn. Sci.* **19**:506–514. [13]

Park, H., and M. Poo. 2012. Neurotrophin Regulation of Neural Circuit Development and Function. *Nat. Rev. Neurosci.* **14**:7–23. [3]

Paşca, S. P., T. Portmann, I. Voineagu, et al. 2011. Using iPSC-Derived Neurons to Uncover Cellular Phenotypes Associated with Timothy Syndrome. *Nat. Med.* **17**:1657–1662. [3]

Pattwell, S. S., K. G. Bath, B. J. Casey, I. Ninan, and F. S. Lee. 2011. Selective Early-Acquired Fear Memories Undergo Temporary Suppression during Adolescence. *PNAS* **108**:1182–1187. [12]

Paul, L. K., W. S. Brown, R. Adolphs, et al. 2007. Agenesis of the Corpus Callosum: Genetic, Developmental and Functional Aspects of Connectivity. *Nat. Rev. Neurosci.* **8**:287–299. [3]

Paus, T., M. Keshavan, and J. N. Giedd. 2008. Why Do Many Psychiatric Disorders Emerge during Adolescence? *Nat. Rev. Neurosci.* **9**:947–957. [11, 12]

Paus, T., A. Zijdenbos, K. Worsley, et al. 1999. Structural Maturation of Neural Pathways in Children and Adolescents: *In Vivo* Study. *Science* **283**:1908–1911. [8]

Penn, A. A., P. A. Riquelme, M. B. Feller, and C. J. Shatz. 1998. Competition in Retinogeniculate Patterning Driven by Spontaneous Activity. *Science* **279**:2108–2112. [2]

Penzes, P., A. Buonanno, M. Passafaro, C. Sala, and R. A. Sweet. 2013. Developmental Vulnerability of Synapses and Circuits Associated with Neuropsychiatric Disorders. *J. Neurochem.* **126**:165–182. [5]

Peper, J. S., H. E. Hulshoff Pol, E. A. Crone, and J. van Honk. 2011. Sex Steroids and Brain Structure in Pubertal Boys and Girls: A Mini-Review of Neuroimaging Studies. *Neuroscience* **191**:28–37. [8]

Perez, J. D., N. D. Rubinstein, and C. Dulac. 2016. New Perspectives on Genomic Imprinting, an Essential and Multifaceted Mode of Epigenetic Control in the Developing and Adult Brain. *Annu. Rev. Neurosci.* **39**:347–384. [6]

Perrin, J. S., P. Y. Herve, G. Leonard, et al. 2008. Growth of White Matter in the Adolescent Brain: Role of Testosterone and Androgen Receptor. *J. Neurosci.* **28**:9519–9524. [8]

Perrin, J. S., G. Leonard, M. Perron, et al. 2009. Sex Differences in the Growth of White Matter during Adolescence. *NeuroImage* **45**:1055–1066. [8]

Perroud, N., E. Rutembesa, A. Paoloni-Giacobino, et al. 2014. The Tutsi Genocide and Transgenerational Transmission of Maternal Stress: Epigenetics and Biology of the HPA Axis. *World J. Biol. Psychiatry* **15**:334–345. [6]

Pesaresi, M., R. Soon-Shiong, L. French, et al. 2015. Axon Diameter and Axonal Transport: *In Vivo* and *in Vitro* Effects of Androgens. *NeuroImage* **115**:191–201. [8]

Petanjek, Z., M. Judas, G. Simic, et al. 2011. Extraordinary Neoteny of Synaptic Spines in the Human Prefrontal Cortex. *PNAS* **108**:13281–13286. [10]

Petersen, C. C. H., M. Brecht, T. T. G. Hahn, and B. Sakmann. 2004. Synaptic Changes in Layer 2/3 Underlying Map Plasticity of Developing Barrel Cortex. *Science* **304**:739–742. [2]

Petronis, A., I. I. Gottesman, P. Kan, et al. 2003. Monozygotic Twins Exhibit Numerous Epigenetic Differences: Clues to Twin Discordance? *Schizophr. Bull.* **29**:169–178. [6]

Pfeffer, C., M. Xue, M. He, Z. Huang, and M. Scanziani. 2013. Inhibition of Inhibition in Visual Cortex: The Logic of Connections between Molecularly Distinct Interneurons. *Nat. Neurosci.* **16**:1068–1076. [5]

Pfister, B. J. 2004. Extreme Stretch Growth of Integrated Axons. *J. Neurosci.* **24**:7978–7983. [3]

Phoenix, C., R. Goy, A. Gerall, and W. Young. 1959. Organizing Action of Prenatally Administered Testosterone Propionate on the Tissues Mediating Mating Behavior in the Female Guinea Pig. *Endocrinology* **65**:369–382. [8]

Pi, H. J., B. Hangya, D. Kvitsiani, et al. 2013. Cortical Interneurons That Specialize in Disinhibitory Control. *Nature* **503**:521–524. [5]

Piaget, J. 1980. Introduction. In: Les Formes Élémentaires de la Dialectique, ed. J. Piaget, pp. 9–13. Paris: Éditions Gallimard. [13]

Pienkowski, M., and J. J. Eggermont. 2011. Cortical Tonotopic Map Plasticity and Behavior. *Neurosci. Biobehav. Rev.* **35**:2117–2128. [13]

Pizzorusso, T., P. Medini, N. Berardi, et al. 2002. Reactivation of Ocular Dominance Plasticity in the Adult Visual Cortex. *Science* **298**:1248–1251. [5]

Pizzorusso, T., P. Medini, S. Landi, et al. 2006. Structural and Functional Recovery from Early Monocular Deprivation in Adult Rats. *PNAS* **103**:8517–8522. [5]

Plenz, D., and T. C. Thiagarajan. 2007. The Organizing Principles of Neuronal Avalanches: Cell Assemblies in the Cortex? *Trends Neurosci.* **30**:99–110. [2]

Poeppel, D., W. J. Idsardi, and V. van Wassenhove. 2008. Speech Perception at the Interface at Neurobiology and Linguistic. *Phil. Trans. R. Soc. B* **363**:1071–1086. [7]

Polackwich, R. J., D. Koch, R. McAllister, H. M. Geller, and J. S. Urbach. 2015. Traction Force and Tension Fluctuations in Growing Axons. *Front Cell Neurosci* **9**:417. [3]

Poliak, S., D. Morales, L.-P. Croteau, et al. 2015. Synergistic Integration of Netrin and Ephrin Axon Guidance Signals by Spinal Motor Neurons Sebastian. *eLife* **4**:1–26. [3]

Pont-Lezica, L., W. Beumer, S. Colasse, et al. 2014. Microglia Shape Corpus Callosum Axon Tract Fasciculation: Functional Impact of Prenatal Inflammation. *Eur. J. Neurosci.* **39**:1551–1557. [3]

Pooh, R. K., K. Shiota, and A. Kurjak. 2011. Imaging of the Human Embryo with Magnetic Resonance Imaging Microscopy and High-Resolution Transvaginal 3-Dimensional Sonography: Human Embryology in the 21st Century. *Am. J. Obstet. Gynecol.* **204**:e1–77. [3]

Portales-Casamar, E., A. A. Lussier, M. J. Jones, et al. 2016. DNA Methylation Signature of Human Fetal Alcohol Spectrum Disorder. *Epigenetics Chromatin* **9**:25. [6]

Poulain, F. E., and C.-B. Chien. 2013. Proteoglycan-Mediated Axon Degeneration Corrects Pretarget Topographic Sorting Errors. *Neuron* **78**:49–56. [3]

Prerau, M. J., R. E. Brown, M. T. Bianchi, J. M. Ellenbogen, and P. L. Purdon. 2017. Sleep Neurophysiological Dynamics through the Lens of Multitaper Spectral Analysis. *Physiology (Bethesda)* **32**:60–92. [10]

Prinz, A. A., D. Bucher, and E. Marder. 2004. Similar Network Activity from Disparate Circuit Parameters. *Nat. Neurosci.* **7**:1345–1352. [4]

Provençal, N., M. J. Suderman, C. Guillemin, et al. 2012. The Signature of Maternal Rearing in the Methylome in Rhesus Macaque Prefrontal Cortex and T Cells. *J. Neurosci.* **32**:15626–15642. [6]

Prusky, G. T., and R. M. Douglas. 2003. Developmental Plasticity of Mouse Visual Acuity. *Eur. J. Neurosci.* **17**:167–173. [5]

Purdon, P. L., K. J. Pavone, O. Akeju, et al. 2015a. The Ageing Brain: Age-Dependent Changes in the Electroencephalogram during Propofol and Sevoflurane General Anaesthesia. *Br. J. Anaesth.* **115(Suppl 1)**:i46–i57. [10]

Purdon, P. L., E. T. Pierce, E. A. Mukamel, et al. 2013. Electroencephalogram Signatures of Loss and Recovery of Consciousness from Propofol. *PNAS* **110**:E1142–1151. [10]

Purdon, P. L., A. Sampson, K. J. Pavone, and E. N. Brown. 2015b. Clinical Electroencephalography for Anesthesiologists: Part I: Background and Basic Signatures. *Anesthesiology* **123**:937–960. [10]

Putignano, E., G. Lonetti, L. Cancedda, et al. 2007. Developmental Downregulation of Histone Posttranslational Modifications Regulates Visual Cortical Plasticity. *Neuron* **53**:747–759. [5, 6]

Quadrato, G., J. Brown, and P. Arlotta. 2016. The Promises and Challenges of Human Brain Organoids as Models of Neuropsychiatric Disease. *Nat. Med.* **22**: [3]

Quallo, M. M., C. J. Price, K. Ueno, et al. 2009. Gray and White Matter Changes Associated with Tool-Use Learning in Macaque Monkeys. *PNAS* **106**:18379–18384. [13]

Rabbitt, S. M., A. E. Kazdin, and B. Scassellati. 2015. Integrating Socially Assistive Robotics into Mental Healthcare Interventions: Applications and Recommendations for Expanded Use. *Clinical Psychology Review* **35**:35–46. [15]

Radtke, K. M., M. Ruf, H. M. Gunter, et al. 2011. Transgenerational Impact of Intimate Partner Violence on Methylation in the Promoter of the Glucocorticoid Receptor. *Transl. Psychiatry* **1**:e21. [6]

Rago, L., R. Beattie, V. Taylor, and J. Winter. 2014. miR379-410 Cluster miRNAs Regulate Neurogenesis and Neuronal Migration by Fine-Tuning N-Cadherin. *EMBO J.* **33**:906–920. [6]

Rajagopalan, S., L. Deitinghoff, D. Davis, et al. 2004. Neogenin Mediates the Action of Repulsive Guidance Molecule. *Nat. Cell Biol.* **6**:756–762. [3]

Rakic, P., Bourgeois, J. P., and P. S. Goldman-Rakic. 1994. Synaptic Development of the Cerebral Cortex: Implications for Learning, Memory, and Mental Illness. *Prog. Brain Res.* **102**:227–243. [9]

Rakic, P., I. Suñer, and R. W. Williams. 1991. A Novel Cytoarchitectonic Area Induced Experimentally within the Primate Visual Cortex. *PNAS* **88**:2083–2087. [4]

Rakic, P., and P. I. Yakovlev. 1968. Development of the Corpus Callosum and Cavum Septi in Man. *J. Comp. Neurol.* **132**:45–72. [3]

Ramón y Cajal, S. 1892. La Rétine Des Vertébrés. *Cellule* **1**:121–247. [3]

———. 1894. The Croonian Lecture: La Fine Structure Des Centres Nerveux. *Proc. Roy. Soc. Lond.* **55**:444–468. [13]

Rando, O. J. 2012. Combinatorial Complexity in Chromatin Structure and Function: Revisiting the Histone Code. *Curr. Opin. Genet. Dev.* **22**:148–155. [6]

Raver, S. M., S. P. Haughwout, and A. Keller. 2013. Adolescent Cannabinoid Exposure Permanently Suppresses Cortical Oscillations in Adult Mice. *Neuropsychopharmacology* **38**:2338–2347. [12]

Recanzone, G. H., C. E. Schreiner, and M. M. Merzenich. 1993. Plasticity in the Frequency Representation of Primary Auditory Cortex Following Discrimination Training in Adult Owl Monkeys. *J. Neurosci.* **13**:87–103. [13]

Reichman, S., A. Terray, A. Slembrouck, et al. 2014. From Confluent Human iPS Cells to Self-Forming Neural Retina and Retinal Pigmented Epithelium. *PNAS* **111**:8518–8523. [3]

Reinagel, P., and R. C. Reid. 2002. Precise Firing Events Are Conserved across Neurons. *J. Neurosci.* **22**:6837–6841. [2]

Renier, N., M. Schonewille, F. Giraudet, et al. 2010. Genetic Dissection of the Function of Hindbrain Axonal Commissures. *PLoS Biol.* **8**:e1000325. [3]

Renier, N., Z. Wu, D. J. Simon, et al. 2014. Idisco: A Simple, Rapid Method to Immunolabel Large Tissue Samples for Volume Imaging. *Cell* **159**:1–15. [4]

Reyna, V., and F. Farley. 2006. Risk and Rationality in Adolescent Decision Making: Implications for Theory, Practice, and Public Policy. *Psychol. Sci. Publ. Int.* **7**:1–44. [14]

Rhines, R., and W. F. Windle. 1941. The Early Development of the Fasciculus Longitudinalis Medialis and Associated Secondary Neurons in the Rat, Cat and Man. *J. Comp. Neurol.* **75**:165–189. [3]

Ribary, U. 2005. Dynamics of Thalamo-Cortical Network Oscillations and Human Perception. *Prog. Brain Res.* **150**:127–142. [7]

Ribary, U., S. M. Doesburg, and L. M. Ward. 2014. Thalamocortical Network Dynamics: A Framework for Typical/Atypical Cortical Oscillations and Connectivity. Magnetoencephalography: From Signals to Dynamic Cortical Networks. Heidelberg: Springer-Verlag. [15]

———. 2017. Unified Principles of Thalamo-Cortical Processing: The Neuronal Switch. *Biomed. Eng. Lett.* **7**:229–235. [15]

Ribary, U., A. A. Ioannides, K. D. Singh, et al. 1991. Magnetic Field Tomography of Coherent Thalamocortical 40-Hz Oscillations in Humans. *PNAS* **88**:11037–11041. [7]

Ribeiro, P. F., L. Ventura-Antunes, M. Gabi, et al. 2013. The Human Cerebral Cortex Is Neither One nor Many: Neuronal Distribution Reveals Two Quantitatively Different Zones in the Gray Matter, Three in the White Matter, and Explains Local Variations in Cortical Folding. *Front. Neuroanat.* **7**:28. [4]

Rico, B., and O. Marín. 2011. Neuregulin Signaling, Cortical Circuitry Development and Schizophrenia. *Curr. Opin. Genet. Dev.* **21**:262–270. [5]

Ripke, S., C. O'Dushlaine, K. Chambert, et al. 2013. Genome-Wide Association Analysis Identifies 13 New Risk Loci for Schizophrenia. *Nat. Genet.* **45**:1150–1159. [3]

Rivera-Gaxiola, M., J. Silva-Pereyra, and P. K. Kuhl. 2005. Brain Potentials to Native and Non-Native Speech Contrasts in 7- and 11-Month-Old American Infants. *Dev. Sci.* **8**:162–172. [7]

Rivolta, D., T. Heidegger, B. Scheller, et al. 2015. Ketamine Dysregulates the Amplitude and Connectivity of High-Frequency Oscillations in Cortical-Subcortical Networks in Humans: Evidence from Resting-State Magnetoencephalography-Recordings. *Schizophr. Bull.* **41**:1105–1114. [10]

Robbe, D., S. M. Montgomery, A. Thome, et al. 2006. Cannabinoids Reveal Importance of Spike Timing Coordination in Hippocampal Function. *Nat. Neurosci.* **9**:1526–1533. [12]

Robertson, C. E., E. M. Ratai, and N. Kanwisher. 2016. Reduced GABAergic Action in the Autistic Brain. *Curr. Biol.* **26**:80–85. [10]

Robinson, T. E., and B. Kolb. 2004. Structural Plasticity Associated with Drugs of Abuse. *Neuropharmacology* **47 (Suppl 1)**:33–46. [9]

Röder, B., O. Stock, S. Bien, H. Neville, and F. Rösler. 2002. Speech Processing Activates Visual Cortex in Congenitally Blind Humans. *Eur. J. Neurosci.* **16**:930–936. [11]

Rodriguez, E., N. George, J. P. Lachaux, et al. 1999. Perception's Shadow: Long-Distance Synchronization of Human Brain Activity. *Nature* **397**:430–433. [7, 11]

Root, C. M., N. A. Velázquez-Ulloa, G. C. Monsalve, E. Minakova, and N. C. Spitzer. 2008. Embryonically Expressed GABA and Glutamate Drive Electrical Activity Regulating Neurotransmitter Specification. *J. Neurosci.* **28**:4777–4784. [4]

Rose, S. A., J. F. Feldman, and J. J. Jankowski. 2004a. Dimensions of Cognition in Infancy. *Intelligence* **32**:245–262. [7]

———. 2004b. Infant Visual Recognition Memory. *Dev. Rev.* **24**:74–100. [7]

Rose, S. A., J. F. Feldman, J. J. Jankowski, and R. A. Van Rossem. 2008. Cognitive Cascade in Infancy: Pathways from Prematurity to Later Mental Development. *Intelligence* **36**:367–378. [7]

Rose, T., J. Jaepel, M. Hübener, and T. Bonhoeffer. 2016. Cell-Specific Restoration of Stimulus Preference after Monocular Deprivation in the Visual Cortex. *Science* **352**:1319–1322. [7]

Rubinstein, R., C. A. Thu, M. Goodman, et al. 2015. Molecular Logic of Neuronal Self-Recognition through Protocadherin Domain Interactions Article Molecular Logic of Neuronal Self-Recognition through Protocadherin Domain Interactions. *Cell* **163**:1–14. [3]

Rucklidge, J. J., and B. J. Kaplan. 2013. Broad-Spectrum Micronutrient Formulas for the Treatment of Psychiatric Symptoms: A Systematic Review. *Expert. Rev. Neurother.* **13**:49–73. [9]

Ruiz de Almodovar, C., P. J. Fabre, E. Knevels, et al. 2011. Vegf Mediates Commissural Axon Chemoattraction through Its Receptor Flk1. *Neuron* **70**:966–978. [3]

Sabatier, C., A. S. Plump, Le Ma, et al. 2004. The Divergent Robo Family Protein Rig-1/Robo3 Is a Negative Regulator of Slit Responsiveness Required for Midline Crossing by Commissural Axons. *Cell* **117**:157–169. [3]

Sadato, N., A. Pascual-Leone, J. Grafman, et al. 1996. Activation of the Primary Visual Cortex by Braille Reading in Blind Subjects. *Nature* **380**:526–528. [11]

Sahin, M., and M. Sur. 2015. Genes, Circuits and Precision Therapies for Autism and Related Neurodevelopmental Disorders. *Science* **350**:926. [7]

Sajo, M., G. Ellis-Davies, and H. Morishita. 2016. Lynx1 Limits Dendritic Spine Turnover in the Adult Visual Cortex. *J. Neurosci.* **36**:9472–9478. [5]

Sakaguchi, H., T. Kadoshima, M. Soen, et al. 2015. Generation of Functional Hippocampal Neurons from Self-Organizing Human Embryonic Stem Cell-Derived Dorsomedial Telencephalic Tissue. *Nat. Commun.* **6**:8896. [3]

Sale, A., J. F. Maya Vetencourt, P. Medini, et al. 2007. Environmental Enrichment in Adulthood Promotes Amblyopia Recovery through a Reduction of Intracortical Inhibition. *Nat. Neurosci.* **10**:679–681. [5, 11]

Sampaio-Baptista, C., A. A. Khrapitchev, S. Foxley, et al. 2013. Motor Skill Learning Induces Changes in White Matter Microstructure and Myelination. *J. Neurosci.* **33**:19499–19503. [9]

Sanes, D. H., and S. M. Woolley. 2011. A Behavioral Framework to Guide Research on Central Auditory Development and Plasticity. *Neuron* **72**:912–929. [7]

Santiago, M., C. Antunes, M. Guedes, N. Sousa, and C. J. Marques. 2014. TET Enzymes and DNA Hydroxymethylation in Neural Development and Function: How Critical Are They? *Genomics* **104**:334–340. [6]

Sato, M., and M. P. Stryker. 2008. Distinctive Features of Adult Ocular Dominance Plasticity. *J. Neurosci.* **28**:10278–10286. [5, 7]

Sawtell, N. B., M. Y. Frenkel, B. D. Philpot, et al. 2003. NMDA Receptor-Dependent Ocular Dominance Plasticity in Adult Visual Cortex. *Neuron* **38**:977–985. [5, 7]

Saxe, R. R., and N. Kanwisher. 2003. People Thinking About Thinking People: The Role of the Temporo-Parietal Junction in "Theory of Mind." *NeuroImage Clin.* **19**:1835–1842. [11]

Saxe, R. R., S. Whitfield-Gabrieli, J. Scholz, and K. A. Pelphrey. 2009. Brain Regions for Perceiving and Reasoning About Other People in School-Aged Children. *Child Dev.* **80**:1197–1209. [11]

Saxonov, S., P. Berg, and D. L. Brutlag. 2006. A Genome-Wide Analysis of CpG Dinucleotides in the Human Genome Distinguishes Two Distinct Classes of Promoters. *PNAS* **103**:1412–1417. [6]

Schäfer, C. B., B. R. Morgan, A. X. Ye, M. J. Taylor, and S. M. Doesburg. 2014. Oscillations, Networks, and Their Development: MEG Connectivity Changes with Age. *Hum. Brain Mapp.* **35**:5249–5261. [11]

Schafer, D. P., E. K. Lehrman, A. G. Kautzman, et al. 2012. Microglia Sculpt Postnatal Neural Circuits in an Activity and Complement-Dependent Manner. *Neuron* **74**:691–705. [7]

Schick, B., P. de Villiers, J. de Villiers, and R. Hoffmeister. 2007. Language and Theory of Mind: A Study of Deaf Children. *Child Dev.* **78**:376–396. [11]

Schillen, T. B., and P. König. 1991. Stimulus-Dependent Assembly Formation of Oscillatory Responses: II. Desynchronization. *Neural Comput.* **3**:167–178. [2]

Schlegel, A. 2001. The Global Spread of Adolescent Culture. In: Negotiating Adolescence in Times of Social Change, ed. L. Crocket and R. Silbereisen, pp. 63–86. New York: Cambridge University Press. [14]

Schmiedek, F., M. Lövdén, and U. Lindenberger. 2010. Hundred Days of Cognitive Training Enhance Broad Cognitive Abilities in Adulthood: Findings from the COGITO Study. *Frontiers in Aging Neuroscience* **2**:27. [13]

———. 2014. Younger Adults Show Long-Term Effects of Cognitive Training on Broad Cognitive Abilities over Two Years. *Dev. Psychol.* **50**:2304–2310. [13]

Scholz, J., Y. Niibori, P. W. Frankland, and J. P. Lerch. 2015. Rotarod Training in Mice Is Associated with Changes in Brain Structure Observable with Multimodal MRI. *NeuroImage* **107**:182–189. [13]

Schultz, A., U. Grouven, I. Zander, et al. 2004. Age-Related Effects in the EEG during Propofol Anaesthesia. *Acta anaesthesiologica Scandinavica* **48**:27–34. [10]

Schultz, W., P. Dayan, and P. R. Montague. 1997. A Neural Substrate of Prediction and Reward. *Science* **275**:1593–1599. [15]

Schulz, K. M., H. A. Molenda-Figueira, and C. L. Sisk. 2009a. Back to the Future: The Organizational-Activational Hypothesis Adapted to Puberty and Adolescence. *Horm. Behav.* **55**:597–604. [8, 11]

Schulz, K. M., H. N. Richardson, J. L. Zehr, et al. 2004. Gonadal Hormones Masculinize and Defeminize Reproductive Behaviors during Puberty in the Male Syrian Hamster. *Horm. Behav.* **45**:242–249. [8, 11]

Schulz, K. M., and C. L. Sisk. 2016. The Organizing Actions of Adolescent Gonadal Steroid Hormones on Brain and Behavioral Development. *Neurosci. Biobehav. Rev.* **70**:148–158. [8, 11]

Schulz, K. M., J. L. Zehr, K. Y. Salas-Ramirez, and C. L. Sisk. 2009b. Testosterone Programs Adult Social Behavior before and during, but Not after, Adolescence. *Endocrinology* **150**:3690–3698. [8]

Scott, N., M. Prigge, O. Yizhar, and T. Kimchi. 2015. A Sexually Dimorphic Hypothalamic Circuit Controls Maternal Care and Oxytocin Secretion. *Nature* **525**:519–522. [8]

Segalowitz, S. J., D. L. Santesso, and M. K. Jetha. 2010. Electrophysiological Changes during Adolescence: A Review. *Brain Cogn.* **72**:86–100. [10]

Seidenbecher, T., T. R. Laxmi, O. Stork, and H.-C. Pape. 2003. Amygdala and Hippocampal Theta Rhythm Synchronization during Fear Memory Retrieval. *Science* **301**:846–850. [2]

Sejnowski, T. J., C. Koch, and P. S. Churchland. 1990. Computational Neuroscience. In: Connectionist Modeling and Brain Function: The Developing Interface, ed. S. J. Hanson and C. R. Olson, pp. 5–35. Cambridge, MA: MIT Press. [13]

Sengpiel, F. 2014. Plasticity of the Visual Cortex and Treatment of Amblyopia. *Curr. Biol.* **24**:R936–940. [5]

Shafer, B., K. Onishi, C. Lo, G. Colakoglu, and Y. Zou. 2011. Vangl2 Promotes Wnt/Planar Cell Polarity-Like Signaling by Antagonizing Dvl1-Mediated Feedback Inhibition in Growth Cone Guidance. *Dev. Cell* **20**:177–191. [3]

Shapero, B. G., S. K. Black, R. T. Liu, et al. 2014. Stressful Life Events and Depression Symptoms: The Effect of Childhood Emotional Abuse on Stress Reactivity. *J. Clin. Psychol.* **70**:209–223. [6]

Sharma, J., A. Angelucci, and M. Sur. 2000. Induction of Visual Orientation Modules in Auditory Cortex. *Nature* **404**:841–847. [7]

Shatz, C. J., and M. P. Stryker. 1978. Ocular Dominance in Layer IV of the Cat's Visual Cortex and the Effects of Monocular Deprivation. *J. Physiol. (Lond.)* **281**:267–283. [5]

Shi, Y., H. Inoue, J. C. Wu, and S. Yamanaka. 2017. Induced Pluripotent Stem Cell Technology: A Decade of Progress. *Nat. Rev. Drug Discov.* **16**:115–130. [3]

Shing, Y. L., M. Werkle-Bergner, S.-C. Li, and U. Lindenberger. 2008. Associative and Strategic Components of Episodic Memory: A Life-Span Dissociation. *J. Exp. Psychol. Gen.* **127**:495–513. [11]

Shinmyo, Y., M. Asrafuzzaman Riyadh, G. Ahmed, et al. 2015. Draxin from Neocortical Neurons Controls the Guidance of Thalamocortical Projections into the Neocortex. *Nat. Commun.* **6**:10232. [3]

Shooner, C., L. E. Hallum, R. D. Kumbhani, et al. 2015. Population Representation of Visual Information in Areas V1 and V2 of Amblyopic Macaques. *Vis Res* **114**:56–67. [5]

Shrager, J., and M. H. Johnson. 1996. Dynamic Plasticity Influences the Emergence of Function in a Simple Cortical Array. *Neural Netw.* **9**:1119–1129. [13]

Silingardi, D., M. Scali, G. Belluomini, and T. Pizzorusso. 2010. Epigenetic Treatments of Adult Rats Promote Recovery from Visual Acuity Deficits Induced by Long-Term Monocular Deprivation. *Eur. J. Neurosci.* **31**:2185–2192. [5, 7]

Silveri, M. M., A. D. Dager, J. E. Cohen-Gilbert, and J. T. Sneider. 2016. Neurobiological Signatures Associated with Alcohol and Drug Use in the Human Adolescent Brain. *Neurosci. Biobehav. Rev.* **70**:244–259. [9]

Simerly, R. B. 2002. Wired for Reproduction: Organization and Development of Sexually Dimorphic Circuits in the Mammalian Forebrain. *Annu. Rev. Neurosci.* **25**:507–536. [8]

Simons, D. J., W. R. Boot, N. Charness, et al. 2016. Do "Brain-Training" Programs Work? *Psychol. Sci. Publ. Int.* **17**:103–186. [13]

Simons, K. 2005. Amblyopia Characterization, Treatment, and Prophylaxis. *Sur Ophthalmol* **50**:123–166. [5]

Simpson, J. S., S. G. Crawford, E. T. Goldstein, et al. 2011. Systematic Review of Safety and Tolerability of a Complex Micronutrient Formula Used in Mental Health. *BMC Psychiatry* **11**:62. [9]

Singer, W. 1995. Development and Plasticity of Cortical Processing Architectures. *Science* **270**:758–764. [2]

———. 1999. Neuronal Synchrony: A Versatile Code for the Definition of Relations? *Neuron* **24**:49–65. [2, 7]

———. 2013. Cortical Dynamics Revisited. *Trends Cogn. Sci.* **17**:616–626. [2]

Singer, W., and C. M. Gray. 1995. Visual Feature Integration and the Temporal Correlation Hypothesis. *Annu. Rev. Neurosci.* **18**:555–586. [2]

Singer, W., and A. Lazar. 2016. Does the Cerebral Cortex Exploit High-Dimensional, Non-Linear Dynamics for Information Processing? *Front. Comput. Neurosci.* **10**:1–10. [2]

Singer, W., and J. P. Rauschecker. 1982. Central Core Control of Developmental Plasticity in the Kitten Visual Cortex: II. Electrical Activation of Mesencephalic and Diencephalic Projections. *Exp. Brain Res.* **47**:223–233. [2]

Singer, W., F. Tretter, and U. Yinon. 1982. Central Gating of Developmental Plasticity in Kitten Visual Cortex. *J. Physiol. (Lond.)* **324**:221–237. [2]

Sisk, C. L., and D. L. Foster. 2004. The Neural Basis of Puberty and Adolescence. *Nat. Neurosci.* **7**:1040–1047. [14]

Sloan, T. F. W., M. A. Qasaimeh, D. Juncker, P. T. Yam, and F. Charron. 2015. Integration of Shallow Gradients of Shh and Netrin-1 Guides Commissural Axons. *PLoS Biol.* **13**:e1002119. [3]

Smyser, C. D., T. E. Inder, J. S. Shimony, et al. 2010. Longitudinal Analysis of Neural Network Development in Preterm Infants. *Cereb. Cortex* **20**:2852–2862. [11]

Snowdon, D. A., S. J. Kemper, J. A. Mortimer, et al. 1996. Linguistic Ability in Early Life and Cognitive Function and Alzheimer's Disease in Late Life: Findings from the Nun Study. *JAMA* **275**:528–532. [4]

Sohal, V. S., F. Zhang, O. Yizhar, and K. Deisseroth. 2009. Parvalbumin Neurons and Gamma Rhythms Enhance Cortical Circuit Performance. *Nature* **459**:698–702. [10]

Southwell, D. G., R. C. Froemke, A. Alvarez-Buylla, M. P. Stryker, and S. P. Gandhi. 2010. Cortical Plasticity Induced by Inhibitory Neuron Transplantation. *Science* **327**:1145–1148. [4]

Spatazza, J., H. H. Lee, A. A. Di Nardo, et al. 2013. Choroid-Plexus-Derived Otx2 Homeoprotein Constrains Adult Cortical Plasticity. *Cell Rep.* **3**:1815–1823. [5]

Spear, L. P. 2000. The Adolescent Brain and Age-Related Behavioral Manifestations. *Neurosci. Biobehav. Rev.* **24**:417–463. [9, 14]

Spear, L. P. 2016. Consequences of Adolescent Use of Alcohol and Other Drugs: Studies Using Rodent Models. *Neurosci. Biobehav. Rev.* **70**:228–243. [9]

Spitzer, N. C. 2017. Neurotransmitter Switching in the Developing and Adult Brain. *Annu. Rev. Neurosci.* **40**:1–19. [4]

Spitzer, N. C., and J. E. Lamborghini. 1976. The Development of the Action Potential Mechanism of Amphibian Neurons Isolated in Culture. *PNAS* **73**:1641–1645. [4]

Squarzoni, P., G. Oller, G. Hoeffel, et al. 2014. Microglia Modulate Wiring of the Embryonic Forebrain. *Cell Rep.* **8**:1271–1279. [3]

Squarzoni, P., M. S. Thion, and S. Garel. 2015. Neuronal and Microglial Regulators of Cortical Wiring: Usual and Novel Guideposts. *Front. Neurosci.* **9**:1–16. [3]

Srour, M., J.-B. B. Rivière, J. M. Pham, et al. 2010. Mutations in DCC Cause Congenital Mirror Movements. *Science* **328**:592. [3]

Stagg, C. J., J. G. Best, M. C. Stephenson, et al. 2009. Polarity-Sensitive Modulation of Cortical Neurotransmitters by Transcranial Stimulation. *J. Neurosci.* **29**:5202–5206. [13]

Steinberg, L. 2005. Cognitive and Affective Development in Adolescence. *Trends Cogn. Sci.* **9**:69–74. [11]

———. 2010. A Dual Systems Model of Adolescent Risk-Taking. *Dev. Psychobiol.* **52**:216–224. [14]

———. 2014. Age of Opportunity: Lessons from the New Science of Adolescence. Boston: Houghton Mifflin Harcourt. [14]

———. 2016. Commentary on Special Issue on the Adolescent Brain: Redefining Adolescence. *Neurosci. Biobehav. Rev.* **70**:343–346. [9]

Steinberg, L., G. Icenogle, E. P. Shulman, et al. 2018. Around the World, Adolescence Is a Time of Heightened Sensation Seeking and Immature Self-Regulation. *Dev. Sci.* **21**:e12532. [14]

Steinschneider, M., K. V. Nourski, H. Kawasaki, et al. 2011. Intracranial Study of Speech-Elicited Activity on the Human Posterolateral Superior Temporal Gyrus. *Cereb. Cortex* **21**:2332–2347. [7]

Stephany, C. É., L. L. Chan, S. N. Parivash, et al. 2014. Plasticity of Binocularity and Visual Acuity Are Differentially Limited by Nogo Receptor. *J. Neurosci.* **34**:11631–11640. [5]

Steriade, M., A. McCormick, and T. J. Sejnowski. 1993. Thalamocortical Oscillations in the Sleeping and Aroused Brain. *Science* **262**:679–685. [2]

Stiefel, K. M., F. Tennigkeit, and W. Singer. 2005. Synaptic Plasticity in the Absence of Backpropagating Spikes of Layer II Inputs to Layer V Pyramidal Cells in Rat Visual Cortex. *Eur. J. Neurosci.* **21**:2605–2610. [2]

Stodieck, S. K., F. Greifzu, B. Goetze, K. F. Schmidt, and S. Löwel. 2014. Brief Dark Exposure Restored Ocular Dominance Plasticity in Aging Mice and after a Cortical Stroke. *Exp Gerontol* **60**:1–11. [5]

Strahl, B. D., and C. D. Allis. 2000. The Language of Covalent Histone Modifications. *Nature* **403**:41–45. [6]

Stuart, G. J., and M. Häusser. 2001. Dendritic Coincidence Detection of EPSPs and Action Potentials. *Nat. Neurosci.* **4**:63–71. [2]

Suárez, R., I. Gobius, and L. J. Richards. 2014. Evolution and Development of Interhemispheric Connections in the Vertebrate Forebrain. *Front. Hum. Neurosci.* **8**:497. [3]

Suberbielle, E., B. Djukic, M. Evans, et al. 2015. DNA Repair Factor BRCA1 Depletion Occurs in Alzheimer Brains and Impairs Cognitive Function in Mice. *Nat. Commun.* **6**:8897. [7]

Subramaniam, K., T. L. Luks, M. Fisher, et al. 2012. Computerized Cognitive Training Restores Neural Activity within the Reality Monitoring Network in Schizophrenia. *Neuron* 73:842–853. [15]

Suda, S., K. Iwata, C. Shimmura, et al. 2011. Decreased Expression of Axon-Guidance Receptors in the Anterior Cingulate Cortex in Autism. *Mol. Autism* 2:14. [3]

Suderman, M., N. Borghol, J. J. Pappas, et al. 2014. Childhood Abuse Is Associated with Methylation of Multiple Loci in Adult DNA. *BMC Med. Genomics* 7:13. [6]

Suderman, M., P. O. McGowan, A. Sasaki, et al. 2012. Conserved Epigenetic Sensitivity to Early Life Experience in the Rat and Human Hippocampus. *PNAS* **109 (Supp 2)**:17266–17272. [6]

Sugiyama, S., A. A. Di Nardo, S. Aizawa, et al. 2008. Experience-Dependent Transfer of Otx2 Homeoprotein into the Visual Cortex Activates Postnatal Plasticity. *Cell* **134**:508–520. [3, 5]

Suleiman, A. B., A. Galván, P. Harden, and R. E. Dahl. 2017. Becoming a Sexual Being: The "Elephant in the Room" of Adolescent Brain Development. *Dev. Cogn. Neurosci.* **25**:209–220. [14]

Sultan, K. T., Z. Han, X.-J. Zhang, et al. 2016. Clonally Related GABAergic Interneurons Do Not Randomly Disperse but Frequently Form Local Clusters in the Forebrain. *Neuron* **92**:31–44. [3]

Sun, L. 2010. Early Childhood General Anaesthesia Exposure and Neurocognitive Development. *Br. J. Anaesth.* **105(Suppl 1)**:i61–68. [10]

Sun, Y., T. Ikrar, M. F. Davis, et al. 2016. Neurogulin-1/Erbb4 Signaling Regulates Visual Cortical Plasticity. *Neuron* **92**:160–173. [5]

Sur, M., I. Nagakura, N. Chen, and S. H. 2013. Mechanisms of Plasticity in the Developing and Adult Visual Cortex. *Prog. Brain Res.* **207**:243–254. [7]

Sur, M., and J. Rubenstein. 2005. Patterning and Plasticity of the Cerebral Cortex. *Science* **310**:805–810. [7]

Suzuki, I. K., and P. Vanderhaeghen. 2015. Is This a Brain Which I See before Me? Modeling Human Neural Development with Pluripotent Stem Cells. *Development* **142**:3138–3150. [3]

Swartz, J. R., A. R. Hariri, and D. E. Williamson. 2016. An Epigenetic Mechanism Links Socioeconomic Status to Changes in Depression-Related Brain Function in High-Risk Adolescents. *Mol. Psychiatry* [6]

Swithers, S. E., M. McCurley, E. Hamilton, and A. Doerflinger. 2008. Influence of Ovarian Hormones on Development of Ingestive Responding to Alterations in Fatty Acid Oxidation in Female Rats. *Horm. Behav.* **54**:471–477. [8]

Syken, J., T. Grandpre, P. O. Kanold, and C. J. Shatz. 2006. PirB Restricts Ocular-Dominance Plasticity in Visual Cortex. *Science* **313**:1795–1800. [5]

Syme, M. R., J. W. Paxton, and J. A. Keelan. 2004. Drug Transfer and Metabolism by the Human Placenta. *Clin. Pharmacokinet.* **43**:487–514. [3]

Tagawa, Y., P. O. Kanold, M. Majdan, and C. J. Shatz. 2005. Multiple Periods of Functional Ocular Dominance Plasticity in Mouse Visual Cortex. *Nat. Neurosci.* **8**:380–388. [7]

Taha, S., and M. P. Stryker. 2002. Rapid Ocular Dominance Plasticity Requires Cortical but Not Geniculate Protein Synthesis. *Neuron* **34**:425–436. [5]

Takano, T., and T. Ogawa. 1998. Characterization of Developmental Changes in EEG-Gamma Band Activity during Childhood Using the Autoregressive Model. *Pediatr. Intl.* **40**:446–452. [7]

Takesian, A. E., and T. K. Hensch. 2013. Balancing Plasticity/Stability across Brain Development. *Prog. Brain Res.* **207**:3–34. [5–7, 9, 10, 15]

Taki, Y., H. Hashizume, B. Thyreau, et al. 2012. Sleep Duration during Weekdays Affects Hippocampal Gray Matter Volume in Healthy Children. *NeuroImage* **60**:471–475 [14]

Tallon-Baudry, C., and O. Bertrand. 1999. Oscillatory Gamma Activity in Humans and Its Role in Object Representation. *Trends Cogn. Sci.* **3**:151–162. [7]

Tallon-Baudry, C., O. Bertrand, and C. Fischer. 2001. Oscillatory Synchrony between Human Extrastriate Areas during Visual Short-Term Memory Maintenance. *J. Neurosci.* **21 (RC177)**:1–5. [2]

Tallon-Baudry, C., S. Mandon, W. A. Freiwald, and A. K. Kreiter. 2004. Oscillatory Synchrony in the Monkey Temporal Lobe Correlates with Performance in a Visual Short-Term Memory Task. *Cereb. Cortex* **14**:713–720. [2]

Tamnes, C. K., M. M. Herting, A.-L. Goddings, et al. 2017. Development of the Cerebral Cortex across Adolescence: A Multisample Study of Inter-Related Longitudinal Changes in Cortical Volume, Surface Area, and Thickness. *J. Neurosci.* **37**:3402–4312. [9]

Tang, E., C. Giusti, G. L. Baum, et al. 2017. Developmental Increases in White Matter Network Controllability Support a Growing Diversity of Brain Dynamics. *Nat. Commun.* **8**:1252. [13]

Tarokh, L., J. M. Saletin, and M. A. Carskadon. 2016. Sleep in Adolescence: Physiology, Cognition and Mental Health. *Neurosci. Biobehav. Rev.* **70**:182–188. [14]

Tashijian, S. N., D. Goldenberg, and A. Galván. 2017. Neural Connectivity Moderates the Association between Sleep and Impulsivity in Adolescents. *Dev. Cogn. Neurosci.* **27**:35–44. [14]

Tashiro, Y., A. Oyabu, Y. Imura, et al. 2011. Morphological Abnormalities of Embryonic Cranial Nerves after *in Utero* Exposure to Valproic Acid: Implications for the Pathogenesis of Autism with Multiple Developmental Anomalies. *Int. J. Dev. Neurosci.* **29**:359–364. [3]

Tchernichovski, O., P. P. Mitra, T. Lints, and F. Nottebohm. 2001. Dynamics of the Vocal Imitation Process: How a Zebra Finch Learns Its Song. *Science* **291**:2564–2569. [2]

Teh, A. L., H. Pan, L. Chen, et al. 2014. The Effect of Genotype and *in Utero* Environment on Interindividual Variation in Neonate DNA Methylomes. *Genome Res.* **24**:1064–1074. [6]

Telzer, E. H., A. J. Fuligni, M. D. Lieberman, and A. Galván. 2013a. The Effects of Poor Quality Sleep on Brain Function and Risk Taking in Adolescence. *NeuroImage* **71**:275–283. [14]

———. Meaningful Family Relationships: Neurocognitive Buffers of Adolescent Risk Taking. *J. Cogn. Neurosci.* **25**:374–387. [14]

Telzer, E. H., D. Goldenberg, A. J. Fuligni, M. D. Lieberman, and A. Galván. 2015. Sleep Variability in Adolescence Is Associated with Altered Brain Development. *Dev. Cogn. Neurosci.* **14**:16–22. [14]

ten Donkelaar, H. J., M. Lammens, P. Wesseling, et al. 2004. Development and Malformations of the Human Pyramidal Tract. *J. Neurol.* **251**:1429–1442. [3]

Tessier-Lavigne, M., and C. S. Goodman. 1996. The Molecular Biology of Axon Guidance. *Science* **274**:1123–1133. [3]

Theunissen, T. W., M. Friedli, Y. He, et al. 2016. Molecular Criteria for Defining the Naive Human Pluripotent State. *Cell Stem Cell* **19**:502–515. [4]

Thompson-Schill, S. L., M. Ramscar, and E. G. Chrysikou. 2009. Cognition without Control: When a Little Frontal Lobe Goes a Long Way. *Curr. Dir. Psychol.* **18**:259–263. [11]

Thut, G., P. G. Schyns, and J. Gross. 2011. Entrainment of Perceptually Relevant Brain Oscillations by Non-Invasive Rhythmic Stimulation of the Human Brain. *Front. Psychol.* **2**:170. [12]

Tobi, E. W., L. H. Lumey, R. P. Talens, et al. 2009. DNA Methylation Differences after Exposure to Prenatal Famine Are Common and Timing- and Sex-Specific. *Hum. Mol. Genet.* **18**:4046–4053. [6]

Tobler, P. N., C. D. Fiorillo, and W. Schultz. 2005. Adaptive Coding of Reward Value by Dopamine Neurons. *Science* **307**:1642–1645. [2]

Tolner, E. A., A. Sheikh, A. Y. Yukin, K. Kaila, and P. O. Kanold. 2012. Subplate Neurons Promote Spindle Bursts and Thalamocortical Patterning in the Neonatal Rat Somatosensory Cortex. *J. Neurosci.* **36**:692–702. [4]

Tonegawa, S., M. Pignatelli, D. S. Roy, and T. J. Ryan. 2015. Memory Engram Storage and Retrieval. *Curr. Opin. Neurobiol.* **35**:101–109. [13]

Torres, O. V., H. A. Tejeda, L. A. Natividad, and L. E. O'Dell. 2008. Enhanced Vulnerability to the Rewarding Effects of Nicotine during the Adolescent Period of Development. *Pharmacol. Biochem. Behav.* **90**:658–663. [14]

Tottenham, N., and A. Galván. 2016. Stress and the Adolescent Brain: Amygdala-Prefrontal Cortex Circuitry as Developmental Targets. *Neurosci. Biobehav. Rev.* **70**:217–227. [9]

Toyoizumi, T., H. Miyamoto, Y. Yazaki-Sugiyama, et al. 2013. A Theory of the Transition to Critical Period Plasticity: Inhibition Selectively Suppresses Spontaneous Activity. *Neuron* **80**:51–63. [5, 7]

Trachtenberg, J. T., and M. P. Stryker. 2001. Rapid Anatomical Plasticity of Horizontal Connections in the Developing Visual Cortex. *J. Neurosci.* **21**:3476–3482. [5]

Trachtenberg, J. T., C. Trepel, and M. P. Stryker. 2000. Rapid Extragranular Plasticity in the Absence of Thalamocortical Plasticity in the Developing Primary Visual Cortex. *Science* **287**:2029–2032. [5]

Tritsch, N. X., E. Yi, J. E. Gale, E. Glowatzki, and D. E. Bergles. 2007. The Origin of Spontaneous Activity in the Developing Auditory System. *Nature* **450**:50–55. [4]

Tropea, D., A. K. Majewska, R. Garcia, and M. Sur. 2010. Structural Dynamics of Synapses *in Vivo* Correlate with Functional Changes during Experience-Dependent Plasticity in Visual Cortex. *J. Neurosci.* **30**:11086–11095. [5, 7]

Tsankova, N. M., O. Berton, W. Renthal, et al. 2006. Sustained Hippocampal Chromatin Regulation in a Mouse Model of Depression and Antidepressant Action. *Nat. Neurosci.* **9**:519–525. [6]

Tsao, F. M., H. M. Liu, and P. K. Kuhl. 2006. Perception of Native and Non-Native Affricative-Fricative Contrasts: Cross-Language Tests on Adults and Infants. *J. Acoust. Soc. Am.* **120**:2285–2294. [7]

Tseng, K. Y., and P. O'Donnell. 2007. Dopamine Modulation of Prefrontal Cortical Interneurons Changes during Adolescence. *Cereb. Cortex* **17**:1235–1240. [12]

Tsirlin, I., L. Colpa, H. C. Goltz, and A. M. Wong. 2015. Behavioral Training as New Treatment for Adult Amblyopia: A Meta-Analysis and Systemic Review. *Invest Ophthalmol Vis Sci* **56**:4061–4075. [5]

Tung, J., L. B. Barreiro, Z. P. Johnson, et al. 2012. Social Environment Is Associated with Gene Regulatory Variation in the Rhesus Macaque Immune System. *PNAS* **109**:6490–6495. [6]

Turecki, G., M. J. Meaney, E. Evans, et al. 2016. Effects of the Social Environment and Stress on Glucocorticoid Receptor Gene Methylation: A Systematic Review. *Biol. Psychiatry* **79**:87–96. [6]

Tyzio, R., R. Nardou, D. C. Ferrari, et al. 2014. Oxytocin-Mediated GABA Inhibition during Delivery Attenuates Autism Pathogenesis in Rodent Offspring. *Science* **343**:675–679. [4]

Tyzio, R., A. Represa, I. Jorquera, et al. 1999. The Establishment of GABAergic and Glutamatergic Synapses on CA1 Pyramidal Neurons Is Sequential and Correlates with the Development of the Apical Dendrite. *J. Neurosci.* **19**:10372–10382. [4]

U.S. Census Bureau. Decennial Censuses 1790 to 1940 and the Current Population Survey: Annual Social and Economic Supplements, 1947 to 2015. https://www.census.gov/prod/www/decennial.html; https://www.census.gov/topics/health/health-insurance/guidance/cps-asec.html. (accessed Mar. 6, 2018). [14]

Uemura, M., S. Nakao, S. T. Suzuki, M. Takeichi, and S. Hirano. 2007. OL-Protocadherin Is Essential for Growth of Striatal Axons and Thalamocortical Projections. *Nat. Neurosci.* **10**:1151–1159. [3]

Uesaka, N., M. Uchigashima, T. Mikuni, et al. 2015. Retrograde Signaling for Climbing Fiber Synapse Elimination. *Cerebellum* **14**:4–7. [7]

Uhlhaas, P. J., G. Pipa, B. Lima, et al. 2009a. Neural Synchrony in Cortical Networks: History, Concept and Current Status. *Front. Integrat. Neurosci.* **3**:1–19. [2]

Uhlhaas, P. J., G. Pipa, S. Neuenschwander, M. Wibral, and W. Singer. 2011. A New Look at Gamma? High- (> 60 Hz) Gamma-Band Activity in Cortical Networks: Function, Mechanisms and Impairment. *Prog. Biophys. Mol. Biol.* **105**:14–28. [7]

Uhlhaas, P. J., F. Roux, E. Rodriguez, A. Rotarska-Jagiela, and W. Singer. 2010. Neural Synchrony and the Development of Cortical Networks. *Trends Cogn. Sci.* **14**:72–80. [7, 11–13]

Uhlhaas, P. J., F. Roux, W. Singer, et al. 2009b. The Development of Neural Synchrony Reflects Late Maturation and Restructuring of Functional Networks in Humans. *PNAS* **106**:9866–9871. [11, 12]

Uhlhaas, P. J., and W. Singer. 2010. Abnormal Neural Oscillations and Synchrony in Schizophrenia. *Nat. Rev. Neurosci.* **11**:100–113. [10, 12]

———. 2011. The Development of Neural Synchrony and Large-Scale Cortical Networks during Adolescence: Relevance for the Pathophysiology of Schizophrenia and Neurodevelopmental Hypothesis. *Schizophr. Bull.* **37**:514–523. [4]

———. 2012. Neuronal Dynamics and Neuropsychiatric Disorders: Toward a Translational Paradigm for Dysfunctional Large-Scale Networks. *Neuron* **75**:963–980. [12]

———. 2013. High-Frequency Oscillations and the Neurobiology of Schizophrenia. *Dialogues in clinical neuroscience* **15**:301–313. [10]

Uy, J. P., and A. Galván. 2017. Sleep Duration Moderates the Association between Insula Activation and Risky Decisions under Stress in Adolescents and Adults. *Neuropsychologia* **95**:119–129. [14]

Vaisvaser, S., S. Modai, L. Farberov, et al. 2016. Neuro-Epigenetic Indications of Acute Stress Response in Humans: The Case of MicroRNA-29c. *PLoS One* **11**:e0146236. [6]

Vakorin, V. A., and S. M. Doesburg. 2016. Development of Human Neurophysiological Activity and Network Dynamics. In: Multimodal Oscillation-Based Connectivity Theory, ed. S. Palva, pp. 107–122. Heidelberg: Springer-Verlag. [11]

Vakorin, V. A., S. M. Doesburg, R. C. Leung, et al. 2017. Developmental Changes in Neuromagnetic Rhythms and Network Synchrony in Autism. *Ann. Neurol.* **81**:199–211. [10, 11]

Valiente, M., and O. Marín. 2010. Neuronal Migration Mechanisms in Development and Disease. *Curr. Opin. Neurobiol.* **20**:68–78. [3]

Vallortigara, G., V. A. Sovrano, and C. Chiandetti. 2009. Doing Socrates Experiment Right: Controlled Rearing Studies of Geometrical Knowledge in Animals. *Curr. Opin. Neurobiol.* **19**:20–26. [11]

van Battum, E. Y., S. Brignani, and R. J. Pasterkamp. 2015. Axon Guidance Proteins in Neurological Disease. *The Lancet Neurology* **14**:532–546. [3]

van den Bos, W., M. X. Cohen, T. Kahnt, and E. A. Crone. 2012. Striatum-Medial Prefrontal Cortex Connectivity Predicts Developmental Changes in Reinforcement Learning. *Cereb. Cortex* **22**:1247–1255. [15]

Van Duijvenvoorde, A. C., B. R. Jansen, J. C. Bredman, and H. M. Huizenga. 2012. Age-Related Changes in Decision Making: Comparing Informed and Noninformed Situations. *Dev. Psychol.* **48**:192–203. [15]

Van Speybroeck, L. 2002. From Epigenesis to Epigenetics: The Case of C. H. Waddington. *Ann. NY Acad. Sci.* **981**:61–81. [6]

van Versendaal, D., and C. N. Levelt. 2016. Inhibitory Interneurons in Visual Cortical Plasticity. *Cell. Mol. Life Sci.* **73**:3677–3691. [5]

Varela, F., J. P. Lachaux, E. Rodriguez, and J. Martinerie. 2001. The Brain Web: Phase Synchronization and Large-Scale Integration. *Nat. Rev. Neurosci.* **2**:229–239. [7]

Varlinskaya, E., and L. Spear. 2008. Social Interactions in Adolescent and Adult Sprague-Dawley Rats: Impact of Social Deprivation and Test Context Familiarity. *Behav. Brain Res.* **188**:398–405. [14]

Vassoler, F. M., E. M. Byrnes, and R. C. Pierce. 2014. The Impact of Exposure to Addictive Drugs on Future Generations: Physiological and Behavioral Effects. *Neuropharmacology* **76 (Pt B)**:269–275. [9]

Venkatesh, S., and J. L. Workman. 2015. Histone Exchange, Chromatin Structure and the Regulation of Transcription. *Nat. Rev. Mol. Cell Biol.* **16**:178–189. [6]

Verhage, M., A. S. Maia, J. J. Plomp, et al. 2000. Synaptic Assembly of the Brain in the Absence of Neurotransmitter Secretion. *Science* **287**:864–869. [4]

Victor, E. C., and A. R. Hariri. 2015. A Neuroscience Perspective on Sexual Risk Behavior in Adolescence and Emerging Adulthood. *Dev. Psychopathol.* **28**:471–487. [14]

Vinck, M., and C. A. Bosman. 2016. More Gamma More Predictions: Gamma-Synchronization as a Key Mechanism for Efficient Integration of Classical Receptive Field Inputs with Surround Predictions. *Front. Syst. Neurosci.* **10**:1–27. [2]

Voelkle, M. C., and J. H. L. Oud. 2015. Relating Latent Change Score and Continuous Time Models. *Struct. Equ. Modeling* **22**:366–381. [15]

Vogel, A. C., J. D. Power, S. E. Petersen, and B. L. Schlaggar. 2010. Development of the Brain's Functional Network Architecture. *Neuropsychol. Rev.* **20**:10.1007/s11065–11010–19145–11067. [9]

von der Malsburg, C., W. A. Phillips, and W. Singer. 2010. Dynamic Coordination in the Brain: From Neurons to Mind, vol. 5. Strüngmann Forum Reports, J. Lupp, series ed. Cambridge, MA: MIT Press. [2]

Wahlstrohm, K., B. Dretzke, M. Gordon, et al. 2014. Examining the Impact of Later School Start Times on the Health and Academic Performance of High School Students: A Multi-Site Study. Center for Applied Research and Educational Improvement. St Paul: Univ. of Minnesota. [14]

Walhovd, K. B., A. M. Fjell, J. Giedd, and A. M. Dale. 2017. Through Thick and Thin: A Need to Reconcile Contradictory Results on Trajectories in Human Cortical Development. *Cereb. Cortex* **27**:1472–1481. [11]

Wallace, M. L., A. C. Burette, R. J. Weinberg, and B. D. Philpot. 2012. Maternal Loss of Ube3a Produces an Excitatory/Inhibitory Imbalance through Neuron Type-Specific Synaptic Defects. *Neuron* **74**:793–800. [6]

Wallen, K. 2005. Hormonal Influences on Sexually Differentiated Behavior in Nonhuman Primates. *Front. Neuroendocrin.* **26**:7–26. [8]

Wang, B. S., R. Sarnaik, and J. Cang. 2010. Critical Period Plasticity Matches Binocular Orientation Preference in the Visual Cortex. *Neuron* **65**:246–256. [5, 7]

Wang, D., M. Szyf, C. Benkelfat, et al. 2012. Peripheral SLC6A4 DNA Methylation Is Associated with *in Vivo* Measures of Human Brain Serotonin Synthesis and Childhood Physical Aggression. *PLoS One* **7**:e39501. [6]

Wang, H. C., C. C. Lin, R. Cheung, et al. 2015. Spontaneous Activity of Cochlear Hair Cells Triggered by Fluid Secretion Mechanism in Adjacent Support Cells. *Cell* **163**:1348–1359. [4]

Wang, H. X., and W. J. Gao. 2009. Cell Type-Specific Development of NMDA Receptors in the Interneurons of Rat Prefrontal Cortex. *Neuropsychopharmacology* **34**:2028–2040. [12]

Wang, Y., P. M. Smallwood, M. Cowan, et al. 1999. Mutually Exclusive Expression of Human Red and Green Visual Pigment-Reporter Transgenes Occurs at High Frequency in Murine Cone Photoreceptors. *PNAS* **96**:5251–5256. [4]

Wang, Z., B. Tang, Y. He, and P. Jin. 2016. DNA Methylation Dynamics in Neurogenesis. *Epigenomics* **8**:401–414. [6]

Ward, L. M. 2003. Synchronous Neural Oscillations and Cognitive Processes. *Trends Cogn. Sci.* **7**:553–559. [7]

Warner, C. E., W. C. Kwan, D. Wright, et al. 2015. Preservation of Vision by the Pulvinar Following Early-Life Primary Visual Cortex Lesions. *Curr. Biol.* **25**:424–434. [3]

Weaver, I. C. G., N. Cervoni, F. A. Champagne, et al. 2004. Epigenetic Programming by Maternal Behavior. *Nat. Neurosci.* **7**:847–854. [6]

Weiss, L. A., D. E. Arking, M. J. Daly, and A. Chakravarti. 2009. A Genome-Wide Linkage and Association Scan Reveals Novel Loci for Autism. *Nature* **461**:802–808. [3]

Weiss, P. 1941. Nerve Patterns: The Mechanics of Nerve Growth. The Third Growth Symposium. *Growth* **5**:163–203. [3]

Weisstanner, C., G. Kasprian, G. M. Gruber, P. C. Brugger, and D. Prayer. 2015. MRI of the Fetal Brain. *Clin. Neuroradiol.* **25**:189–196. [3]

Welborn, B. L., M. D. Lieberman, D. Goldenberg, et al. 2016. Neural Mechanisms of Social Influence in Adolescence. *Soc. Cogn. Affect. Neurosci.* **11**:100–109. [14]

Wenger, E., C. Brozzoli, U. Lindenberger, and M. Lövdén. 2017a. Expansion and Renormalization of Human Brain Structure during Skill Acquisition. *Trends Cogn. Sci.* **21**:930–939 [13]

Wenger, E., S. Kühn, J. Verrel, et al. 2017b. Repeated Structural Imaging Reveals Nonlinear Progression of Experience-Dependent Volume Changes in Human Motor Cortex. *Cereb. Cortex* **27**:2911–2925. [13, 15]

Wenger, E., S. Schaefer, H. Noack, et al. 2012. Cortical Thickness Changes Following Spatial Navigation Training in Adulthood and Aging. *NeuroImage* **59**:3389–3397 [13]

Werker, J. F., and T. K. Hensch. 2015. Critical Periods in Speech Perception: New Directions. *Annu. Rev. Psychol.* **66**:173–196. [6, 7]

Werker, J. F., and R. C. Tees. 1984. Cross-Language Speech Perception: Evidence for Perceptual Reorganization during the First Year of Life. *Infant. Behav. Dev.* **7**:49–63. [7]

Werker, J. F., H. H. Yeung, and K. A. Yoshida. 2012. How Do Infants Become Experts at Native-Speech Perception? *Curr. Dir. Psychol. Sci.* **21**:221–226. [7]

White, L. E., D. M. Coppola, and D. Fitzpatrick. 2001. The Contribution of Sensory Experience to the Maturation of Orientation Selectivity in Ferret Visual Cortex. *Nature* **411**:1049–1052. [7]

Whittington, M. A., R. D. Traub, N. Kopell, B. Ermentrout, and E. H. Buhl. 2000. Inhibition-Based Rhythms: Experimental and Mathematical Observations on Network Dynamics. *Int. J. Psychophysiol.* **38**:315–336. [2, 10]

Wiesel, T. N., and D. H. Hubel. 1963. Single-Cell Responses in Striate Cortex of Kittens Deprived of Vision in One Eye. *J. Neurophysiol.* **26**:1003–1017. [5]

———. 1970. The Period of Susceptibility to the Physiological Effects of Unilateral Eye Closure in Kittens. *J. Physiol. (Lond.)* **206**:419–436. [5]

Willing, J., and J. M. Juraska. 2015. The Timing of Neuronal Loss across Adolescence in the Medial Prefrontal Cortex of Male and Female Rats. *Neuroscience* **301**:268–275. [8]

Willshaw, D. J., and C. von der Malsburg. 1976. How Patterend Neural Connections Can Be Set up by Self-Organization. *Proc. R. Soc. Lond. B* **194**:431–445. [2]

Wright, K. M., K. A. Lyon, H. Leung, et al. 2012. Dystroglycan Organizes Axon Guidance Cue Localization and Axonal Pathfinding. *Neuron* **76**:931–944. [3]

Wu, H., and Y. Zhang. 2014. Reversing DNA Methylation: Mechanisms, Genomics, and Biological Functions. *Cell* **156**:45–68. [6]

Xu, T., X. Yu, A. J. Perlik, et al. 2009. Rapid Formation and Selective Stabilization of Synapses for Enduring Motor Memories. *Nature* **462**:915–919 [13]

Yamagata, M., and J. R. Sanes. 2008. Dscam and Sidekick Proteins Direct Lamina-Specific Synaptic Connections in Vertebrate Retina. *Nature* **451**:465–469. [3]

Yamamoto, J., J. Suh, D. Takeuchi, and S. Tonegawa. 2014. Successful Execution of Working Memory Linked to Synchronized High-Frequency Gamma Oscillations. *Cell* **157**:845–857. [2]

Yang, E.-J., E. W. Lin, and T. K. Hensch. 2012. Critical Period for Acoustic Preference in Mice. *PNAS* **109(Suppl)**:17213–17220. [3, 5, 11]

Yang, G., F. Pan, and W. B. Gan. 2009a. Stably Maintained Dendritic Spines Are Associated with Lifelong Memories. *Nature* **462**:920–924. [13]

Yang, J. W., S. An, J. J. Sun, et al. 2013. Thalamic Network Oscillations Synchronize Ontogenetic Columns in the Newborn Rat Barrel Cortex. *Cereb. Cortex* **23**:1299–1316. [4]

Yang, J. W., I. L. Hanganu-Opatz, J.-J. Sun, and H. J. Luhmann. 2009b. Three Patterns of Oscillatory Activity Differentially Synchronize Developing Neocortical Networks *in Vivo*. *J. Neurosci.* **29**:9011–9025. [2, 4]

Yang, Y., T. Yamada, K. K. Hill, et al. 2016. Chromatin Remodeling Inactivates Activity Genes and Regulates Neural Coding. *Science* **353**:300–305. [6]

Yao, B., K. M. Christian, C. He, et al. 2016. Epigenetic Mechanisms in Neurogenesis. *Nat. Rev. Neurosci.* **17**:537–549. [6]

Yashiro, K., R. Corlew, and B. D. Philpot. 2005. Visual Deprivation Modifies Both Presynaptic Glutamate Release and the Composition of Perisynaptic/Extrasynaptic NMDA Receptors in Adult Visual Cortex. *J. Neurosci.* **25**:11684–11692. [5]

Yazaki-Sugiyama, Y., S. Kang, H. Câteau, T. Fukai, and T. K. Hensch. 2009. Bidirectional Plasticity in Fast-Spiking GABA Circuits by Visual Experience. *Nature* **462**:218–221. [5]

Ye, Y., H. Xu, X. Su, and X. He. 2016. Role of MicroRNA in Governing Synaptic Plasticity. *Neural Plast.* **2016**:1–13. [6]

Yoon, K.-J., H. N. Nguyen, G. Ursini, et al. 2014. Modeling a Genetic Risk for Schizophrenia in iPSCs and Mice Reveals Neural Stem Cell Deficits Associated with Adherens Junctions and Polarity. *Cell Stem Cell* **15**:79–91. [3]

Yu, A. J., and P. Dayan. 2005. Uncertainty, Neuromodulation, and Attention. *Neuron* **46**:681–692. [15]

Yu, H., A. K. Majewska, and M. Sur. 2011a. Rapid Experience-Dependent Plasticity of Synapse Function and Structure in Ferret Visual Cortex *in Vivo*. *PNAS* **108**:21235–21240. [5]

Yu, N.-K., S.-H. Baek, and B.-K. Kaang. 2011b. DNA Methylation-Mediated Control of Learning and Memory. *Mol. Brain* **4**:5. [6]

Yuan, P., and N. Raz. 2014. Prefrontal Cortex and Executive Functions in Healthy Adults: A Meta-Analysis of Structural Neuroimaging Studies. *Neurosci. Biobehav. Rev.* **42**:180–192. [13]

Zeanah, C. H., M. R. Gunnar, R. B. McCall, J. M. Kreppner, and N. A. Fox. 2011. Sensitive Periods. *Monogr. Soc. Res. Child Dev.* **76**:147–162. [6]

Zelina, P., H. Blockus, Y. Zagar, et al. 2014. Signaling Switch of the Axon Guidance Receptor Robo3 during Vertebrate Evolution. *Neuron* **84**:1–15. [3]

Zhang, M., R. Katzman, D. Salmon, et al. 1990. The Prevalence of Dementia and Alzheimer's Disease in Shanghai, China: Impact of Age, Gender, and Education. *Ann. Neurol.* **27**:428–437. [4]

Zhou, L., I. Bar, Y. Achouri, et al. 2008. Early Forebrain Wiring: Genetic Dissection Using Conditional Celsr3 Mutant Mice. *Science* **320**:946–949. [3]

Zhu, Y., T. Matsumoto, S. Mikami, T. Nagasawa, and F. Murakami. 2009. SDF1/CXCR4 Signalling Regulates Two Distinct Processes of Precerebellar Neuronal Migration and Its Depletion Leads to Abnormal Pontine Nuclei Formation. *Development* **136**:1919–1928. [3]

Zhu, Y., T. Yu, X.-C. Zhang, et al. 2002. Role of the Chemokine SDF-1 as the Meningeal Attractant for Embryonic Cerebellar Neurons. *Nat. Neurosci.* **5**:719–720. [3]

Zipursky, S. L., and J. R. and Sanes. 2010. Chemoaffinity Revisited: Dscams, Protocadherins, and Neural Circuit Assembly. *Cell* **143**:343–353. [3]

Zovkic, I. B., B. S. Paulukaitis, J. J. Day, D. M. Etikala, and J. D. Sweatt. 2014. Histone H2A.Z Subunit Exchange Controls Consolidation of Recent and Remote Memory. *Nature* **515**:582–586. [6]

Subject Index

Further Titles in the Strüngmann Forum Report Series[1]

Better Than Conscious? Decision Making, the Human Mind, and Implications For Institutions
edited by Christoph Engel and Wolf Singer, ISBN 978-0-262-19580-5

Clouds in the Perturbed Climate System: Their Relationship to Energy Balance, Atmospheric Dynamics, and Precipitation
edited by Jost Heintzenberg and Robert J. Charlson, ISBN 978-0-262-01287-4

Biological Foundations and Origin of Syntax
edited by Derek Bickerton and Eörs Szathmáry, ISBN 978-0-262-01356-7

Linkages of Sustainability
edited by Thomas E. Graedel and Ester van der Voet, ISBN 978-0-262-01358-1

Dynamic Coordination in the Brain: From Neurons to Mind
edited by Christoph von der Malsburg, William A. Phillips and Wolf Singer, ISBN 978-0-262-01471-7

Disease Eradication in the 21st Century: Implications for Global Health
edited by Stephen L. Cochi and Walter R. Dowdle, ISBN 978-0-262-01673-5

Animal Thinking: Contemporary Issues in Comparative Cognition
edited by Randolf Menzel and Julia Fischer, ISBN 978-0-262-01663-6

Cognitive Search: Evolution, Algorithms, and the Brain
edited by Peter M. Todd, Thomas T. Hills and Trevor W. Robbins, ISBN 978-0-262-01809-8

Evolution and the Mechanisms of Decision Making
edited by Peter Hammerstein and Jeffrey R. Stevens, ISBN 978-0-262-01808-1

Language, Music, and the Brain: A Mysterious Relationship
edited by Michael A. Arbib, ISBN 978-0-262-01962-0

Cultural Evolution: Society, Technology, Language, and Religion
edited by Peter J. Richerson and Morten H. Christiansen, ISBN 978-0-262-01975-0

Schizophrenia: Evolution and Synthesis
edited by Steven M. Silverstein, Bita Moghaddam and Til Wykes, ISBN 978-0-262-01962-0

Rethinking Global Land Use in an Urban Era
edited by Karen C. Seto and Anette Reenberg, ISBN 978-0-262-02690-1

Trace Metals and Infectious Diseases
edited by Jerome O. Nriagu and Eric P. Skaar, ISBN 978-0-262-02919-3

Translational Neuroscience: Toward New Therapies
edited by Karoly Nikolich and Steven E. Hyman, ISBN: 9780262029865

---

[1] available at https://mitpress.mit.edu/books/series/str%C3%BCngmann-forum-reports-0

The Pragmatic Turn: Toward Action-Oriented Views in Cognitive Science
edited by Andreas K. Engel, Karl J. Friston and Danica Kragic
ISBN: 978-0-262-03432-6

Complexity and Evolution: Toward a New Synthesis for Economics
edited by David S. Wilson and Alan Kirman, ISBN: 9780262035385

Computational Psychiatry: New Perspectives on Mental Illness
edited by A. David Redish and Joshua A. Gordon, ISBN: 9780262035422

Investors and Exploiters in Ecology and Economics: Principles and Applications
edited by Luc-Alain Giraldeau, Philipp Heeb and Michael Kosfeld
Hardcover: ISBN: 9780262036122, eBook: ISBN: 9780262339797

The Cultural Nature of Attachment: Contextualizing Relationships and Development
Edited by Heidi Keller and Kim A. Bard
Hardcover: ISBN: 9780262036900, ebook: ISBN: 9780262342865

Printed in the United States
by Baker & Taylor Publisher Services